AF794858

Agitator Design Technology for Biofuels and Renewable Chemicals

# Agitator Design Technology for Biofuels and Renewable Chemicals

*Gregory T. Benz*

*Benz Technology International, Inc.,*
*Clarksville, OH, USA*

WILEY

AIChE
The Global Home of Chemical Engineers

This edition first published 2022
© 2022 John Wiley & Sons, Inc.

All rights reserved. No part of this publication may be reproduced, stored in a retrieval system, or transmitted, in any form or by any means, electronic, mechanical, photocopying, recording or otherwise, except as permitted by law. Advice on how to obtain permission to reuse material from this title is available at http://www.wiley.com/go/permissions.

The right of Gregory T. Benz to be identified as the author of this work has been asserted in accordance with law.

*Registered Office*
John Wiley & Sons, Inc., 111 River Street, Hoboken, NJ 07030, USA

*Editorial Office*
111 River Street, Hoboken, NJ 07030, USA

For details of our global editorial offices, customer services, and more information about Wiley products visit us at www.wiley.com.

Wiley also publishes its books in a variety of electronic formats and by print-on-demand. Some content that appears in standard print versions of this book may not be available in other formats.

*Limit of Liability/Disclaimer of Warranty*
The contents of this work are intended to further general scientific research, understanding, and discussion only and are not intended and should not be relied upon as recommending or promoting scientific method, diagnosis, or treatment by physicians for any particular patient. In view of ongoing research, equipment modifications, changes in governmental regulations, and the constant flow of information relating to the use of medicines, equipment, and devices, the reader is urged to review and evaluate the information provided in the package insert or instructions for each medicine, equipment, or device for, among other things, any changes in the instructions or indication of usage and for added warnings and precautions. While the publisher and authors have used their best efforts in preparing this work, they make no representations or warranties with respect to the accuracy or completeness of the contents of this work and specifically disclaim all warranties, including without limitation any implied warranties of merchantability or fitness for a particular purpose. No warranty may be created or extended by sales representatives, written sales materials or promotional statements for this work. The fact that an organization, website, or product is referred to in this work as a citation and/or potential source of further information does not mean that the publisher and authors endorse the information or services the organization, website, or product may provide or recommendations it may make. This work is sold with the understanding that the publisher is not engaged in rendering professional services. The advice and strategies contained herein may not be suitable for your situation. You should consult with a specialist where appropriate. Further, readers should be aware that websites listed in this work may have changed or disappeared between when this work was written and when it is read. Neither the publisher nor authors shall be liable for any loss of profit or any other commercial damages, including but not limited to special, incidental, consequential, or other damages.

*Library of Congress Cataloging-in-Publication Data applied for:*

Hardback ISBN: 9781119815495

Cover Design: Wiley
Cover Image: © photosbyjim/Getty Images

Set in 9.5/12.5pt STIXTwoText by Straive, Pondicherry, India

# Contents

# Foreword

In his new book, Gregory T. Benz covers the full range of agitator design technology theory and applications in the biofuels and renewable chemicals industry. His specialty in fermentation and bioreactors for the biofuels and renewable chemicals industry comes through loud and clear. He also has experience in continuous flow and multistage reactor design. Greg is well-known for his expertise in agitation, mixing and bioreactor design consulting, including equipment specification and bid evaluation. He has designed more than 150 fermentor agitators, as well as many more polymer and biofuel agitation systems, including bioethanol, biobutanol, corn and cellulosic ethanol, biodiesel, and other value-added renewable chemicals and biochemicals.

As evidence of Greg's expertise described in this book, he covers areas such as the purpose behind agitator design, the principles of how to design agitators, and the various types of impellors and how to use each type. But as a bioprocess development, scaleup, and commercialization professional, Greg's treatment of various applications including starch and sugar-based ethanol applications, cellulosic ethanol and renewable chemicals applications, biodiesel applications, and pyrolysis applications are my personal favorites. He even discusses the features of cellulosic feedstocks and their unique agitation requirements. Application is where the rubber meets the road and R&D is translated into commercial processes. These topics provide real-life applications and insights into these processes. Furthermore, Greg discusses the keys to successful scaleup of agitator systems, principles of rheology, the use of computational fluid dynamics (CFD) to agitator design, mechanical issues, and finally, the evaluation of vendors.

Many, if not most, of the well-known biofuel and biotechnology technology providers have used Greg as a consultant during development, scaleup, and commercial implementation. I am one of them. I have known Greg for over 15 years. Ten of those years were as corporate Director of R&D at ICM, Inc. where Greg was my go-to expert for agitator design, testing, and commercial implementation. Greg's agitator expertise was invaluable in the many successes of our fully integrated pilot, demonstration, and commercial scale facilities with individual reactors ranging in size from 5000 gallons to 500000+ gallons in capacity. He brings his expertise out in this book. I consider this new book the go-to resource for bioprocess agitator applications.

*Douglas B. Rivers, PhD*
*Executive Vice President*
*Lee Enterprises Consulting, Inc.*
*Birmingham, AL, USA*
*23 August 2021*

# Foreword to *Agitator Design Technology for Biofuels and Renewable Chemicals*, Gregory T. Benz

Biofuels and renewable chemicals by definition use feedstocks that are solid and have been recently grown and harvested from the land. These materials may typically be masticated by some mechanical means to get them into a form amenable to a water solution where chemical/biological reactions can be performed on them. Still in the solid form, materials such as ground corn, pulverized wood chips, ground corn stover or grass will be suspended in a solution that ultimately will end up in vessel that needs to be mixed to enable the chemical or biological reactions to lead to desirable fuel and chemical products. Fuel is an inherently low-margin product. Ethanol from corn struggled for 20 years while being shored up by various government subsidies until it achieved process designs at a scale that would make it profitable. After the turn of the twenty-first century, this industry exploded, limited only by the maximum amount that could be blended into standard gasoline.

One area that led to the ultimate profitability of that industry was extremely large fermentations. Fermenters as large as 500 000 gallons or even larger became common. Mixing a slurry of corn mash in these tanks was critical to sustaining productivity that allowed for the shortest turn around for these vessels. The high rate of production per tank is important to return on investment of these high capital cost items.

With the production of various other commodity chemicals via fermentation, the profit margin continues to be low and the complexity of the fermentation has increased. Production of some chemicals involves the careful addition of very specific amounts of oxygen (air). Too much air and some undesired by-product might be the result. Too little air and the biological organism responsible for the production of your product might die, leaving a tank of raw material at the mercy of foreign organisms waiting to use your raw material and make some useless mix of stuff that has no value.

In the last 10 years, there has been a shift from corn starch to raw cellulosic materials. These cellulosic materials are more difficult to get into a solution that is mostly liquid continuous. They will stay as solids until nearly all are converted into simple molecules and then there will be residues of unreacted material.

The unknowns of how to mix these in extremely large tanks, up to as much as 1 200 000 gallons, are made greater by extremely viscous solutions. Viscosity in saccharification, with or without fermentation for cellulose feedstocks, generally limits the percent solids in the tank. Higher solids enable more economical use of these large vessels. Fermentation might take place simultaneously to the saccharification or in a separate vessel, but again, the addition of air might prove critical.

As a chemical engineer involved in biofuels and renewable chemicals, I first met Gregory T. Benz about 20 years ago while working on a fermentation that had a critical air need. It was amazing to me that the large agitation companies had decided to forgo technical support in house, because we as engineers needed that support. Greg was there to provide it. At that time, he had about 25 years of experience working at a large agitator company. As a chemical engineer, he could easily understand the problem we were facing, and, with his experience in the agitator industry, was able to design and recommend equipment that would work. In addition, as an independent party no longer affiliated with a single company, he could help us evaluate competing designs of multiple companies, making sure we got the best product for our application.

Over the years, I have had the occasion to work with Greg on the most difficult agitation problem I know of saccharification of pretreated corn stover and other cellulosic feedstocks. To build a saccharification and fermentation vessel that is 1 200 000 gallons in size with data from tanks that are, at most, only a few thousand gallons, can be quite a risk for the project and probably one's job. Greg was able to design experiments at small scale that gave the proper scaleup information to enable confidence in designing such large systems.

Greg has now brought together his combined 45 years of experience designing tough agitation systems and describes how you can use the theory and experiments to confidently design the agitation systems that are required for this nascent industry.

I am very glad to see that Greg has put together a volume dedicated to Biofuels and Renewable Chemical Agitator design. This is a high capital area of a plant making low-margin materials and this text will help designers get it right. Thank you, Greg.

*Robert J. Wooley, PhD, PE*
*Consulting Engineer*
*Cedar Key, FL, USA*
*24 November 2021*

## Foreword by Dan Beacom (Beacom Consulting)

Early in my career I had the good fortune to attend a two-day mixing course prepared and taught by Gregory T. Benz. As a microbiologist, I appreciated how he could describe the concepts in engineering terms, mathematically, and in plain language. Over the years I've had opportunity to work with Greg as THE go-to expert on multitude of projects from lab fermentations to large-scale plants. I'm really glad we now have this written reference to access his know-how out our fingertips. I'm sure you will find this book as valuable as I've found Greg's consultations over the years.

*Jenkintown, PA*
*7 December 2021*

# Preface

This is a book about fluid agitation technology, as applied to processes used to make biofuels and renewable chemicals. Though there are industrial processes (such as pyrolysis/biogas routes) that do not involve agitation, except perhaps for some ancillary tanks, ours is on mechanically agitated systems, consisting of a closed vessel with a rotating shaft and impellers.

Though side entering agitators have been used in some biofuel applications, their use is discouraged by this author, as they require many times as much power as top entering units. Moreover, because they have a submerged seal, they have higher maintenance costs. Thus, we only devote one chapter to side entering agitator design, though they are also mentioned in the application guides for specific processes.

Many books have been written about fluid agitation. Much, though not all, of the material in this book has been covered in such books. A number of books have also been written about biofuels and renewable chemicals, but little specific information is given on agitation in such books. All books on fluid agitation cover much more than agitator design for biofuels and chemicals. For example, typical books on agitation cover topics such as solids suspension (almost never an issue in biofuels), highly viscous systems (>100 000 cP), specialized impellers such as helical ribbons, anchors, augers, and others that have no use in biofuels, etc. Little has been published in such books about how to acquire the proper pilot data for agitator design, or how to minimize energy consumption. And none that the author is aware of cover the very unique agitation requirements of cellulosic reactor applications. This book fills that void.

The main purpose of this book is to be a single-source reference on all the major issues related to agitator design for biofuels. It is intended to save the reader time by avoiding the need to consult multiple references or sift through many pages of text to find what is needed specifically for agitator design for these industries. This book will also cover important related topics such as power cost, basic agitator mechanical design, and vendor bid evaluation.

Though some introductory fundamental theory is included, the main focus is on practical application of theory to real-world agitator design. This book is more of a how-to book than an academic treatise. The relative brevity of the book is intentional. It is hoped that the brevity will encourage people to actually read the entire book, not just skim an occasional page or chapter.

This book is intended to be useful for a variety of people. Since it is primarily a technical document, most readers will have a science or engineering degree. Many will be chemical engineers. Some will be chemists or microbiologists tasked with operating facilities in a way that can produce scalable data. Academic degrees among readers will vary from Bachelor up through Post-Doc. Most readers will be employed by companies using bioprocessing of renewable feedstocks to make biofuels or valuable products as well as many making commodity products. Some will work for agitator manufacturers. If used as a course supplement, some will be college students or professors. Top level managers may want to skim the contents to make sure their teams are properly staffed and have a high level view of what their team should be doing. They will find the overview and flowchart described in Chapter 2 especially useful. Information on energy use optimization will also be of interest to business unit managers. Information on bid evaluation should be of interest to procurement professionals. Although written primarily for users of agitation equipment and operators of production facilities, engineers employed by agitator manufacturers will likely find it of interest as it provides a deeper window into the details of these applications than they are accustomed to, as well as how their bids may be viewed in a competitive environment.

A note about symbols: rather than making the reader refer to a list of symbols in the appendix, each chapter has the symbols used in that chapter listed and defined at the end. That should save the reader some time. Also, it lets the author use the same symbol for different purposes in different contexts, reducing the number of symbols needed. For example, T means tank diameter in most cases, but in the context of mechanical calculations, it can be used to mean torque.

Considerable space is devoted to cellulosic hydrolysis reactor design, as this is the most complex application described in this book. Space is given on how to define and measure its rheology, pilot protocol and scaleup.

Application guides are given by major industry, and include simple solution tanks, slurry tanks, compounding tanks and fermenters/bioreactors. As used in this book, "fermenter" will mean an anaerobic bioreactor. Though there are aerobic processes used in some biofuel processes, the topic is complex enough to require at least an entire book devoted to it. Such a book has also been written by this author, and is titled *Agitator Design for Gas–Liquid Fermenters and Bioreactors* published by Wiley InterScience (Ref. [1]).

Those who read and study this book will learn to:

- Choose the right impellers for the job
- Understand how power draw and pumping are calculated

- Be familiar with biofuel/biomass agitator sizing guidelines
- Understand the principles and limitations of scaleup
- Understand the most common non-Newtonian rheology applicable to biofuel applications
- Use an agitator as a viscometer
- Properly set up a facility for piloting biofuel agitation requirements
- Design lab tests and scaleup cellulosic hydrolysis agitation
- Understand the uses and limitations of computational fluid dynamics
- Understand basic mechanical design
- Appreciate the difference between top-tier and lower-tier agitator suppliers

Historically, agitator specifications for the biofuel industry have been written by the end user or by engineering firms, but the personnel employed by such companies are generally not experts in fluid agitation. The result is that many mistakes have been made, though they are perhaps a bit less frequent these days. Among those I have personally seen specific to the fuel ethanol industry are as follows:

- Impeller diameters and speeds specified that would not result in even minimal solids suspension or liquid motion.
- Combinations of motor size, impeller system and speed specified that result in less than 1% of motor power being invested. (Yes, you read that right; 1% is not a typo!)
- Bizarre impeller designs, with poorly understood power, performance and hydraulic/mechanical characteristics.
- Higher power specified than needed in some applications.
- Motor power specified without regard to impeller type or shaft speed.
- Inefficient impeller designs specified, with poor hydraulic stability.
- Radial turbines called out with up-and-down mixing specified.
- Expensive horizontal ribbon blenders specified where turbines will suffice.
- Shaft diameters specified that would break if full motor power were applied.
- Smaller shafts specified through the seal than the extension shaft, resulting in a built-in weak point in an expensive location to repair.
- Seal type left up to the equipment vendor, when fugitive emissions regulations may dictate seal type.
- Grain slurry make-down at temperatures greater than 65 °C with no consideration given to breaking up the pasted starch balls that form.
- No differentiation of sizing based on stage in fermentation.
- Two impellers specified where one will suffice.
- One impeller specified where two are required.
- Upper impeller located above normal liquid level.
- Impeller so high off bottom that flow reversal occurs, resulting in no solids suspension.
- Pumping capacity specified without defining equations to be used, making it open to liberal interpretation.

**Figure P.1** The author processing biomass.

- Steady bearing materials specified which are subject to rapid wear in slurry service.
- No thought given to energy costs or maintenance costs.

Hopefully, the readers of this book will not make such mistakes!

As an aside, the author himself practices one time-honored way of converting biomass to energy, as illustrated in Figure P.1.

## Reference

**1** Benz, G. (2021). *Agitator Design for Gas–Liquid Fermenters and Bioreactors*, 1e. Wiley InterScience. ISBN-10: 1119650496, ISBN-13: 978-1119650492.

# Author Biography

Gregory T. Benz is President of Benz Technology International, Inc., located at 2305 S Clarksville Road, Clarksville, Ohio 45113. Phone 937-289-4504; fax 937-289-3914; e-mail g.benz@benz-tech.com.com. He received his BSChE from the University of Cincinnati in 1976, and has taken a course on Fermentation Biotechnology from The Center for Professional Advancement. A registered professional engineer in Ohio, he has over 46 years' experience in the design of agitation systems, specializing in polymer reactors, biofuel applications, fermentation, and bioreactors. Currently, his company does agitation, mixing, and bioreactor design consultation, including designing pilot protocol, scaling up, equipment specification, and bid evaluation. He has designed more than 150 fermenter agitators, as well as many polymer and biofuel agitation systems. He is a member of AIChE, ISPE, SIMB, SBE, American Mensa, and the American Chamber of Commerce in Shanghai. He is affiliated with Lee Enterprises Consulting. He has authored more than 38 technical articles. He has specific experience doing business in China, having managed his former employer's business there for years, and having consulted there for multiple clients. He has traveled to China dozens of times and is a student of its language and culture.

When not busy agitating vessels or people, he enjoys cutting and splitting firewood on his tree farm, photography, hiking, listening to music, and eating spicy food.

# 1

# Purpose of Agitator Design

The purpose of using the agitator design principles in this book is to ensure, to the extent possible, that the user of agitation equipment achieves the process objectives, and does so in a reliable and economical manner.

Agitators are employed in many different industries. The process results/objectives desired from the agitators vary by industry and by application within each industry. Since an agitator is ultimately nothing more than a kind of pump, and the agitated tank is essentially a deadheaded pump, it would be ideal if the objectives could be stated in purely physical terms, mostly related to flow and head. For example, some would describe agitation in terms of pumping capacity, characteristic fluid velocity [1], G-value [2], or other physical terms.

Some process results correlate well with simple physical measurements of agitation. For example, the ability to overcome density differences or viscosity ratios correlates well with characteristic fluid velocities [1]. However, many other process objectives do not correlate well with such simple measures. Examples of process results that have complex relationships to agitation and do not correlate well with pumping capacity, fluid velocity, or other simple measures would include blend time, mass transfer rate, heat transfer rate, off-bottom solids suspension, solids suspension degree of uniformity, solids suspension cloud height, rate of particle attrition or shear damage, dissolved oxygen spatial distribution, reaction rate, reaction product distribution, and many others.

Since this book is about agitator design for the biofuels and renewable chemicals industries, we will focus on the attributes of agitator design most important for those applications. Most biofuel applications are liquid motion controlled, so we will focus on such physical attributes as pumping capacity, velocity, ability to overcome physical property differences, and avoidance of dead spots in the tank. Though some processes in this industry use aerobic bioreactors, and thus have mass transfer as an important process result, I have written a whole book on the subject [3]. Readers interested in that subject are encouraged to refer to that book. We will not go into detail on aerobic processes in this book.

*Agitator Design Technology for Biofuels and Renewable Chemicals*, First Edition. Gregory T. Benz.
© 2022 John Wiley & Sons, Inc. Published 2022 by John Wiley & Sons, Inc.

Readers will be given a brief introduction to agitation fluid dynamics theory, but the focus will be on practical application. Guides are given for each major application, organized by industry. Several chapters are devoted to lignocellulosic agitation requirements, as they are more complex than others.

As we will see, in most cases, the desired process result can be accomplished over a range of power inputs by varying impeller size, type, and shaft speed. Therefore, there are opportunities to optimize the combination of capital and power costs.

In processes where the agitation requirements vary at different times during the operating cycle, there is an opportunity to reduce power costs by using a variable speed drive judiciously.

Finally, a good agitator design must be mechanically reliable. The process does not run when the agitator is broken!

In summary, there are three main purposes to agitator design:

1) Make sure agitation is sufficient to achieve the process result desired.
2) Try to minimize energy cost in both design and operation.
3) Make sure the design will provide many years of service with minimum downtime.

## References


1 Hicks, R.W., Morton, J.R., and Fenic, J.G. (1976). How to design agitators for desired process response. *Chemical Engineering Magazine* 22–30.

2 Benz, G.T. (2007). The G-value for agitator design: time to retire it? *Chemical Engineering Progress* 103 (3): 43–47.

3 Benz, G.T. *Agitator Design for Gas-Liquid Fermenters and Bioreactors.* Wiley InterScience. ISBN: 978-1-119-65049-2.

# 2

# Major Steps in Successful Agitator Design

This chapter presents an overview of the main steps and logic required to achieve the best agitation system design. Subsequent chapters will provide more technical details and fundamental concepts to facilitate the execution of each step. Figure 2.1 provides a graphic summary of these steps. We will describe each one in more detail in the following paragraphs. The flow chart concept used here was inspired by the procedures in Ref. [1], but is expanded upon in more detail here specifically for biofuel and renewable chemical agitator design.

## Define Process Results

The first step in agitator design, or, for that matter, the design of any kind of process equipment, is to define the expected process result. For agitators, that could be a number of different things, such as degree of solids suspension, blend time to some specified degree of uniformity, characteristic fluid velocity, heat transfer coefficient, etc. Some or all of these process results may be needed or applicable to biofuel and renewable chemicals process design. The most common requirement for the biofuel industry is to keep slowly-settling solids in suspension, which is treated as a liquid motion problem. Some applications require liquid–liquid dispersion. Some blend liquids with physical property differences, which must be overcome to allow mixing. And some require assurance of complete tank motion with no dead spots. The key is to clearly define the process results. However, they must be stated in physical terms in order to relate them to agitation parameters. For example, it would not be appropriate to specify a reaction rate or a degree of conversion. Agitators do not control reaction kinetics; all they can do is bring the materials together in a certain pattern. The rest is up to the chemistry and operating conditions.

*Agitator Design Technology for Biofuels and Renewable Chemicals*, First Edition. Gregory T. Benz.
© 2022 John Wiley & Sons, Inc. Published 2022 by John Wiley & Sons, Inc.

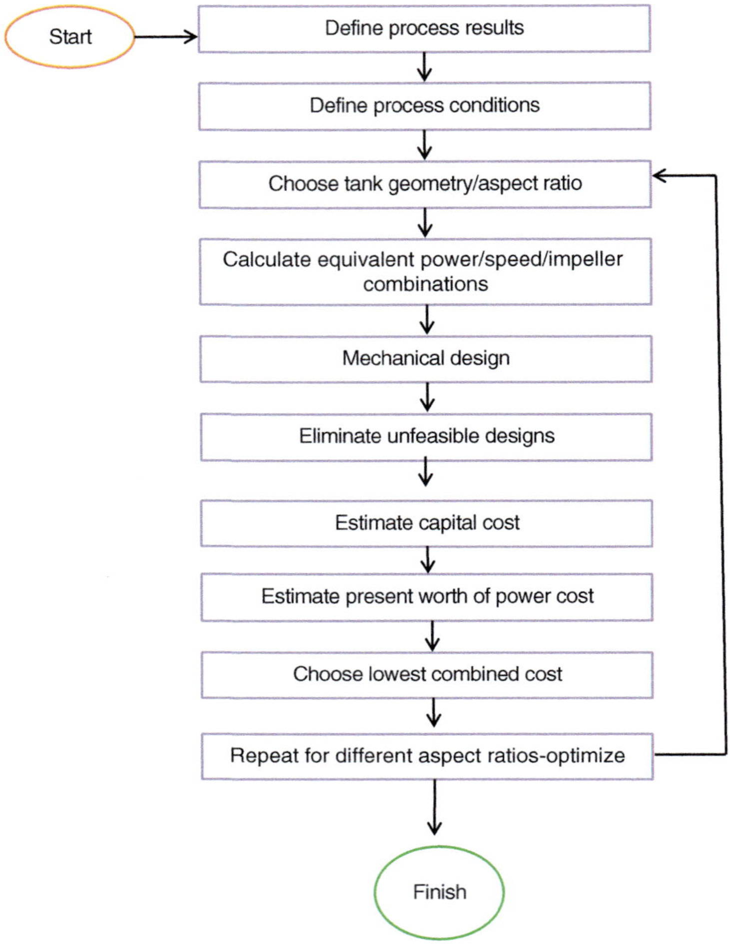


**Figure 2.1** Agitator design flow chart.

For the special case of a cellulosic hydrolysis reactor, the main process result is to avoid dead spots, especially at the beginning of the reaction sequence. We will delve into this in more detail in Chapter 13.

## Define Process Conditions

All conditions impacting the agitator design, the effect on process results and ancillary functions such as heat transfer must be delineated. (Thermal properties are not needed if there are no heat transfer surfaces within the tank). A partial list follows:

- Fluid density (e.g., specific gravity, or density such as $kg/m^3$); if two or more fluids are to be mixed, supply properties of each one.
- Fluid viscosity (e.g., cP or Pa-s). If the fluid is non-Newtonian, the model and parameters describing it must be included. If two or more fluids are to be blended, provide these properties for each one.
- Total volume to be agitated.
- Volume and properties at each processing step.
- Operating temperature.
- Mean barometric pressure (not needed if the tank has no seal).
- Back pressure in vapor space (not needed if the tank has no seal).
- Heat capacity of the process fluid at constant pressure, $C_P$. Sample units J/g-C or BTU/lb-F.
- Thermal conductivity of the process fluid, k. Sample units J/h-m-C or BTU/h-ft-F.
- For the ancillary heat transfer calculations, need the above thermal properties for the heat transfer medium as well.
- For the heat transfer calculations, need the available flow and temperature of the heat transfer medium.
- Any other process constraints affecting design.

## Choose Tank Geometry

In principle, many tank shapes can be used. That can include cylindrical, rectangular, and spherical tanks. However, odd-shaped tanks may be hard to baffle and agitate properly. Rectangular tanks may be harder to clean and sterilize if they have sharp corners. Most tanks used in this industry are cylindrical. Most are mounted with their axis vertical and centered. Any other geometry greatly increases hydraulic forces on the impeller, complicating agitator mechanical design, and increasing cost.

For the purposes of this book, we will limit the discussion to vertical cylindrical tanks. With that restriction, the geometry to be decided is the ratio of liquid height to tank diameter (Z/T), often referred to as aspect ratio. Agitators can be designed for any aspect ratio within reason. For very large, field-fabricated tanks, the geometry tends to be rather squat; Z/T ratios are usually between 0.25 and 1. For indoor tanks, the aspect ratio is usually between 2 and 4. This is often influenced by a transition between shop fabricated vessels and field fabricated vessels. The maximum diameter shop vessel that can usually be transported is about 14 ft. or 4.26 m.

There can be restrictions on aspect ratio due to building constraints. If the vessel must fit within a given floor space, that may place restrictions on the diameter, forcing a certain minimum aspect ratio. Sometimes local building codes carry height restrictions. So, the allowable aspect ratio range may be bound by such constraints, among others.

The main effect aspect ratio has on agitator design is on how many impellers are needed and possible shaft speeds. In most cases, the agitator cost increases as aspect ratio increases. Net power consumption also generally increases. Vessel cost may or may not increase, depending on various factors such as wall thickness, shop vs. field fabrication, etc.

Tank bottom geometry can be very important. Best results are obtained with dished heads, especially 2 : 1 elliptical. However, for noncritical applications such as starch or grain slurry tanks and fermenters, it is not unusual to use flat, sloped, or shallow cone-bottomed tanks. ("Shallow" as used here means no more than a 15° slope from horizontal.) For cellulosic hydrolysis reactors, one should *NEVER* use flat or sloped bottom tanks. With reasonable levels of agitation, you will get dead spots that will not react. It is possible to overcome these with about five times the power input, overagitating the bulk of the vessel just to get the last 2% of the volume to move and react. But a dished bottom or a permanent tank fillet can solve this problem. We will discuss this in more detail in Chapter 13.

## Calculate Equivalent Power/Shaft Speed/Impeller Systems

Every agitation requirement has more than one possible solution. Different impeller styles may have different power and torque requirements to achieve the same process result. Even within the same impeller style, there are different combinations of diameter and shaft speed that will yield equivalent results. In general, smaller impellers require higher shaft speeds than larger ones, and the higher shaft speeds are accompanied by higher power requirements for most process results. Therefore, the designer may look into several equivalent designs, which will differ in capital cost, power cost, and mechanical feasibility. This will allow a degree of optimization, which will be covered in later steps.

The calculation steps depend on the problem classification. For this book, almost all problems are classified as liquid motion, which will be covered in Chapter 5. Some special notes will be given for liquid–liquid dispersions as used in biodiesel production in Chapter 9, and for lignocellulosic applications in Chapters 12 and 13.

Though not all gear drives match the American Gear Manufacturer's Association (AGMA) standard speeds, they are a good place to start, prior to engaging in detailed mechanical design. The speeds relevant for agitator design, in rpm, are 5, 6, 7.5, 9, 11, 13.5, 16.5, 20, 25, 30, 45, 56, 68, 84, 100, 125, 155, 190, 230, 280, and 350 rpm. Lab units may have considerably higher speeds than this range.

## Basic Mechanical Design

Mechanical design may be done in two stages. The first stage is done to determine such things as gear drive selection and shaft design. Detailed mechanical design will be needed to actually manufacture the equipment, and this will be done by the agitator manufacturer. But for basic design, we will calculate shaft diameter based on stress, estimate impeller weight and then check the shaft for critical speed issues, changing the shaft diameter if necessary. With this information, one can determine feasibility and estimate cost with the help of the agitator supplier. More information on mechanical design will be found in Chapter 15.

## Eliminate Unfeasible Designs

There are a number of reasons why a design may not be feasible, or only marginally feasible. For example, although larger impellers and slower shaft speeds are often desirable to save power cost, years of research and experimentation have revealed that a D/T ratio greater than 1 is not likely to be successful. ☺

On a more serious note, if the D/T ratio is too large, poor flow patterns, such as flow inversion, can result.

Similarly, a design can be unfeasible for mechanical reasons. For example, if the shaft design requires that $d > D$, it clearly is not feasible. In general, the shaft should not be more than 50% of the impeller diameter, even if hollow shafts are used. Fortunately, this is rarely a problem.

In the feasible but not preferred category, we may have some issues with gear drive selection. For example, high power/rpm selections may exceed the thermal limit of a gearbox, and require gear drive cooling. This adds capital and maintenance costs and should be avoided if possible.

## Estimate Capital Costs

This step will normally have to be done by the agitator manufacturer unless the user has a database of similar units that he can use to estimate cost. Regardless of who does it, it is an important step in the overall optimization.

## Estimate Present Worth of Power Cost

Most agitators are powered by electricity. Electrical costs vary by region. The idea here is to calculate the present worth of electric power cost over the planned useful life of the equipment or facility. Essentially, it is how much money you need right now to pay for the total power usage over the life of the equipment. To do this, one needs to know the power cost, the number of years the equipment will operate, the duty cycle (hours per year), the actual power consumed, and the interest rate for the time value of money.

Two popular ways for determining this are shown below: the discreet method, and the continuous compounding method.

Equation (2.1) discreet method:

$$PW = R\left(\left(1+i\right)^{n}-1\right)/\left(i\left(1+i\right)^{n}\right), \tag{2.1}$$

where PW is the present worth, R is the payment per period, i is the interest rate per period and n is the number of periods.

Equation (2.2) continuous method:

$$PW = R_T\left(1/rn\right)\left(1-e^{-rn}\right), \tag{2.2}$$

where $R_T$ is the total payment over the life of the equipment, n is the number of interest periods (usually in years) and r is the nominal interest rate per period.

The value for R is based on the cost per kW-h and the hours per period operated. For example, if the motor output is 10 kW and the period is one month, the hours per month operated is 684 and the cost per kW-h is $0.10 (USD), R is 684 * 10 * 0.1 = $684/month.

Likewise, if the same motor were operated 684 hours per month for an evaluation period of 15 years, $R_T$ would have a value of 684 * 12 * 10 * 0.1 * 15 = $123 120, with no adjustment for inflation. (We can include inflation effects in the interest rate.)

A useful website for doing these calculations is https://www.ajdesigner.com/phppresentworth/present_worth_value_equation.php.

The above website was accessed by the author on 14 November 2021.

If we use a nominal interest rate of 1% per month; APR of 12.6%, and a 15 year evaluation period, the above website gives us a PW of $56 992 using the discrete method on a monthly basis, and we would get $55 301 using the continuous method calculating by hand. It should be obvious that it may be worthwhile to consider paying a little more for a design that saves power cost. In my experience, vendors will usually quote, and customers will usually buy, the design with the lowest capital cost. However, in many cases, a design that cuts the required power in half will only cost about 15% more. For the above example, that would mean cutting the power draw from 10 to 5 kW would save a present worth cost in excess of $25 000, which is far more than the capital cost difference.

## Choose Lowest Combined Cost

If there are no constraints prohibiting doing so, choose the design among the remaining feasible designs that has the lowest combined capital and present worth of operating cost. We have discussed the need to calculate the capital cost of the agitator and the present worth of power cost. Equally important is the present worth of such operating costs as maintenance and downtime. Unfortunately, these figures are difficult to obtain. They will vary by design and equipment vendor. Past experience may provide some guidance. A lot of this is part of the vendor selection process, described in Chapter 17.

## Repeat for Different Aspect Ratios

All of the previous steps were within the confines of the starting aspect ratio. So, up to this point, hopefully, we have an optimum design for that ratio. However, that ratio may not be optimum overall. So, ideally the entire process should be repeated over the range of aspect ratios that are not constrained by other factors, such as site restrictions, shop fabricated vs. field fabricated issues, etc. Only by doing this will we find the economically optimum design. The capex and opex of the agitator, vessel, and ancillary equipment should ideally be included.

## Finish

When all of the steps are completed as many times as it takes to get the final optimum, the capex and opex per unit of capacity will be optimized. As you may have surmised, that is a lot of work. However, the savings could be quite significant. Moreover, a very experienced agitator designer can quickly go through the optimization for a given set of conditions and aspect ratio by instinctively avoiding designs that his experience indicates are poor or infeasible.

The balance of the book provides background information and details needed to complete these steps to the degree possible. Outside resources will be needed for cost data. The individual chapters are not organized as extensions of the step-by-step procedure, but, rather, as sources of information and calculation methods, as well as providing enough fundamental understanding to use the procedures described herein.

## Summary of Chapter

This chapter has presented a series of steps to arrive at optimum design and operation. All of these steps will be covered in this book, in varying degrees of detail. The book will not follow this logic chapter by chapter, as a lot of background information and principles must be established before optimization can begin. The next several chapters will do that. A sizeable portion of the book is devoted to industry-specific application guidelines, within which optimization can be applied.

## List of Symbols

| | |
|---|---|
| $C_P$ | Heat capacity at constant pressure |
| d | Shaft diameter |
| D | Impeller diameter |
| i | Interest rate per period |
| k | Thermal conductivity |
| n | Number of periods in interest calculation |
| r | Nominal interest rate per period (usually one year.) |
| R | Power cost per period |
| $R_T$ | Total power cost over the life of the equipment |
| T | Tank diameter |
| Z | Liquid height |

## Reference

**1** Hicks, R.W., Morton, J.R., and Fenic, J.G. (1976). How to design agitators for desired process response. *Chemical Engineering Magazine* 22–30.

# 3

# Agitator Design Basics

Before delving into details of agitation specific to biofuels and renewable chemicals, we must establish a common framework of terminology and principles common to all agitation systems. This chapter will cover basic terminology, how experimental data are usually correlated, and some basic viscosity models commonly used in agitator design, including those that work well with lignocellulosic slurries.

## Agitated Tank Terminology

A very simplified view of an agitated tank may be found in Figure 3.1. Though simplified, all of the main elements of an agitator and tank may be seen there. A schematic view with components labeled and a few major nomenclature symbols may be found in Figure 3.2.

An agitated tank consists of a number of elements and is dimensionally described by a number of symbols. We will go through these more or less in the order of power flow, referring to the nomenclature of Figure 3.2.

## Prime Mover

Motive power is provided to an agitator by means of a prime mover, which provides power in a rotary fashion. Usually, it is an electric motor, as shown in Figure 3.2. For some applications, a variable speed drive is often provided, usually by means of a variable frequency drive (VFD), though other technologies are possible. Variable speed can save total power costs when the agitation requirements are not constant, or to fine tune a process that may not be well understood initially. In principle, many other rotary power sources could be used. Some that

*Agitator Design Technology for Biofuels and Renewable Chemicals*, First Edition. Gregory T. Benz.
© 2022 John Wiley & Sons, Inc. Published 2022 by John Wiley & Sons, Inc.

**Figure 3.1** Agitated tank. *Source:* Photo courtesy Chemineer, a brand of NOV. Permission granted by NOV.

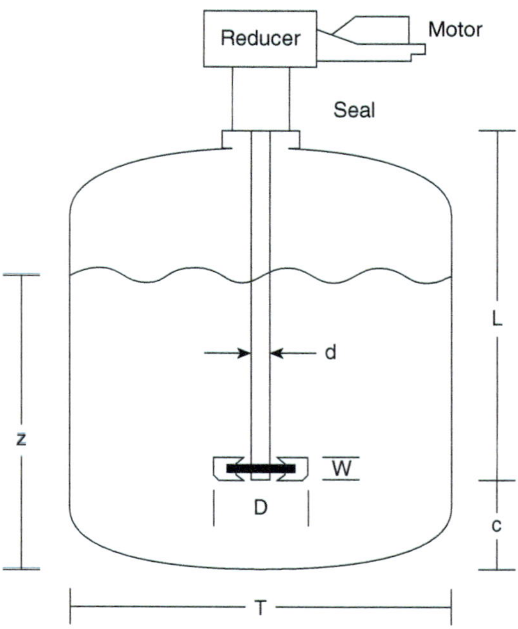


**Figure 3.2** Agitated tank sketch.

the author has seen used include air motors, DC motors, hydraulic motors, and even diesel engines. But, probably more than 99% of the time, the prime mover will be an electric motor.

## Reducer

Most agitator designs do not operate at direct motor speed, except in very small tanks. The reducer decreases the shaft speed to a value less than motor speed and increases torque. In most agitator designs, the reducer must also support the weight of the shaft and impellers, the thrust due to tank pressure or vacuum, and the bending moment created by random fluid forces acting on the impellers. In some cases, those forces are supported by a separate set of bearings, and the shaft is flexibly coupled to the reducer.

The two most common reducer designs in industry are belt drive and gear drive. Most modern top-entering agitators use gear drives. More discussion of drive types will be found in Chapter 15.

## Shaft Seal

Although not all agitators have shaft seals (some are mounted on open-top tanks or basins), those used in biofuel production almost always do. The purpose of the seal, in addition to maintaining tank pressure or vacuum, is to isolate tank contents from the outside environment. This may be done to keep foreign matter from contaminating the tank contents or to protect plant personnel from exposure to potentially harmful organisms or gases. The shaft seal area may be heated to create a sterile barrier. Most tanks used in the biofuels industry use simple packing for a shaft seal. Vessels under pressure greater than 1 bar-g usually use some kind of mechanical seal, though it is possible to use packing at higher pressures than that.

## Wetted Parts

The power and torque from the reducer are transmitted to the tank contents by means of a shaft with diameter d, extending a distance L from the mounting flange. On the shaft are mounted one or more impellers of diameter D and actual blade width or height, w, located off the bottom at a distance C. For this book, we measure C from the bottom edge of the impeller. Some other sources use the centerline of the impeller. We also define D as the flat-to-flat dimension of the blades in plan view, rather than the swept circle, called $D_S$. This makes a difference of

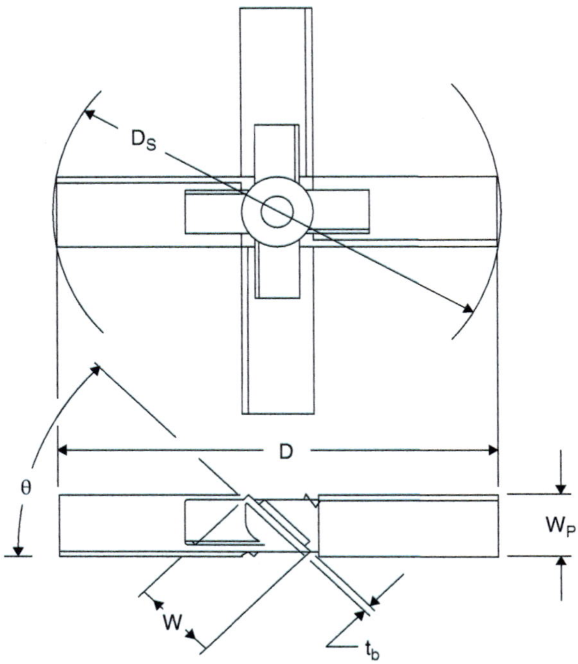


**Figure 3.3** Swept diameter.

about 1% in the calculated diameter for a pitched blade turbine of standard design, for example. This is illustrated in Figure 3.3, along with a few other relevant dimensions, such as the blade thickness, $t_b$.

For multiple impellers, we would use subscripts such as $D_1$, $D_2$, $C_1$, $C_2$, etc.

## Tank Dimensions

The tank diameter is designated as T. The liquid level is designated as Z. Other tank dimensions, not shown on the sketch, could include head depths, straight side, nozzle projections, baffles (width, length, and offset from wall), and any relevant internals.

## How Agitation Parameters Are Calculated

Agitation systems, just as any other system producing or modifying fluid flow, must obey the laws of physics. In terms of mathematical models, they obey the equations of continuity and the Navier–Stokes equations. Unfortunately, those equations can usually only be used to solve problems analytically in relatively simple geometries, such as flow in a pipe, and, often, only in laminar flow. Such equations can be supplemented by various turbulence models, which give approximate solutions.

An agitated tank, however, is a very complex geometry. Most would agree that it is all but impossible to solve the equations of motion for an agitated tank by analytical methods. In modern times, there have been many successful attempts to model agitated tanks by using numerical methods, which in essence convert differential equations into a series of algebraic approximations. Those approximations can be very good, depending on the skill of the modeler and the computational power used. These methods are often called CFD (computational fluid dynamics) and sometimes called CFM (computational fluid mixing). Chapter 14 describes some of the uses of CFD as applied to agitator design.

The traditional way of solving agitation problems is quite different. The approach that has been used in most studies, and which is still the staple of agitator design, is to use the equations of motion to derive dimensionless number groups, and then correlate experimental data in terms of those dimensionless numbers. That is the approach we will take for the majority of this book.

We will not show the derivation of the dimensionless numbers in this book, but will describe the ones important for our use in designing agitators, and how they are used, especially for the applications described in this book.

Some readers may be unfamiliar with the concept of dimensionless numbers, so we will give a brief description here, prior to getting into the most commonly used ones.

A dimensionless number is a ratio of quantities such that the dimensions and units in the numerator exactly match the dimensions and units in the denominator, thereby canceling all dimensions and units. The resulting dimensionless number has no units or dimensions; it is just a scalar number. It also does not depend on what units are used, though converting dissimilar units to a consistent set of units will assist with the math.

A rather trivial example is the concept of the aspect ratio of a cylinder, which equals its height or length divided by its diameter. A 5-ft. tall cylinder with a 12″ diameter has an aspect ratio of 5. That is because 5 ft. is five times as much, in terms of its dimension (length), as 12″. But the math would be more obvious and less prone to error if we first converted the diameter to feet by dividing by 12, or, alternatively, converting the height of the cylinder to inches by multiplying by 12. Obviously, the important point is that it is the ratio of the actual physical dimensions, and is not unit dependent. We could have stated the dimensions as meters, microns, furlongs, or cubits; the dimensionless number we are calling aspect ratio would still be 5.

In the next several sections, we will cover the major dimensionless numbers used in agitator design, and then we will show how some of them are used.

## Reynolds Number

The Reynolds number is the most widely used number in fluid agitation and most other fluid flow calculations. Many other dimensionless numbers are functions of it, as we will see in several subsequent chapters. Conceptually, the Reynolds number represents a ratio of inertial forces to viscous forces. When the Reynolds number is high, inertial forces dominate. This is the turbulent flow regime. When the Reynolds number is low, viscous forces dominate. This is the laminar regime. Intermediate Reynolds numbers constitute the transition flow regime and may exhibit attributes of both turbulent and laminar flow. In fact, in an agitated tank, there can be regions of laminar flow and regions of turbulent flow in the same tank when operating in the transition flow range of Reynolds numbers. Generally, in transition flow, the area near impellers is turbulent and the area near the walls is laminar.

Mathematically, the Reynolds number is the product of a reference dimension times a reference velocity times the fluid density, divided by the fluid viscosity. The chosen reference dimension and velocity depend on the system being studied. For example, for a pipe, the typical reference dimension is the pipe inside diameter, and the reference velocity is the bulk velocity in the pipe.

For an agitated tank, it is customary to use the impeller diameter, D, as the reference length dimension. Likewise, it is customary to use a form of the impeller tip speed, $\pi$ND, as the reference velocity. However, to avoid building a constant into a dimensionless number, $\pi$ is dropped, so ND is used for velocity. The resulting expression is:

$$N_{Re} = D^2 N \rho / \mu \tag{3.1}$$

Because all units must cancel, it is best to use a consistent set of units. SI (Systeme Internationale) units work well. Normally, this is no problem, except for viscosity. Most of the time, viscosity is stated as cP (centipoise). However, to use SI units, viscosity should be in units of kg/m-s, also known as Pa-s. Fortunately, the conversion is simple. 1 cP = 1 mPa-s = 0.001 Pa-s = 0.001 kg/m-s. English units become very peculiar. If the lengths are in feet and the density is in lb/ft.$^3$, with time in seconds, the viscosity unit to use is pound mass/foot-second. The conversion from cP is 1 cP = 6.72E-4 pound mass/foot-second. (I have yet to see a viscometer that reads in units of pound mass/foot-second.)

For those who prefer to use inches, rpm, specific gravity and cP, we can build in a conversion factor using those units:

$$N_{Re} = 10.7 * D^2 * N * sg / \mu \tag{3.2}$$

Caution: do not use Eq. (3.2) with any other than the prescribed units!

## Power Number

Power number is conceptually the ratio of power draw to impeller diameter, speed, and density, raised to appropriately derived exponents. It is defined as:

$$N_P = P / \rho N^3 D^5 \tag{3.3}$$

The main use of power number is to calculate power draw. It is a function of impeller type, Reynolds number, and various geometric factors. When using SI units, the power will be expressed in watts. If one chooses to use inches, rpm, Hp, and specific gravity, we can insert the conversion factor and re-arrange for power:

$$P(Hp) = 6.555E - 14 * N_P * sg * N^3 * D^5 \tag{3.4}$$

Same caution as for Eq. (3.2); one must use the above English units.

Note that power draw is quite sensitive to both impeller diameter and shaft speed.

## Pumping Number

Pumping number is the ratio of impeller discharge rate (aka primary pumping capacity) to the cube of its diameter and the shaft speed:

$$N_Q = Q / ND^3 \tag{3.5}$$

It is used to calculate the flow created by the impeller, which can be used to determine a characteristic velocity in the tank. It is a function of impeller type, Reynolds number, and geometric parameters.

The units used for N and D determine the resultant units for Q. For example, if D is in feet, and N is in rpm, Q will be expressed as cubic feet per minute. Likewise, if D is in m and N is in revolutions per second, Q will be in cubic meters per second.

There are many different ways to measure impeller pumping, and they do not all give the same results. The most widely accepted methods define a discharge area around the impeller and measure flow through it, usually by use of either laser-Doppler (LDV) or digital particle image velocimeters (DPIV). It is not within the scope of this book to discuss such methods. The above methods measure what is commonly called primary impeller flow or discharge. There can also be entrained flow, which can be several times as high as the primary flow. Some methods, such as hot-wire anemometry, give higher pumping numbers than LDV or DPIV, probably by including at least some of the effects of entrained flow. Any claimed impeller pumping capacity should state whether it is primary flow or total flow. And any vendor-claimed pumping number substantially higher than those listed in the appendices to Chapter 4 should be challenged.

## Dimensionless Blend Time

Dimensionless blend time is the product of blend time (defined as the time to reach some degree of concentration variance reduction after an assigned starting time) times the shaft speed. In other words:

$$N_B = \tau * N \tag{3.6}$$

This group is used to determine the blend time to some degree of attenuation of concentration differences. It is the product of blend time and shaft speed. Essentially, it is how many revolutions of the impeller are required to achieve a certain degree of blending. It is a function of Reynolds number, impeller type, and geometric factors. Rarely is blend time a limiting factor in common agitator applications. The exception is highly viscous fluids.

Sometimes, people will add a factor of $(D/T)^{\alpha}$ to the right hand side of the equation to correct for geometric effects. The value of $\alpha$ depends on the impeller type, but is usually about 2.3 for pitched blade and straight blade turbines, and about 1.73 for propellers and hydrofoils.

## Aeration Number

Aeration number, also called gas flow number, is the ratio of actual gas flow rate at the impeller (corrected for absolute pressure and temperature) divided by the impeller diameter cubed and the shaft speed:

$$N_A = Q_g / ND^3 \tag{3.7}$$

This group is used for power draw calculations in the gassed condition, along with other dimensionless groups. It can be thought of, in a way, as being proportional to the ratio of gas flow rate to the impeller pumping capacity. We will not be using this number in this book, but it is used for aerobic fermenter design.

## Gassing Factor

When gas is introduced into an impeller or is present in the tank, it affects the impeller power draw. Usually, it reduces the power draw, but under some circumstances, it may increase it. The ratio of gassed power to ungassed power is called the gassing factor, and it does not usually have a special symbol for it. Instead, the ratio is simply expressed as a ratio: $P_g/P_u$.

It is a function of impeller type, Reynolds number, Aeration number, Froude number, and geometric factors. It is used mainly for aerobic fermenter design, which will not be covered in this book. See Ref. [1] in the Preface of this book for information on a book about agitator design for fermenters and bioreactors.

## Nusselt Number

The Nusselt number is the ratio of the convective heat transfer coefficient times a reference length divided by the thermal conductivity of the fluid. The reference length dimension depends on the heat exchange surface. For a vessel jacket, that will normally be the tank diameter. For tubes inside the tank, it would normally be the tube OD. Expressed in terms of tank diameter, we have:

$$N_{Nu} = hT / k \tag{3.8}$$

It is used in heat transfer calculations, to determine the convective coefficient, h. It is a function of impeller type, Reynolds number, Prandtl number, and various geometric factors, as well as the ratio of local viscosity to bulk viscosity.

## Froude Number

Froude number is the ratio of a reference length times a reference velocity divided by the local gravitational acceleration. For similar reasons that we used for Reynolds number, we use D for length and ND for velocity (eliminating $\pi$), resulting in:

$$N_{Fr} = N^2 D / g \tag{3.9}$$

Note that g is not a constant; it is the local gravitational acceleration. While it is almost uniform on the Earth's surface, it will be quite different on other planets. So the same impeller operating at the same speed would have different Froude numbers on Earth, the moon, Mars, and Jupiter. This will be important if we ever install agitators on another planet.

Conceptually, the Froude number is the ratio of inertial forces to gravitational forces. High Froude numbers mean inertia dominates. This is associated with a choppy surface and vortex formation. Low Froude numbers mean gravity dominates, which is associated with a quiet, flat surface. Froude number, in combination with Reynolds number, impeller type, baffle number, and various geometric factors, can be used to predict mean vortex depth.

In gas–liquid contacting, the impeller gassing factor is a function of Froude number, impeller type, Reynolds number, Aeration number, and geometry factors. That is its principal use for bioreactor design. It is also used in most impeller flooding correlations.

## Prandtl Number

Prandtl number is simply a physical property group used in heat transfer correlations:

$$N_{Pr} = C_P \mu / k \tag{3.10}$$

The Nusselt number is a function of the Prandtl number. Care must be taken to make sure all units cancel. The most common units for the fluid properties, especially viscosity, do not cancel. SI units work well. If common English units are used (BTU, h, ft.), the viscosity must be converted to pounds mass/foot-hour. The conversion is 1 cP = 2.419 lb/ft.-h.

## Geometric Ratios

Relative size and placement of impellers in a tank affect power draw, blend time, pumping capacity, solids suspension capability, vortex formation and maybe a few other things. Some parameters are more sensitive to geometry than others. The dimensionless geometric ratios used in calculations include D/T, Z/T, C/T, S/D, O/T, and W/D. As long as the same units are used for numerator and denominator, it does not matter what system of units is used.

## Baffle Number

Although this could be lumped in under geometric ratios, it affects things in different and important ways, so I decided to mention it separately. Essentially, it represents total baffle width (calculated normal to the tank wall) divided by tank diameter:

$$n_b = \#\text{of baffles} * w_b / T \tag{3.11}$$

Power draw and vortex formation depend on baffle number. At a baffle number of zero, the tank is called unbaffled and power draw is at a minimum. Under such a condition, in turbulent flow, there is a lot of swirl, with mostly tangential motion rather than axial or radial. Gas dispersion is essentially impossible to do under such a condition. Likewise, solids suspension is poor. As baffle number increases, power draw increases to a maximum, and then falls off. In the United States, "standard" baffling is 4 baffles at 90° to the tank wall, each one being 1/12 of the tank diameter, resulting in a baffle number of 0.33. In Europe, it is more common to use a baffle width of 1/10 of the tank diameter, giving a baffle number of 0.4. The power draw is pretty constant over this range and is basically at the maximum. D/T and impeller type also interact with baffle number.

Use of standard baffling is recommended for most applications. However, as we will see in Chapters 8 and 13, it is necessary to use an unbaffled tank for high solids lignocellulosic slurries prior to a significant degree of hydrolysis.

## Dimensionless Hydraulic Force

When an impeller operates in turbulent flow, the loads on each blade are fluctuating about a mean. To better visualize this, imagine riding along in a car with your hand outside the window. You will note that your hand is buffeted about, with a highly variable force, rather than a smooth, steady one. This is due to turbulence, involving various shedding of vortices, etc. The same thing happens to an impeller in turbulent flow. The load on each blade varies with time, and the loads on each blade are not synchronous with each other. The result is that there will be a fluctuating net side load on the impeller, creating a bending load on the shaft. This load, sometimes called an imbalanced hydraulic force, is not due to mechanical imbalance or any lack of manufacturing precision. It is entirely due to the nature of turbulence. Therefore, making a geometrically perfect impeller with perfect mechanical balance would have no effect on this force.

The dimensionless hydraulic force equals the product of hydraulic force times impeller diameter divided by impeller torque:

$$F_d = F_h * D / T_q \tag{3.12}$$

Normally the peak value of this ratio is measured and correlated, so that a peak value of hydraulic force may be predicted and used in shaft and impeller design. This group is a function of Reynolds number and impeller type, as well as geometric ratios, Aeration number, and Froude number, and direction of pumping for axial flow impellers. Theoretically, it will have a value of zero in laminar flow, as there is no turbulence.

## Thrust Number

This relates impeller thrust to impeller parameters and density:

$$N_T = F_T / \rho N^2 D^4 \tag{3.13}$$

It is primarily used to determine impeller thrust for mechanical purposes. It also is used to predict cavern size for shear-thinning fluids, as will be discussed in Chapter 12. Qualitatively, it is a constant in laminar and turbulent flow, but goes through a minimum in transition flow.

## Typical Dimensionless Number Curves

Figure 3.4 represents a typical log/log plot of power number as a function of Reynolds number, with D/T as a parameter. (The D/T effect is primarily noticeable on axial impellers. Radial impellers show much less dependency on D/T, as we will see in Chapter 4.)

Examining this curve, we can observe several things. One is that the power number becomes constant for a given D/T under turbulent conditions, i.e., at high Reynolds numbers (typically, above 20 000). When a manufacturer of an impeller states its power number, it is normally the turbulent power number at a D/T of 1/3 and a C/T of 1/3. Axial impellers will tend to have a decreasing power number at increasing D/T.

Also note that the curve becomes a 45° angle at laminar flow conditions (typically, $N_{Re} < 10$). That means the product of Reynolds number and power number is constant in that range. A simple derivation reveals that in the laminar range,

$$P \propto \mu N^2 D^3 \tag{3.14}$$

The range between laminar and turbulent is called the transition range. It is a very broad range. That is because turbulence commences first near the impeller blade tips somewhere in the region of $N_{Re} = 10$, but does not reach all of the far corners of the tank until $N_{Re} > 20\,000$. The power number gradually levels out in this range. For some impellers, it goes through a minimum in the transition range.

Figure 3.5 represents a typical pumping number curve, as a function of Reynolds number and having D/T as a parameter.

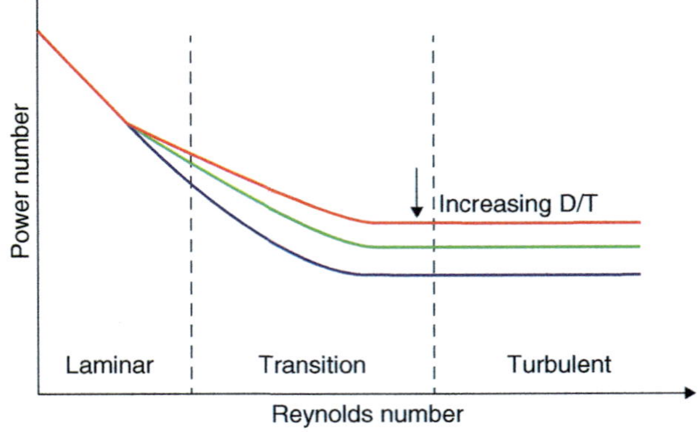


**Figure 3.4** Typical power number curve.

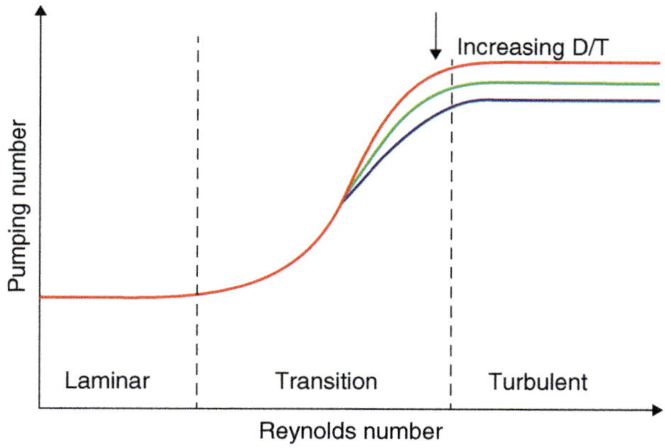


**Figure 3.5** Typical pumping number curve.

In laminar flow, the pumping number becomes a constant. This means that under such conditions, the impeller pumping is independent of viscosity, though the power draw is directly proportional to viscosity.

In turbulent flow, the pumping number again becomes constant at a given D/T, though it is much higher than the laminar pumping number. As D/T increases, the pumping number generally decreases. This is because recirculating flow within the tank creates a velocity opposite the impeller discharge, impeding its flow. Nonetheless, larger impellers pump more for a given power input than smaller ones.

When vendors quote a pumping number, it is usually in turbulent flow and at a D/T of 1/3 and a C/T of 1/3. Many consider this geometry to be "standard," but there is no fundamental reason to adhere to this geometry when designing agitation equipment. In fact, as we will see in Chapter 5, it is usually possible to save energy by using larger impellers than this.

Figure 3.6 represents a typical generic blend time curve.

We can observe that blend time becomes constant in both laminar and turbulent flow. However, the laminar dimensionless blend time is often several orders of magnitude greater than the turbulent blend time. Specialized impellers have been developed for laminar flow mixing. We need not delve into these in this book, as most applications will be in a turbulent flow. Cellulosic hydrolysis will often be in laminar flow, but high viscosity impellers do not work well for such applications, for reasons we will describe in Chapter 13.

Some versions of the dimensionless blend time curve incorporate the D/T effect into the Y-axis expression, as stated under the dimensionless blend time definition given earlier.

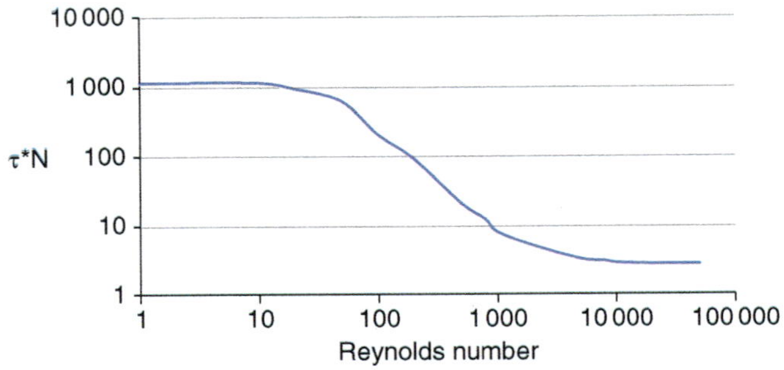


**Figure 3.6** Dimensionless blend time.

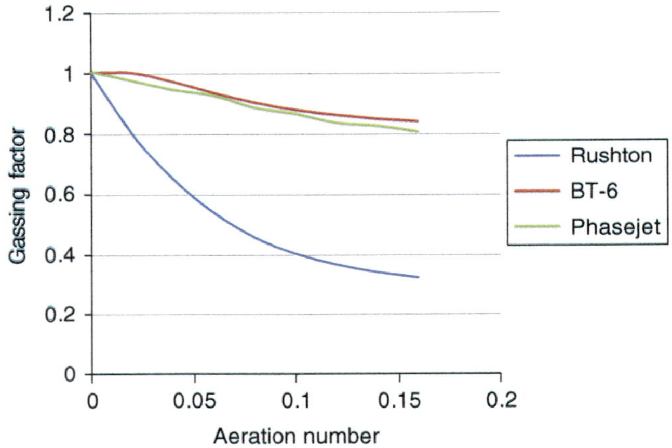


**Figure 3.7** Gassing factors.

Figure 3.7 depicts gassing factor as a function of Aeration number for three different impeller types.

Gassing factor depends on Aeration number, Froude number, D/T, Reynolds number, and impeller type. Therefore it is impossible to show the entire relationship on a simple two dimensional graph. Figure 3.7 is based on turbulent flow, a D/T of 1/3, and an Aeration number of 0.5.

We can readily see that the traditional impeller for gas dispersion, the Rushton (or disk) turbine, has a rather severe drop in power at high gas flow rates. That is why it has been largely replaced by modern, deeply concave impellers such as the BT-6 by Chemineer, the Phasejet by Ekato, and several others that have less effect of gas flow on power draw.

We will cover other dimensionless parameters as needed in subsequent chapters. For now, we will show some example calculations.

**Example 3.1** Power Draw Calculation
A tank has an impeller of 1000 mm diameter, rotating at a shaft speed of 125 rpm, in a fluid with a specific gravity of 1.2. It has a known power number of 0.75. How much power will it draw?

**Answer**
It is best to first convert all parameters to SI units. So, we have a 1 m impeller diameter, turning at a shaft speed of 125 rpm/60, or 2.0833/second, in a fluid with a density of 1200 kg/m$^3$. Since the power number is dimensionless, it remains at 0.75. The definition of power number, from Eq. (3.3), is:

$$N_P = P/\rho N^3 D^5 \tag{3.15}$$

Re-arranging to calculate power:

$$\begin{aligned} P &= N_P \rho N^3 D^5 = 0.75 * \left(1200\ \text{kg/m}^3\right)\left(2.0833/\text{second}\right)^3 \left(1\ \text{m}\right)^5 \\ &= 8138\ \text{kg} - \text{m}^2/\text{s}^3 = 8138\ \text{W} \end{aligned} \tag{3.16}$$

Since it is more common to think of power in kW, this would become 8.14 kW. An 11 kW motor (a standard motor rating) would be ample to power this impeller under these conditions.

**Example 3.2** Pumping Calculation
A tank has a hydrofoil impeller of 6 ft. diameter rotating at 30 rpm. It has a known pumping number of 0.5. How much fluid will it pump?

**Answer**
We do not have to do a unit conversion for this one. The units used merely determine the resultant pumping units. Equation (3.5) defines pumping number, and is reproduced next:

$$N_Q = Q/ND^3 \tag{3.17}$$

Re-arranging to calculate Q (pumping rate), we get:

$$Q = N_Q N D^3 = 0.5 * 30/\text{minute} * \left(6\ \text{ft.}\right)^3 = 3240\ \text{ft.}^3/\text{min} \tag{3.18}$$

Had we used SI units, the calculated pumping capacity would have been expressed as m$^3$/s, but the physical quantity represented would have been the same.

**Example 3.3** Blend Time Calculation
This really only gets interesting in transition flow, where dimensionless blend time varies with Reynolds number. But let's suppose we have an impeller with a dimensionless blend time of 10. How quickly would it blend the tank while operating at 30 rpm?

**Answer**

If $\tau * N = 10$, $\tau = 10/N = 10/30(1/\text{minute}) = 1/3\,\text{minute} = 20\,\text{seconds}$. Note that one can use any units for shaft speed; the resultant blend time units will be determined by this. Note also that as long as the dimensionless blend time is fixed, the result is independent of tank size. This means, for a given D/T and Reynolds number, the blend time at 30 rpm will be the same in any size tank. However, the P/V goes up exponentially when trying to keep the same D/T and the same shaft speed. So, designing for the same blend time in a large tank as in a small one can be problematic.

## A Primer on Rheology

It is not the purpose of this book to be a treatise on rheology. Indeed, many books have been written on the subject. We will get into more detail on rheology specifically for cellulosic applications in Chapter 12.

Therefore, we will focus on a few simple models that do an adequate job of describing the behavior of fluids encountered in the biofuel and renewable chemical industries. We will look at shear stress versus shear rate, and assume that normal (i.e., at right angles to the shear plane) stresses, such as found in viscoelastic fluids, do not apply. We will also confine ourselves to time-independent models because the fluids in our applications are always under shear and in a quasi-steady state.

## Newtonian Model

Newtonian fluids are those which have a shear stress that is directly proportional to shear rate (also called velocity gradient). Mathematically, this can be represented as:

$$\sigma_s = \mu * \gamma \tag{3.19}$$

where $\mu$, the coefficient on shear rate, is also known as viscosity. This can be arranged to:

$$\mu = \sigma_s / \gamma \tag{3.20}$$

Most low viscosity fluids (<50 cP) are Newtonian. Intermediate and even some high viscosity fluids can be Newtonian if they are true solutions, especially if they are solutions of small molecules. For example, honey and corn syrups are often Newtonian even at 30 000 cP. Some silicone solutions are Newtonian even at 20 000 000 cP, but this is quite rare. Most suspensions of more than 5–10% solids exhibit at least some non-Newtonian behavior. High solids slurries can be very non-Newtonian.

Low-solids slurries operating in a water-like fluid are usually Newtonian. High solids slurries may not be, and those containing both an aqueous and an oil phase usually are not. But the overall viscosity for most biosolids slurries is less than 1000 cP. Once again, the exception is fibrous materials.

## Pseudoplastic or Shear Thinning, Model (Aka Power Law Fluid)

These fluids are called shear thinning because the slope of the shear stress vs. shear rate curve decreases as the shear rate increases. Therefore the apparent viscosity decreases as shear increases. This behavior may often be well-described by a power-law model:

$$\sigma_s = M * \gamma^n \tag{3.21}$$

The exponent n is called the power-law exponent. When it has a value of 1, the equation reverts to Newtonian. When it is less than 1, it is shear thinning. While it can be greater than 1, which would mean a shear thickening behavior, this is not found in biofuel applications. Starch slurry above 40% solids can exhibit shear thickening behavior, but this is not a normal situation.

The apparent viscosity can be expressed by dividing by the shear rate on both sides of the equation:

$$\mu_a = M * \gamma^{(n-1)} \tag{3.22}$$

In Eq (3.22), M is the viscosity coefficient, or the viscosity at a shear rate of 1 (usually in units of 1/s).

Many biosolids slurries exhibit this behavior at high percent solids (typically above 20%; above 2 or 3% for fibrous materials.)

## Bingham Plastic

These fluids exhibit a yield stress. That is, until the fluid is exposed to a certain minimum shear stress, it will not move. Examples include some high solids slurries, toothpaste, lignocellulosic slurries, and any fluid that can hold some of its shape after being disturbed. Above the yield stress, Bingham fluids exhibit a constant slope. Mathematically:

$$\sigma_s = \sigma_y + M * \gamma \tag{3.23}$$

and

$$\mu_a = \sigma_y / \gamma + M \tag{3.24}$$

## Herschel–Bulkley

These fluids combine a yield stress with shear thinning behavior above the yield stress:

$$\sigma_s = \sigma_y + M * \gamma^n \tag{3.25}$$

and

$$\mu_a = \sigma_y / \gamma + M * \gamma^{(n-1)} \tag{3.26}$$

One may recognize that all three previous models are special cases of the Herschel–Bulkley model, making this the most versatile model commonly in use.

These four models are described visually in Figures 3.8 and 3.9

## Impeller Apparent Viscosity

In laminar and transition flow, the apparent viscosity at the impeller can be evaluated using the average shear rate:

$$\gamma = dv / dx = K_M * N \tag{3.27}$$

where $K_M$ is the Metzner–Otto constant. There are some disagreements about what that constant is for various impellers. Table 3.1 is an average of some of the published numbers.

In turbulent flow, the shear rates may be higher, so the apparent viscosity may be less for most fluids.

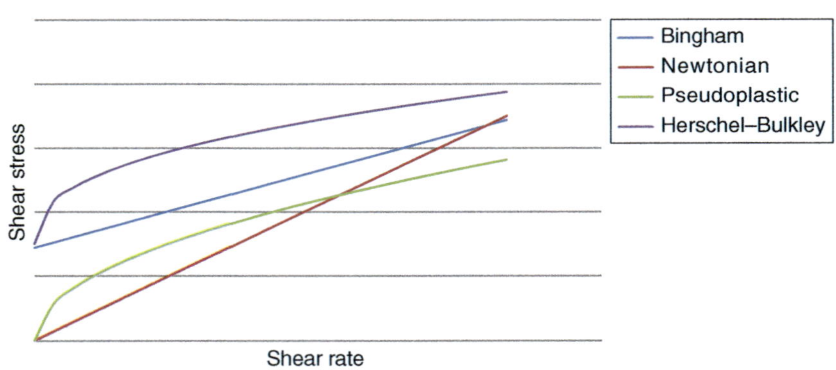


**Figure 3.8** Shear stress curve.

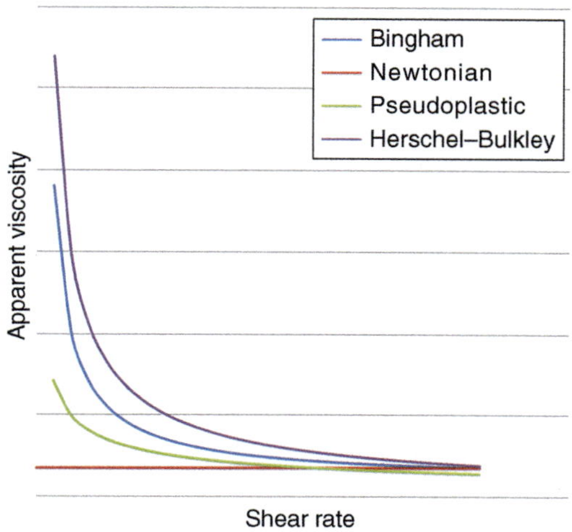


**Figure 3.9** Viscosity curve.

**Table 3.1** Metzner–Otto constants for impellers.

| Impeller type | $K_M$ |
|---|---|
| Propeller | 10 |
| Low solidity hydrofoil | 10 |
| High solidity hydrofoil | 11 |
| Pitched blade turbine | 11 |
| Radial turbines | 11.5 |
| Chaos (reversing pitch) turbines | 40 |

## A Bit of Impeller Physics

Though we will get into impeller details more in Chapter 4, since this chapter is about the fundamentals, we will describe the basic physics of impeller behavior.

Some have said that some impellers invest their power more as flow, and others invest more of it as shear. As we will see next, this is an erroneous description. It incorrectly implies that total power is the sum of power invested in flow plus power invested in shear.

Basically, a rotating impeller moves fluid. Agitation results are side effects of how much fluid is moved and how it moves it. But the macro relationship is very simple: power = flow times head, same as for pumps. Theoretical velocity head is proportional to tip speed squared or discharge velocity squared, though actual head produced depends on many other factors. Head is related to shear as well, as maximum shear depends on tip speed. So, power is more like a product of flow times shear squared, rather than a sum of flow and shear.

Some talk about the ratio of flow to power as an impeller characteristic. This, too, is erroneous, as, for a given impeller size in turbulent flow, flow is directly proportional to shaft speed but power is proportional to shaft speed cubed. Therefore, the ratio of flow to power diminishes as shaft speed increases, even though it is the same impeller. It approaches infinity as shaft speed approaches zero. Likewise, in laminar flow, power is proportional to shaft speed squared. So, there is no attribute of a flow/power ratio for any particular impeller. It depends on the specific application.

So, if we were to want to primarily create flow and minimize power, how would we do it? Let's start with impeller design, based on the flow vs. head relationship. We want to minimize the head term. We will look at several design attributes and the limitations imposed on them.

One attribute is the number of blades. Increasing the number of blades increases the flow, but it also increases the head, as the wake of each blade creates some degree of additivity for the net pressure, because the discharge flow will be moving at a higher velocity as we add blades. So, from a flow-for-power maximization standpoint, the optimum impeller design would have only one blade. But there is a reason why we seldom see one-bladed impellers: we prefer not to break the shaft! The hydraulic imbalance would be huge! Hydraulic forces are still quite large for two-bladed impellers, so they are quite uncommon except in transition flow regimes, and even then, there will usually be several impellers, with the orientation staggered to minimize the bending moment on the shaft. Most "high flow" or hydrofoil impellers have three blades, because this is the minimum that has reasonable hydraulic stability. Using more than three blades improves hydraulic stability somewhat, but at the price of requiring more power for the flow produced at a given impeller diameter.

Another attribute is blade chord angle from the horizontal plane. If the blades were infinitely thin, a zero-degree angle would produce no flow and draw no power, save that created by skin drag on the blade surface. As the angle increases, flow increases and so does power draw, but the power draw increases faster with blade angle than flow does, so a very shallow angle will create the most flow for the power. However, the absolute flow produced may be too small for a given diameter and speed. So, either the diameter would have to be increased or the shaft speed would have to be increased. But we are confined in diameter by the tank, so we cannot always increase the diameter; a D/T value of greater than 1 does not work very well. So, we would have to increase shaft speed in some cases in order to use a very shallow angle. But increasing shaft speed also increases head and power, negating the desired power savings from a low chord angle design. The end result is that hydrofoil impellers available in the marketplace generally have chord angles between 15° and 30°, with 20–25° being most common. The power number range is about 0.2–0.4, with 0.3 being most common. The chord angle must never be greater than the stall angle, which is about 37° in water, or boundary layer separation will occur, greatly increasing power draw. For example, a 45° pitched blade turbine will require almost three times the power as a hydrofoil to create the same flow at the same shaft speed.

A third attribute is blade width or a related measure: impeller solidity (which will be defined more in Chapter 4). As blade width increases, so does pumping, but power also increases due to higher discharge pressure. So, most high-solidity impellers are used for applications that are not just about creating flow.

One of the most important attributes is shaft speed, and this applies to any impeller type. A small impeller rotating at a high speed can create the same flow as a larger impeller rotating at a slower shaft speed. The smaller impeller has a smaller discharge area, hence its mean discharge velocity will be higher, and thus the discharge head will be higher. Therefore, it requires more power to produce the same flow. For flow-controlled applications, significant power savings can be achieved by using larger impellers at slower shaft speed. Sometimes, the capital cost is higher, but the payback time is usually a matter of months.

Of course, not all agitator design problems are flow controlled. For example, gas–liquid dispersion is closer to being turbulent eddy dissipation controlled than anything else. So, in most gas–liquid cases, the power requirements will be governing, and impeller design will focus on other things besides flow.

## Summary of Chapter

This chapter has presented basic material that all must know and understand in order to continue into the details required of general agitator design. It can be referred to as needed while reading the rest of the book. A compendium of useful reference information for agitator design can be found in Ref. [2].

## List of Symbols

| | |
|---|---|
| C | Off bottom clearance |
| $C_P$ | Heat capacity at constant pressure |
| D | Impeller diameter, edge to edge |
| $D_S$ | Swept diameter, or point-to-point diagonal diameter |
| dv/dx | Shear rate |
| $F_d$ | Dimensionless hydraulic force |

$F_h$ Hydraulic force
$F_T$ Impeller axial thrust
g Gravitational acceleration
h Convective heat transfer coefficient
k Thermal conductivity
$K_M$ Metzner–Otto constant
M Viscosity coefficient
N Shaft speed
$N_A$ Aeration number; also called gas flow number
$N_B$ Dimensionless blend time
$N_P$ Power number
$N_Q$ Flow or pumping number
$N_{Re}$ Reynolds number
$N_T$ Thrust number
n Power law exponent
$n_b$ Baffle number
O Impeller offset from tank center line
P Power draw
$P_g$ Gassed power draw
$P_u$ Ungassed power draw
Q Impeller discharge rate; also called pumping capacity
$Q_g$ Actual gas flow at the impeller
sg Specific gravity
S Impeller spacing
T Tank diameter
$T_q$ Impeller torque
$t_b$ Blade thickness
W Actual blade width
$W_P$ Projected blade width
$w_b$ Baffle width, calculated normal to tank wall
v Liquid velocity
Z Liquid height

## Greek Letters

$\theta$ Blade angle from a horizontal plane
$\tau$ Blend time
$\mu$ Dynamic viscosity
$\mu_a$ Apparent viscosity
$\rho$ Fluid density

$\sigma_s$ Shear stress
$\sigma_y$ Yield stress
$\gamma$ Shear rate; dv/dx

## References

Because the descriptions and use of dimensionless numbers in this chapter are found in many references, we chose not to do a line by line reference. Instead, one can find an introduction to these concepts in the references below, among others.

1 Dickey, D.D. and Fenic, J.G. (1976). Dimensional analysis for fluid agitation systems. *Chemical Engineering Magazine* 7–13.
2 Paul, E., Atiemo-Obeng, V., and Kresta, S. (2004). *Handbook of Industrial Mixing*. Wiley InterScience. (Look up by the name of the dimensionless number).

# 4

# Impeller Types and Flow Patterns

There are many kinds of impellers used on agitators. Some have very specific purposes; some are intended for general purposes. In this chapter, we will focus primarily on those classes of impellers that have had considerable utility in biofuels and renewable chemicals.

Some impellers are generic, and some are proprietary. There are always new impellers being developed, so those covered in this chapter will not necessarily include all of them. Our choice of which impellers to include does not necessarily denote an endorsement of a particular manufacturer or product. Our goal, rather, is to explore the different classes of impellers used, as each manufacturer normally has a variety available, and generally have offerings that are similar to those of their competitors. Manufacturers of proprietary impellers often do not publish their performance characteristics, and even when they do, there is no guarantee that the information is accurate for any particular application. Though we will provide estimated performance attributes where possible for some proprietary impellers in this chapter, it is to be understood that the numbers are based on the best information available to the author, but are not guaranteed. It is best to confirm any quantitative measures of agitator performance with the equipment manufacturer.

Since most applications in the biofuels industry operate in turbulent flow conditions, the quantitative impeller attributes defined in this chapter will be for turbulent flow only. The only major exception I am aware of is agitation of high solids lignocellulosic slurries, which often operate in transition or even laminar flow. Therefore, in Chapters 12 and 13, we will supplement the attributes for some impellers in laminar flow.

We will start by classifying impellers by flow pattern, and then provide examples of each with commentary and data. We will provide a summary chart of performance characteristics by type, except for radial impellers, which are covered in depth in Ref. [1].

*Agitator Design Technology for Biofuels and Renewable Chemicals*, First Edition. Gregory T. Benz.
© 2022 John Wiley & Sons, Inc. Published 2022 by John Wiley & Sons, Inc.

## Impeller Flow Patterns

Generally speaking, turbine impeller flow patterns can be divided into three types: axial, radial, and mixed flow. A fourth flow pattern, tangential, does not exist to any significant degree in a fully baffled tank in turbulent flow. Nor does it exist in an unbaffled tank in laminar flow. Since good design practice dictates the use of baffles in turbulent flow and they are not needed in laminar flow, we do not expect to have a significant degree of tangential flow in any properly designed system. Therefore we will not discuss tangential flow patterns further.

### Axial Flow

An axial flow pattern is one in which the predominant flow direction is parallel to the shaft, at least in the region of the impeller. Obviously, it must turn around and change directions at the tank bottom and liquid surface. The flow direction can be either up or down. A stylized and simplified depiction of an axial flow pattern is found in Figure 4.1.

An axial flow pattern can be produced by impellers with varying degrees of solidity. By solidity, we mean the fraction of a circle that just envelops the impeller in plan view that is covered or occluded by the impeller. This concept is illustrated in Figure 4.2, showing what are typically described as low solidity and high solidity impellers in plan view, with a dotted circle drawn around them which just envelops the impeller swept diameter.

Although in principle impellers can be built with a solidity ranging from nearly zero all the way up to 100% (with blades overlapping each other enough to cover

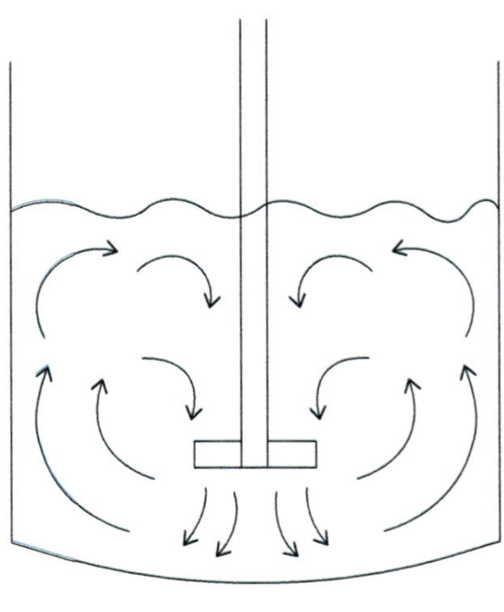

**Figure 4.1** Axial flow pattern.

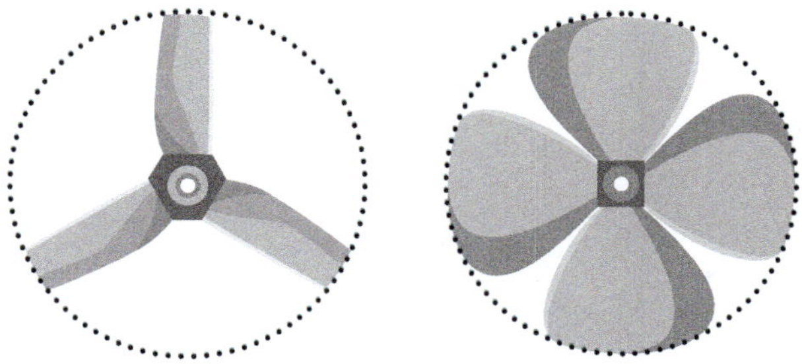


**Figure 4.2** Solidity.

the circle), and anywhere in between, in practice, most impellers on the market are either low solidity, with a solidity value of 10–20%, or high solidity, with a solidity value of 70–100%. This is probably because there are few applications that would benefit from a medium solidity impeller. Flow velocity controlled applications (which includes virtually all nonaerated biofuel applications) will usually use low solidity, as that produces less head and therefore lower power draw for a given amount of flow. High solidity impellers are used where some pressure or a gradual transition of flow and pressure is needed, such as in liquids near the boiling point, some solids suspension applications, and reasonably often for gas–liquid dispersion. They have also been used to reduce abrasive wear in highly abrasive slurries, as they operate at a lower tip speed than low solidity impellers for a given power input. Chord angle is defined as the angle above horizontal of a line drawn from the front blade edge to the back. For many axial impellers, this angle varies from hub to blade tip. For example, it might be 35° at the hub and 20° at the blade tip.

Generally, axial flow impellers will have a chord angle of 37° or less. Chord angles above 37° change the flow pattern into mixed flow and are likely to have boundary layer separation in water-like fluids, resulting in higher power draw.

## Radial Flow

A radial flow pattern is one in which the predominant flow pattern is radially outward from the impeller, i.e. normal to the shaft axis. It is stylistically represented by Figure 4.3.

Such a flow pattern can be beneficial when trying to agitate low liquid levels. In that usage, it is often called a "tickler" or "kicker" impeller. It is also useful for dispersing gasses into liquids, as impellers producing a radial flow pattern

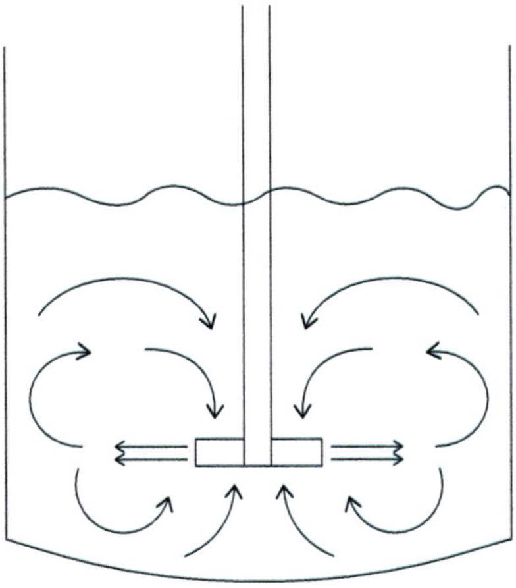

**Figure 4.3** Radial flow pattern.

also produce a strong system of turbulent eddies or vortices which stretch and collapse gas bubbles, creating strings of smaller bubbles, and thereby increasing interfacial area.

Generally, radial flow impellers will have a blade angle of 90° from the horizontal.

## Mixed Flow

For lack of a better description, this flow pattern is between axial and radial flow. Prior to the widespread use of hydrofoils, it was often called axial flow. However, since hydrofoils more nearly approximate a true axial flow pattern, it is appropriate to consider mixed flow a separate category. The mixed flow pattern is stylistically depicted in Figure 4.4.

Mixed flow impellers can be useful at incorporating solids or gas from the liquid surface, as they tend to create deep vortices when operated near the surface, but only in down-pumping mode. They are not as good as true axial impellers at creating bulk motion throughout the tank, at any viscosity.

Mixed flow impellers generally have a chord angle of greater than 37° up to about 75°. Below 37°, the flow pattern is closer to axial, and above 75°, it is closer to radial. The most common angle is 45°.

Impellers with an inner section pitched to pump one way and an outer section pitched to pump the opposite direction actually produce a flow pattern resembling mixed flow when the pattern is time averaged. However, the moment-to-moment flow pattern is extremely variable, even chaotic, especially when looked at in three dimensions. So, we are including a separate category here. It may be the second time such a category has been used in the literature; the first being in Ref. [1].

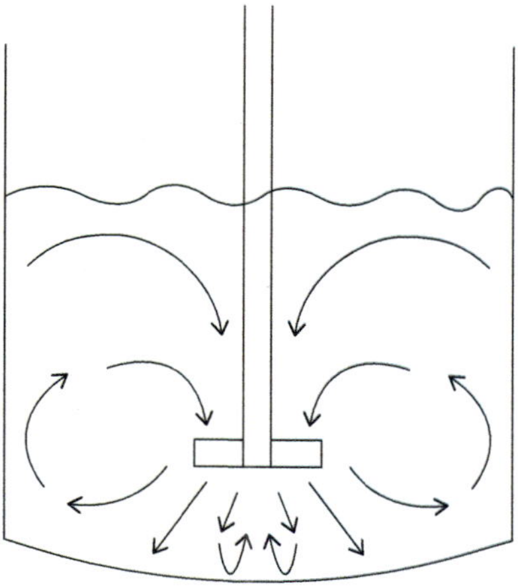

**Figure 4.4** Mixed flow pattern.

### Chaos Flow

This would describe a flow pattern in which the flow vectors change dramatically with time, and in a three-dimensional manner. Mostly this is associated with impellers that have a pitch direction change while traversing the blade from root to tip. Their time-averaged flow pattern may resemble mixed flow, axial flow, or radial flow, depending on the particular impeller design and how it is applied. The flow pattern is very sensitive to D/T. Unfortunately, there is no way to give a good representation of the flow pattern using a still 2D image.

## Examples of Axial Flow Impellers

### Low Solidity

Most axial flow impellers used for blending and solids suspension are low solidity designs. Most have a turbulent power number of about 0.3, though some are higher and some are quite a bit lower. The low power number designs (less than 0.25) have utility for saving power usage in large tanks with gentle agitation. They must be larger than high-power number impellers to invest the same power or create the same flow rate at a given shaft speed. The higher power number designs (greater than 0.4) are often used in tall tanks to save weight and thereby help the shaft design.

Sometimes a vendor will create completely different geometries for their various axial flow impellers, resulting in different power numbers. Sometimes they will just change the angle of attachment at the impeller hub, resulting in a different chord angle at the tip, and also affecting the power number. We will give several specific examples of impellers currently on the market in each category. Note that many manufacturers, if they publish power numbers, do not give information on how those numbers vary with D/T. When only one power number is given, it is probably based on a D/T ratio of about 1/3.

There are many low-solidity hydrofoils on the market. We can only describe a few here. We will divide them up in terms of low power number ($N_P < 0.25$), intermediate ($0.25 \leq N_P < 0.4$) and high ($N_P \geq 0.4$). It is the author's opinion that most of the impellers within each category will have similar performance at the same invested power and shaft speed.

Examples of low power number impellers include the ProQuip HFI (Figure 4.5a), the Lightnin A-510 E12 (not shown; essentially the same as the A-510 E22 and the A-310 except for blade angle) and the Chemineer XE-3 (Figure 4.5b).

Examples of intermediate power number impellers are more numerous and include the Fusion Fluid PF3 (Figure 4.6a), the Lightnin A-510 E22 (Figure 4.6b), the Chemineer HE-3 (Figure 4.6c), and the generic marine propeller (Figure 4.6d). Marine propellers are rarely used in top entering biofuel applications except for small tanks but may be used at times for side entry applications in large tanks, as described in Chapter 6.

High power number examples include the Chemineer SC-3 (Figure 4.7) and the Lightnin A-510 E32 (image not shown; essentially the same as the A-510 E22 but with a steeper blade angle).

Low solidity impellers have the following advantages:

- Better blending performance for the torque input: This advantage increases as the power number decreases. However, there may or may not be power savings, due to limitations on maximum D/T ratio.
- Easier to install in a vessel, due to the narrower blade width.
- Lower hydraulic force, sometimes resulting in a smaller required shaft diameter.

(a) (b)

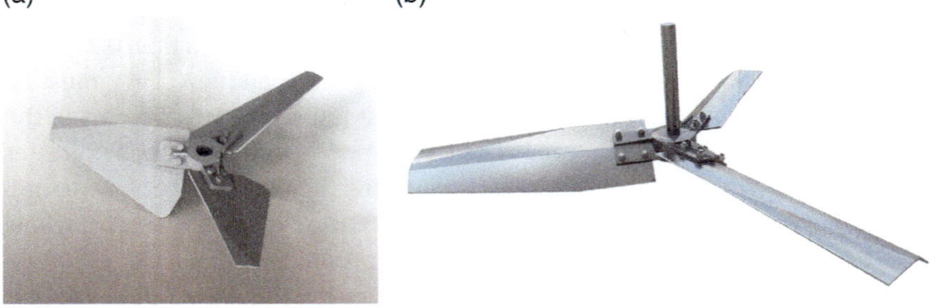

**Figure 4.5** (a) Proquip HFI. *Source:* PRO-QUIP. (b) Chemineer XE-3. *Source:* Used with permission of NOV.

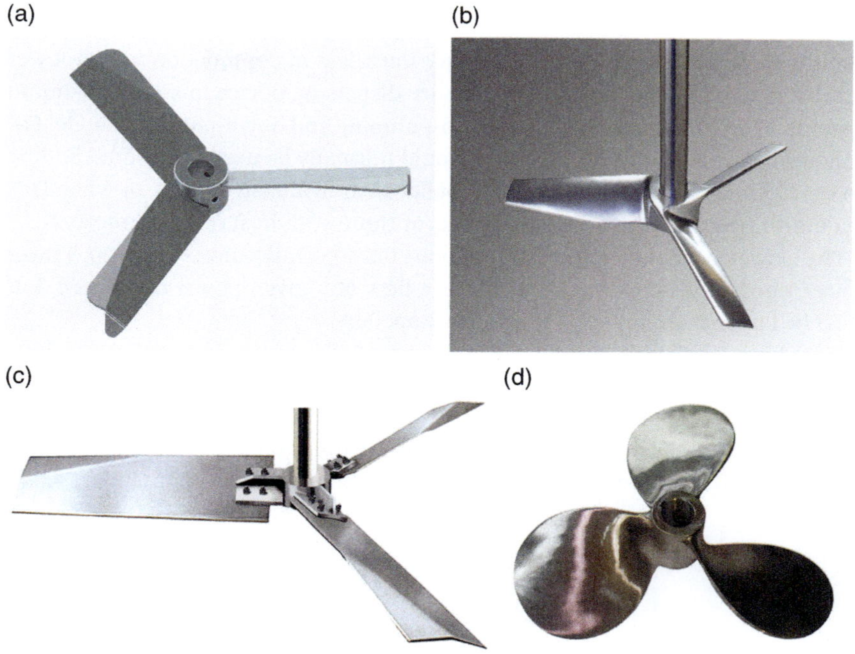


**Figure 4.6** (a) Fusion Fluid PF3. *Source:* Fusion Fluid Equipment. (b) Lightnin A510E22/A-310. *Source:* Courtesy of SPX FLOW, Inc. (c) Chemineer HE-3. *Source:* Used with permission of NOV. (d) Marine propeller. *Source:* Courtesy Fusion Fluid.

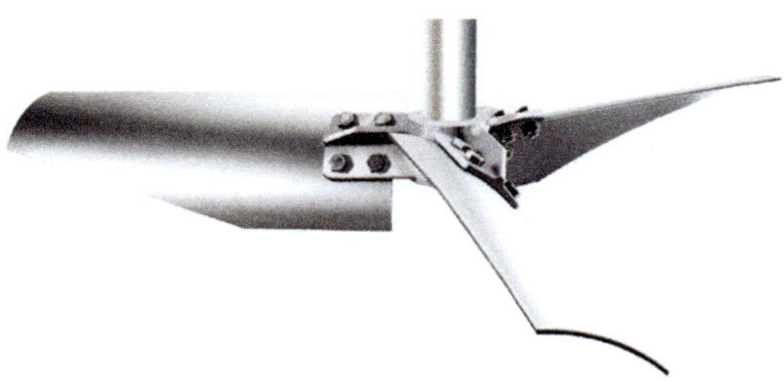

**Figure 4.7** Chemineer SC-3. *Source:* Used with permission of NOV.

But they have the following disadvantages:

- Larger diameter required, or a higher shaft speed, for a given power input. This may cause clearance problems with internals such as heat exchange surfaces.
- Not effective for gas dispersion as a primary impeller.
- Higher weight or higher shaft speed may make shaft design more difficult, due to critical speed issues. See Chapter 15.

## High Solidity

Mainly, these are used both as secondary impellers in conjunction with a lower gas-dispersing impeller and as the primary dispersing device in some gas–liquid systems. They have been used in both up-pumping and down-pumping mode. For non-gas dispersion applications, they would normally be used near liquid boiling points to avoid cavitation; to reduce impeller wear in abrasive solids, or when D/T is a limiting factor, such as in small tanks or those with heat transfer internals.

The power number range is from about 0.5 to a little more than 1.0. Lower power number versions may create more flow at a given power and speed, but must be larger than high power number impellers.

Most have four blades as a means to achieve high solidity, but some have three blades.

Examples of high solidity impellers include the Chemineer Maxflo-W (Figure 4.8a), The Chemineer Maxflo-Y shown in up-pumping mode (Figure 4.8b), the ProQuip SHF High Solidity (Figure 4.8c), and the Lightnin A-315 (Figure 4.8d).

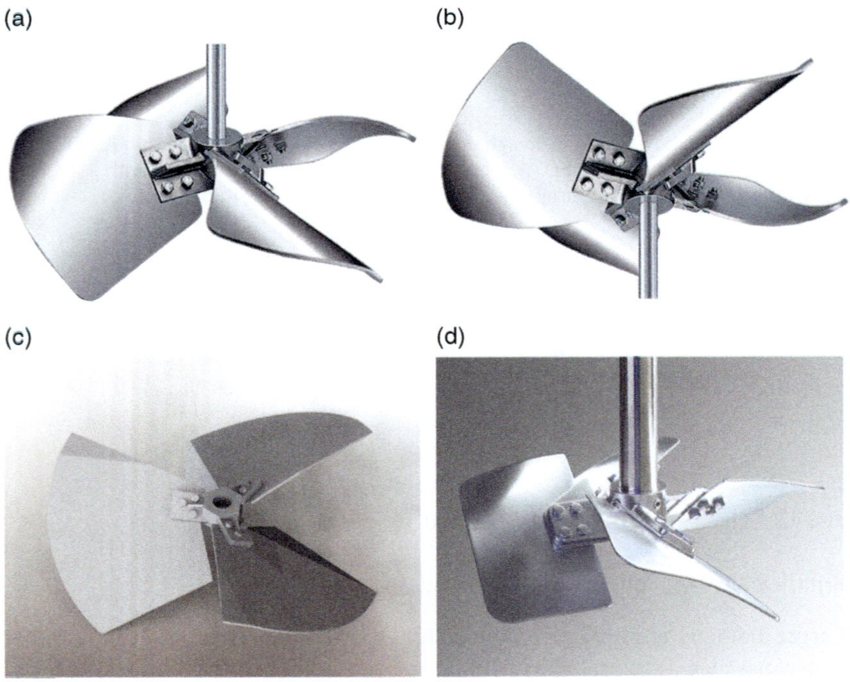

**Figure 4.8** (a) Chemineer Maxflo W. *Source:* Used with permission of NOV. (b) Chemineer Maxflo Y up-pumping. *Source:* Used with permission of NOV. (c) ProQuip SHF High solidity. *Source:* PRO-QUIP. (d) Lightnin A-315. *Source:* Courtesy of SPX FLOW, Inc.

Advantages of high solidity impellers include:

- High power number means smaller impellers are needed to invest the required power.
- Can be used to directly disperse gas, especially in up-pumping mode.

Disadvantages include:

- Lower flow and blending performance for the torque invested.
- Difficulty in fitting wide blades through access openings (there are special split-blade versions to help with this, but they are less sanitary and more expensive).
- Greater weight.
- High hydraulic forces, especially when directly gassed, down-pumping

To summarize this section, we have Table 4.1, using the author's best estimates.

**Example 4.1** What is the approximate power draw of a 1.5 m diameter A-315 impeller, operating at 125 rpm in a 3.75 m diameter vessel, in a fluid with a specific gravity of 1.05?

**Table 4.1** Axial flow power and pumping numbers (D/T ~ 0.33) turbulent flow only.

| Brand or manufacturer | Impeller designation | Solidity class | Power number | Pumping number |
|---|---|---|---|---|
| Chemineer | XE-3 | Low | 0.24 | 0.5 |
| Chemineer | HE-3 | Low | 0.29 | 0.53 |
| Chemineer | SC-3 | Low | 0.5 | 0.71 |
| Chemineer | Maxflo W-4 | High | 1 | 0.77 |
| Chemineer | Maxflo Y-4 | High | 0.5–0.9 | 0.6–0.75 |
| ProQuip | HFI | Low | 0.22 | 0.48 |
| ProQuip | SHF | High | 0.5 | 0.65 |
| Fusion fluid | PF3 | Low | 0.3 | 0.5 |
| Lightnin | A-310 | Low | 0.3 | 0.54 |
| Lightnin | A-510E12 | Low | 0.16 | 0.39 |
| Lightnin | A-510E22 | Low | 0.3 | 0.51 |
| Lightnin | A-510E32 | Low | 0.45 | 0.63 |
| Lightnin | A-315 | High | 0.75 | 0.75 |
| Lightnin | A-320 | High | 0.65 | 0.62 |
| Generic | Marine prop | Low | 0.35 | 0.53 |

**Answer**

From Table 4.1, we can look up a power number of 0.75. Using SI units, we can calculate the power:

$$P = \rho * N_P * N^3 * D^5 = 1050\ \text{kg/m}^3 * 0.75 * \left(125/60\ \text{second}\right)^3 * \left(1.5\ \text{m}\right)^5 \\ = 54073\ \text{W} = 54.1\ \text{kW} \tag{4.1}$$

### Up-Pumping vs. Down-Pumping

When using only a single impeller for blending liquids, it makes little difference whether the impeller pumps up or down, except in very shallow tank geometries, where up-pumping would blend better because it would not draw a vortex into the impeller as readily. In gas–liquid systems, up-pumping is strongly favored. However, we will not be dealing with such systems in this book, as they are covered extensively in Ref. [1]. In most vessel geometries, down-pumping is better at sweeping the bottom and keeping solids in suspension.

## Examples of Radial Flow Impellers

There are many styles of radial flow impellers in the marketplace. Some are generic and some are proprietary. This section will explore some of the attributes of some of the better-known ones. For most proprietary impellers, several manufacturers have essentially similar products, so the characteristics may be close among them.

### Straight Blade Impeller

This style has blades attached to an open-style hub (with no disc) at an angle of 90° to the plane of rotation. The most common style in use today has four blades, with a blade height one-sixth of the impeller diameter. It is illustrated in Figure 4.9.

Mostly this style of impeller is used as a bottom turbine to avoid dead spots and keep solids in suspension, especially at low liquid levels. In that service, it is often referred to as a "tickler" or "kicker" turbine. When placed near the tank bottom (e.g. D/6 or less) with about one-third of the power input of a hydrofoil located above it (say, T/3 off the bottom), the combined system seems to be as effective as if the same power were invested by the hydrofoil alone. This is probably because the flow patterns are synergistic. The combined flow pattern is illustrated in Figure 4.10. Straight blade turbines, especially with fewer than six blades, are not normally used to disperse gas. With four blades, their power number is about 3,

**Figure 4.9** Straight blade turbine by Chemineer. *Source:* Used with permission of NOV.

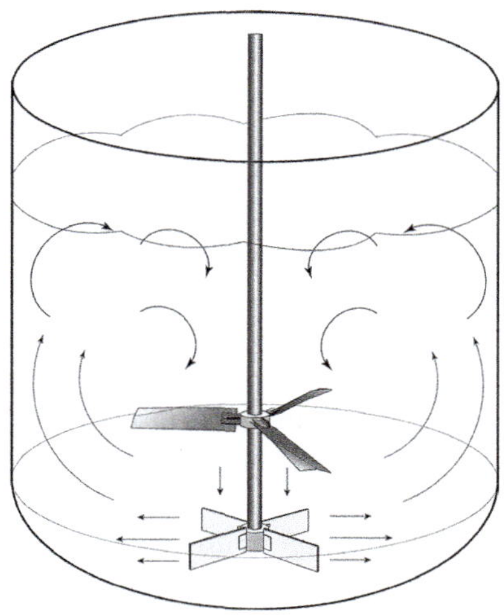

**Figure 4.10** Hydrofoil plus tickler flow pattern.

and the pumping number is about 1. Their gas dispersion characteristics, such as flooding and gassing factor, have not been studied much, so they will not be discussed here.

## Disc, Aka Rushton, Turbines

This impeller style is one of the oldest in industrial use. It has a series of flat blades mounted at 90° to the plane of rotation, attached to a flat disc. Most commonly, there are 6 blades, but impellers have been built with 4–12 blades. The most common six-blade version has a blade height of 20% of the diameter and has a turbulent power number of about 5.5. The turbulent power number is proportional to the w/D ratio for the six-bladed version, and is largely independent of

D/T in the range of 0.2–0.5. This impeller has been extensively studied in the literature. Indeed, many graduate students are still studying it, though not necessarily in optimum uses. It has become a generic impeller, made by virtually every agitator manufacturer in the world. A typical embodiment is shown in Figure 4.11.

This was the impeller style first used in industrial fermentation, notably in the commercialization of penicillin in the 1940s. It remained popular well into the 1980s, but has been mostly replaced by more modern impellers in production-scale equipment, though, curiously, it is still the most common impeller supplied with bench-top and small pilot fermenters. The reasons why it has been largely replaced will become evident in this and the next few impeller sections. In the biofuel industry, it has mainly been used in the yeast propagator tank, as this is an aerobic fermenter.

Rushton impellers go through several changes in gas flow regimes around the impeller blades, as a function of gas loading. These regimes are illustrated in Figure 4.12.

There are roughly three different kinds of gas cavities or pockets that form behind a Rushton impeller blade, as a function of gas loading. The bottom of Figure 4.12 illustrates these. At low gas rates, the gas is collected in two counter-rotating vortex cavities, with rotation direction as shown, and with the vortices rotation shown by the curved arrows. At larger gas flow rates, clinging cavities form. These still have two distinct vortices, but a portion of the cavity clings to the backside of the blade, filling part of the space with gas. At high gas flow rates, large cavities are formed, which fill in the entire back of the blade and extend some distance behind it, after which the bubbles break off and are dispersed by a vortex action.

These transitions do not happen to all blades simultaneously. As the upper part of Figure 4.12 shows, for a six-bladed turbine, it is possible to have alternating

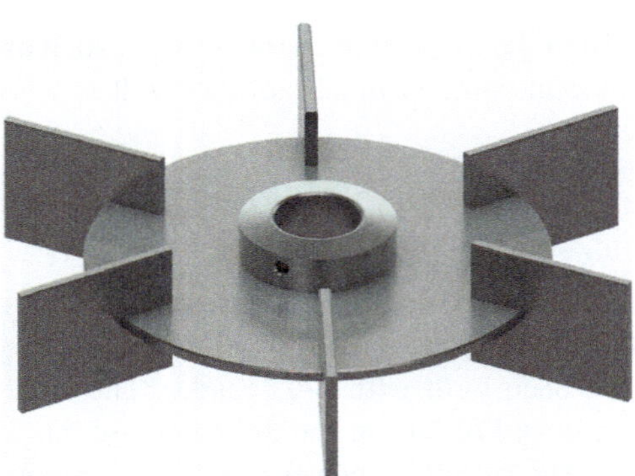

**Figure 4.11** Rushton turbine by Fusion Fluid. *Source:* Fusion Fluid Equipment.

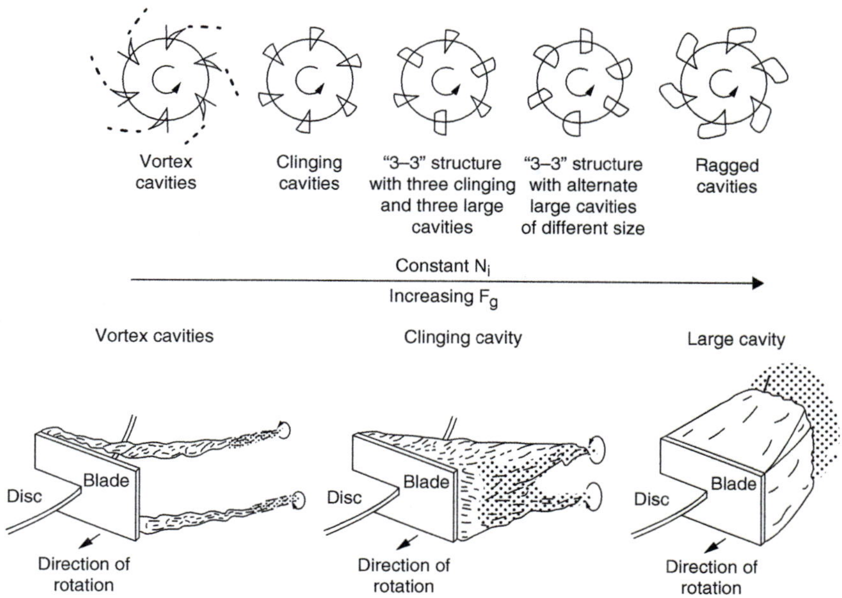


**Figure 4.12** Rushton gas cavities. *Source:* Licensed by Elsevier.

large and clinging cavities, or alternating large cavities of different sizes. These conditions are called a "3–3" structure and are a stable condition. Such a pattern cannot form on an impeller with an odd number of blades. For example, with a five-bladed impeller, one might develop a "3–2" pattern, which can move from blade to blade or change to a "2–3" pattern. This is hydraulically unstable and can create large lateral hydraulic forces. The author recommends avoiding Rushton impellers with an odd number of blades for this reason, at least in gas–liquid systems. The author is unaware of any reason why a Rushton turbine would be the impeller of choice in a single-phase system.

At very high gas loading, "ragged" cavities can form. These indicate that the impeller is close to flooding. In fact, one mechanism of flooding is when the cavities extend from the back of one blade to the front of the other. Essentially the impeller is running in gas. Power draw under this condition varies periodically between a high and a low value, as the impeller "burps" the excess gas as it loses control of the flow pattern. At higher gas rates still, the gas basically rises through the impeller without being dispersed. That condition is called flooding.

The consequence of the existence of these cavities is that they streamline the apparent shape of the impeller for drag creation purposes, reducing both pumping and power draw considerably. It is not unusual for a Rushton impeller in an industrial fermenter to have a gassing factor of 0.4 or less. Since mass transfer depends on gassed power, this represents a considerable loss of mass transfer potential as the gas flow rate increases. Also, if the unit design is based on the gassed power draw, a loss of gas flow can create massive overloading.

Quantitatively, the gassing factors are illustrated in Figure 4.13, based on a D/T ratio of 0.3. The figures end at each Froude number approximately at the flooding point, i.e. the maximum aeration number for that Froude number, or, in some cases where correlation data were not available. Note that the gassing factor can get as low as 0.3, although under typical conditions, it is usually around 0.4. These figures apply when the Rushton turbine is used as a lower impeller. When used as an upper impeller, the gassing factor can be estimated by the following:

$$\left(P_g / P_u\right)_{upper} = 0.9 * \left(P_g / P_u\right)_{lower} + 0.1 \tag{4.2}$$

Rushton turbines have a gas handling ability well correlated by the following equation (Ref. [2]).

$$N_a\left(\text{maximum}\right) = 30 * \left(D / T\right)^{3.5} * N_{Fr} \tag{4.3}$$

As we will see later, that is a much lower gas handling capacity than is afforded by more recent impeller designs. In fact, we are listing the radial impellers in this section in order of increasing gas handling capacity.

Note that the minimum Froude number for any degree of dispersion for a Rushton turbine is about 0.045.

In summary, the Rushton impeller has disadvantages at high gas flow rates due to potential flooding and substantial drop in power draw. But there are cases where it will perform as well as more modern impellers. Basically, these are cases

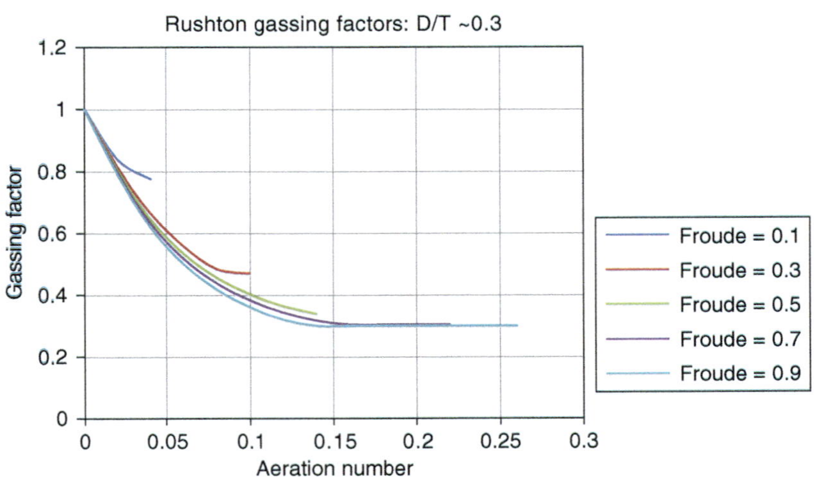


**Figure 4.13** Rushton turbine gassing factors.

where little gas flow is present, such as in microaeration (typical OTR < 10 mmol/l-h) or when a very high utilization of gasses is required due to either the expense of the gas or the cost of disposing of any gas that leaves the reactor. This means low gas flow rates near the end of the process. Also, it has been observed that Rushton impellers are easier to clean with spray balls than concave impellers.

**Example 4.2** How much gas flow can a 1.2 m diameter Rushton impeller handle without flooding when operated in water at 100 rpm in a 4 m diameter tank, and what will its gassed power draw be?

**Answer**

First, let us calculate the ungassed power:

$$\begin{aligned} P_u &= N_P * \acute{A} * N^3 * D^5 = 5.5 * 1\ 000\ \text{kg/m}^3 * (100/60\ \text{second})^3 * (1.2\ \text{m})^5 \\ &= 63\ 360\ \text{W} = 63.4\ \text{kW} \end{aligned} \tag{4.4}$$

Next, let us calculate the Froude number:

$$N_{Fr} = N^2 D/g = (100/60\ \text{second})^2 * 1.2\ \text{m}/(9.8\ \text{m/s}^2) = 0.34 \tag{4.5}$$

Now we can calculate the maximum aeration number:

$$N_a(\text{max}) = 30 * (D/T)^{3.5} * N_{Fr} = 30 * (1.2/4)^{3.5} * 0.34 = 0.15 \tag{4.6}$$

The maximum airflow is thus:

$$Q_g(\text{max}) = N_a(\text{max}) * ND^3 = 0.15 * 100/\text{minute} * (1.2\ \text{m})^3 = 25.92\ \text{m}^3/\text{min}. \tag{4.7}$$

Note that we did not convert N to revolutions per second, as we were looking for a dimensional result. By using rpm, we got the gas flow in cubic meters per minute. Had we used revolutions per second, we would have just gotten gas flow in cubic meters per second.

Next, let us determine the gassing factor. Though Figure 4.13 does not have a recorded value for a Froude number of 0.34 and an aeration number of 0.15, the trend lines would lead one to estimate a value of 0.4 for the gassing factor. Thus, a reasonable estimate of the gassed power would be:

$$P_g = P_u * (P_g/P_u) = 63.4\ \text{kW} * 0.4 = 25.4\ \text{kW} \tag{4.8}$$

## Smith Trbines

One of the pioneer researchers in the field of gas–liquid dispersion is Professor Emeritus John M. Smith. While at the University of Surrey in the UK and Delft Technical University in the Netherlands in the 1980s, his team developed data on the mechanisms of dispersion, power draw, and flooding for the Rushton turbine. In particular, upon observing the gas pockets that form behind the Rushton turbine blades, he reasoned that by using concave blades with the concave side being the leading face, more gas would be discharged from the front of the blade and any gas pockets formed would be smaller, increasing the gas handling capacity and the gassing factor. He tested his ideas and found that both were true. The particular impeller he came up with had blades with circular segments as cross sections, with about a 120° curvature. This design is properly called a Smith Turbine. We do not have an illustration of that design.

In that configuration, with six blades, its typical power number was about 3.8, its gassing factor was about 0.6 under highly aerated conditions, and its gas handling was about 50% more than that of the Rushton. So, it succeeded admirably.

This exact style of turbine has never been produced as a commercial product, to the author's knowledge. Instead, Chemineer, Inc.developed a modification based on Smith's principles, called the CD-6, and sold many of them starting around 1990. Other manufacturers soon started building essentially the same impeller but called it a Smith turbine, even though the geometry is based on the Chemineer CD-6 rather than the actual Smith turbine geometry. Since the original Smith turbine is not what manufacturers offer today, we will provide more detailed commentary on the CD-6 turbine, which will often be referred to as a Smith turbine, as that term is the generic name used today by most manufacturers. Next we describe how the CD-6 is different from the original Smith turbine.

## CD-6 Turbine by Chemineer; Aka Smith Turbine by Many Manufacturers

As stated earlier, Chemineer was likely the first to commercialize the impeller concepts developed by Smith. However, they did additional research before commercializing the impeller, which they named the CD-6 (presumably for Concave Disc-6 bladed turbine). That research included detailed, proprietary power draw algorithms and flooding correlations, as well as some changes to the impeller geometry. The principal change is that the blade cross section is 180°, i.e. a semicircle, rather than the 120° used by Smith. This change lowered the typical power number from 3.8 to about 3.2, further increased the gassing factor from about 0.6 to about 0.65 under highly gassed conditions, and increased gas handling to more than double that of the Rushton at higher D/T and Froude number conditions. The impeller is illustrated in Figure 4.14. This impeller has been widely used since the 1990s and is currently manufactured by multiple vendors under various names.

Unlike the Rushton impeller, the CD-6 does show a dependency of power number on D/T ratio. Myers et al. [3] provided the data shown in Table 4.2. Note that there is a 20% drop in ungassed power going from a D/T of 0.25 to 0.5.

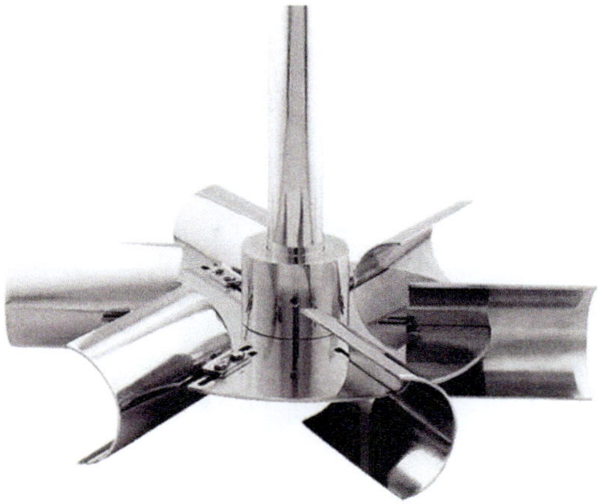

**Figure 4.14** Chemineer CD-6 impeller. *Source:* Used with permission of NOV.

**Table 4.2** Ungassed Power number of the CD-6 impeller.

| D/T | 0.25 | 0.32 | 0.39 | 0.45 | 0.50 |
|---|---|---|---|---|---|
| $N_P$ | 3.21 | 2.89 | 2.77 | 2.56 | 2.55 |

As mentioned previously, the CD-6 unloads much less than the Rushton. This is because the curvature on the rear of the blade allows for less room for gas pockets to form, as shown in Figure 4.15, so that there is less streamlining of the flow. Figure 4.16 summarizes the gassing factor at a D/T of about 0.3. Note that the dependency on Froude number virtually vanishes above an aeration number of 0.15. Note also that the scale is different from the Rushton scale; it covers up to about triple the aeration number. That is because of the higher gas handling characteristics. The curves for each Froude number stop at the flood point or at the limits of the correlation.

Flooding behavior of the CD-6 is quite complex. The maximum aeration number depends on D/T and Froude number. At a given D/T ratio, the maximum aeration number is proportional to Froude number up to a certain Froude number. Above that critical Froude number, the maximum aeration number increases rapidly, sometimes by a factor of 3–5 within a 10–20% increase in Froude number. Reference [3] has considerable analysis of this phenomenon empirically.

If we denote the proportionality constant as $K_{fl}$, we can write this as:

$$N_a(\max) = K_{fl} * N_{Fr} \tag{4.9}$$

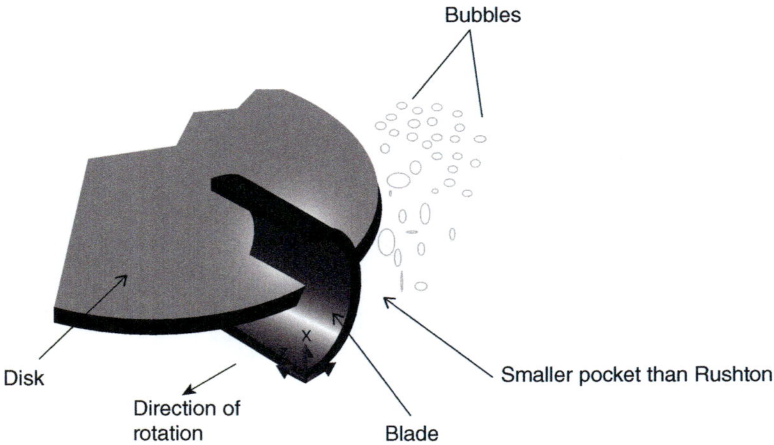


**Figure 4.15** CD-6 gas pockets.

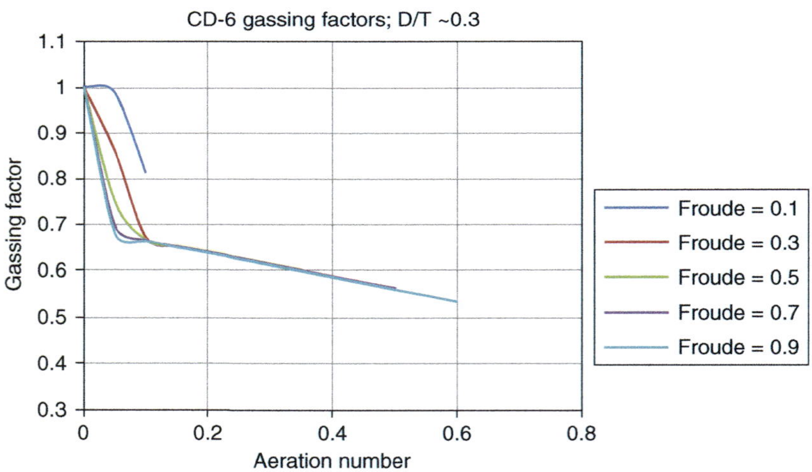


**Figure 4.16** CD-6 gassing factors.

Adding the caveat that there is a critical Froude number above which Eq. (4.9) no longer holds true, we can express this as a function of D/T in tabular form, as shown in Table 4.3.

Rushton turbines do not exhibit such behavior. They retain a basically linear relationship between maximum aeration number and Froude number. The end result is that at small D/T and Froude number values, the CD-6 has little or no gas handling advantage over a Rushton. However, at higher D/T ratios, the CD-6 has considerably more gas handling capacity, even below the critical Froude number.

**Table 4.3** Flooding constants of CD-6.

| D/T | $K_{fl}$ | Critical Froude number |
|---|---|---|
| 0.25 | 0.3 | 0.98 |
| 0.32 | 0.5 | 0.5 |
| 0.39 | 1.0 | 0.3 |
| 0.45 | 2.5 | 0.15 |
| 0.50 | 5.4 | 0.06 |

Above the critical Froude number, the gas handling capacity dwarfs that of the Rushton. And that critical Froude number decreases as the D/T increases. So, by judicious manipulating of D/T ratios, it should be possible to avoid flooding in most cases by using the CD-6 impeller.

**Example 4.3** Use the same problem statement as under Rushton turbines, but substitute a CD-6 impeller.

**Answer**

First, let us calculate the ungassed power. Our D/T ratio is 0.3, so our power number based on Table 4.2 is about 2.87.

$$P_g = N_p * \rho * N^3 * D^5 = 2.87 * 1\ 000\ \text{kg/m}^3 * (100/60\ \text{second})^3 * (1.2\ \text{m})^5$$
$$= 33\ 062\ \text{W} = 33\ \text{kW} \tag{4.10}$$

Since we have the same diameter and speed as before, the Froude number remains at 0.34. From Table 5.3, we can see that we are below the critical Froude number, so we can use a linear assumption. The value of $K_{fl}$, by interpolation, is about 0.6–0.65. We will use 0.6. Thus:

$$N_a(\max) = 0.6 * N_{Fr} = 0.6 * 0.34 = 0.204 \tag{4.11}$$

That is about 36% higher than for the Rushton. The resulting maximum airflow is:

$$Q_g(\max) = N_a(\max) * ND^3 = 0.204 * 100/\text{minute} * (1.2\text{m})^3 = 35.25\text{m}^3/\text{min}. \tag{4.12}$$

Now let us calculate the gassed power. By inspection of Figure 4.16 at an aeration number of 0.204 and a Froude number of 0.34, the gassing factor is about 0.64. Thus:

$$P_g = P_u * (P_g/P_u) = 33\text{kW} * 0.64 = 21.1\text{kW}. \tag{4.13}$$

Though the CD-6 in this case has a ungassed power only 52% of the same-sized Rushton, and a gassed power 83% of that of the Rushton, it can handle 36% more gas flow before flooding. It would handle about double the gas flow at equal gassed power, which would require a larger impeller diameter.

In summary, the CD-6 is a substantial improvement over the Rushton in most cases. The only comparative drawbacks seem to be that the impeller weighs more and is more difficult to clean with spray balls. This is due to several reasons:

- Larger diameter required for equal power input.
- More total metal in the blades (though they can be made of thinner material, as a circular cross section is stronger than a flat plate).
- An additional mechanical load is created: fluctuating loads normal to the disc. This requires a greater disc thickness for stability.
- The curved shape makes spray ball access from above and below insufficient. Immersion in cleaning fluid may be required.

## Deeply Concave Turbines

As noted above, Smith and similar turbines greatly reduce the size of the gas pockets, but they do not eliminate them. Smith himself has said that he was under the impression that strongly curved blades would be most effective. Therefore, several manufacturers embarked on development programs to create impellers that had a deeper degree of concavity than was possible using circular segments as cross-sections. Examples of such impellers include the ICI radial impeller (Figure 4.17a), the SCABA 6SRGT (Figure 4.17b; Ref. [4]), the Lightnin R135 (Figure 4.17c), and the Ekato Phasejet (Figure 4.17d).

All of these impellers are mounted on a disc. All have deeper cross sections than a semicircle, and all are symmetric above and below the disc. As shown, all come to a sharp spine at the trailing edge of the impeller. However, it should be noted that the actual configuration of most Ekato Phasejet turbines sold for fermenters differs from the patent illustration in three ways:

1) The disc is flat, rather than conical or dished.
2) The disc normally does not have serrations.
3) The blades generally end in a curved section, rather than a sharp spine or ridge.

Because all of these are proprietary impellers, it is difficult to obtain reliable performance characteristics. Reference [4] reports the power number of the SCABA 6SRGT impeller as 1.5. Daub [5] reports the power number of the Phasejet as 1.6. Most likely, all are in the range of 1.3–1.8.

Likewise, little has been published about the gassing factor results. Some studies show a value of close to 1 for aeration numbers up to 0.05, but this is

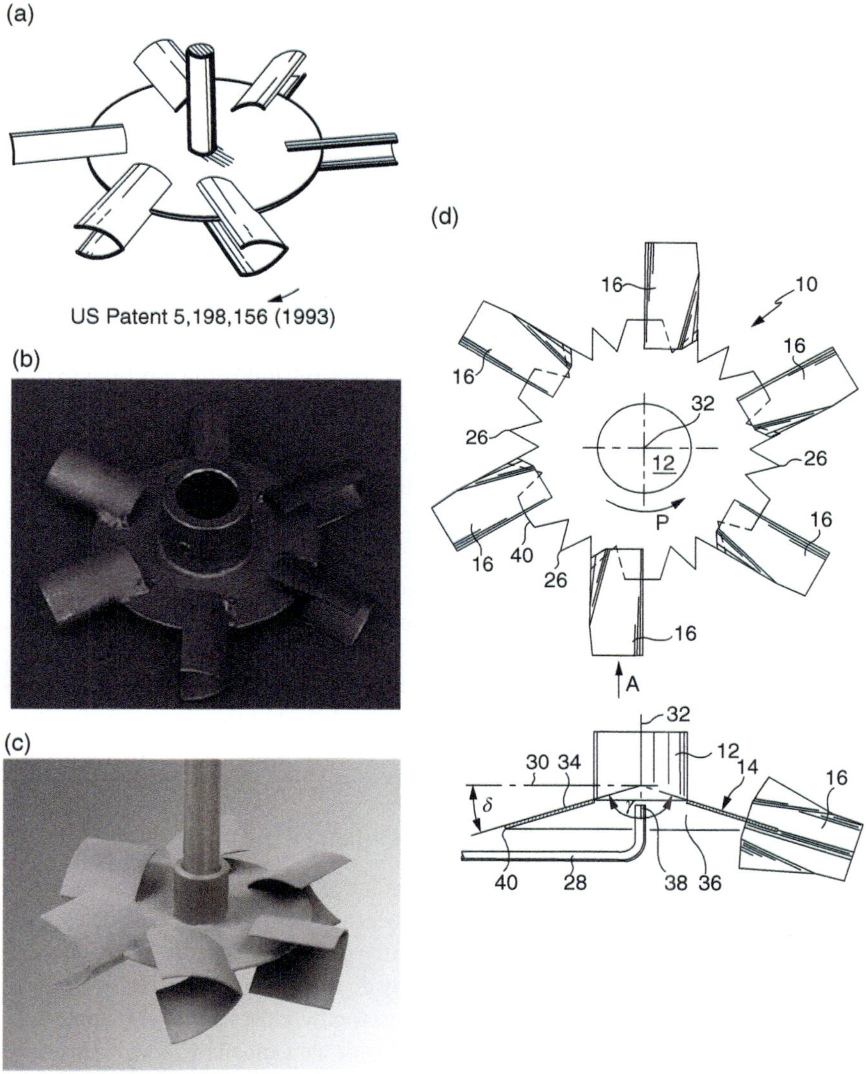


**Figure 4.17** (a) ICI Patent drawing. *Source:* Middleton & Ramshaw 1993/U.S. Patent 5198156A. (b) SCABA 6SRGT. *Source:* Permission of Rob Van der Lans. (c) Lightnin R-135. *Source:* Courtesy of SPX FLOW, Inc. (d) Ekato Phasejet Patent drawing. *Source:* Forschner et al. 1999/U.S. Patent 5904423A.

rather low for impellers operating in fermenters. Some fermenters have been operated with aeration numbers as high as 0.5, and many are in the range of 0.1–0.2. Ekato's literature on the Phasejet claims a gassing factor of 0.8 at an aeration number of 0.16 and a Froude number of 0.5. It is expected that the other impellers in this group will behave similarly. As expected, this gassing factor range is substantially higher than that obtained with Rushton and CD-6/Smith turbines. It is almost as high as just simply accounting for density reduction due to gas holdup.

Gas holdup is expected to be similar to the Rushton at equal gassed power and airflow.

There is also not much in the literature regarding flooding. However, it is expected that these impellers will generally have around twice the gas handling capacity of the CD-6 style impeller at equal invested power and shaft speed, in part because of studies done on the kind of impeller described in the next section.

Summarizing, the impellers in this category have the following advantages over the CD-6 geometry:

- Much higher gas handling capacity before flooding.
- Substantially smaller drop in power due to gassing.

They have the following disadvantages:

- Larger impeller diameter, due to lower power number.
- Greater impeller weight, in part because of greater vertical oscillating loads, requiring a thicker disc.
- Under overload conditions, they can open up (become taller in terms of blade height), which further increases loads and leads to failure. Sometimes a strut (illustrated in the next section) is needed to prevent this.

## Deep Asymmetric Concave Turbine with Overhang (BT-6)

More or less concurrently with the development of the impellers in the previous section, or slightly later in some cases, engineers at Chemineer (notably Dr. André Bakker) asked, since gas flow was inherently asymmetric with respect to the impeller (it always tries to flow upwards), why do the blades need to be symmetric about the disc plane? Their questioning led to the hypothesis that the upper section of the blade could be longer than the lower as a means to force discharge more of the gas on the front side of the impeller, leaving less to form pockets on the rear. They also studied the shape of the gas pockets, and designed the impeller curvature to match it, so that there would effectively be no pockets. The upper part of the blade has a different curvature from that of the lower part of the blade. The resultant impeller was dubbed the BT-6, and was patented, though the patent is now expired and several companies have copied the design. The impeller is depicted in Figure 4.18. A conceptual illustration of the lack of gas pockets is shown in Figure 4.19.

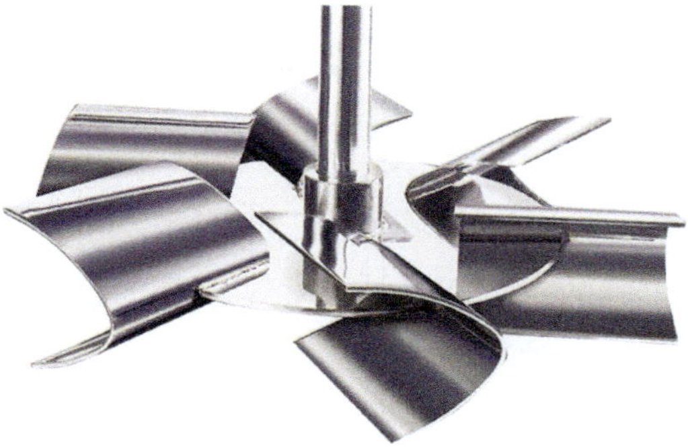

Figure 4.18 BT-6 impeller. *Source:* Photo used with permission of NOV.

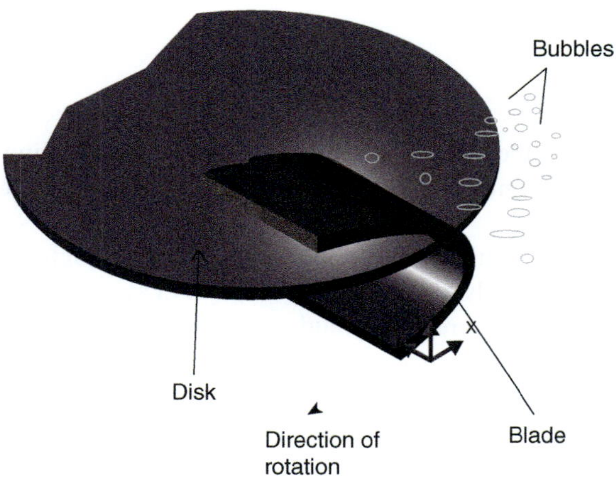


Figure 4.19 BT-6 no gas pockets.

The turbulent power number of the BT-6 is about 2.3 over the range of D/T from 0.25 to 0.5. It can be made lower by reducing the w/D ratio. The standard w/D ratio is 0.2. The power number is proportional to w/D with a six-bladed impeller.

As expected, according to the theory behind its design, its gassing factors are quite high. Figure 4.20 illustrates these for a D/T ratio of 0.3. If it seems that the results are basically independent of Froude number, it is because they are. The main difference Froude number makes is how much airflow can be handled before flooding. Note that the lowest gassing factor reported was about 0.82. One notable feature of these curves is that the gassing factor does not exceed 1 at low gas flows and high Froude numbers, as it can for some radial turbines

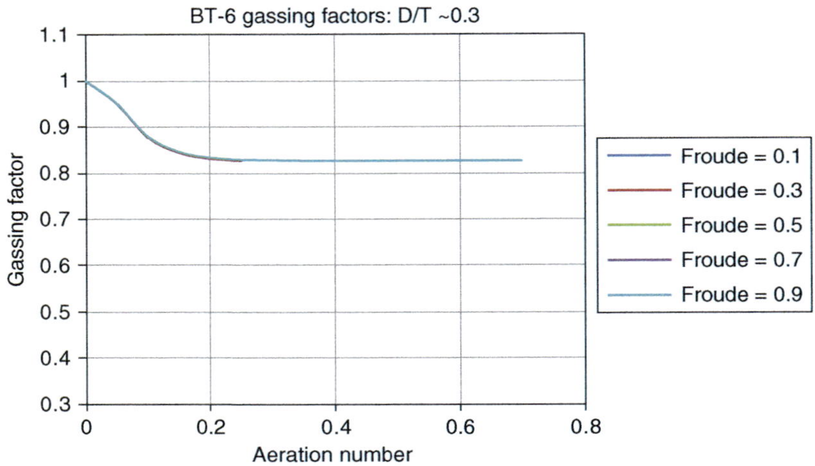


**Figure 4.20** BT-6 gassing factors.

such as the Rushton and CD-6 impellers. This reduces the risk of overloading when designed for ungassed conditions but operated at lightly gassed conditions.

How does this compare with the results in the previous category? Although we do not have gassing factors for all deeply concave turbines listed, Ekato was kind enough to publish values in their literature for a Froude number of 0.5 and aeration numbers from 0 to 0.16. These are plotted along with gassing factors for the BT-6 over the same range in Figure 4.21 plus extending it to an aeration number of 0.25 for the BT-6. On average, except for zero gas flow, the BT-6 shows a 3% higher gassing factor up to an aeration number of 0.16, and a larger difference above that seems probable. In the absence of data for the other deeply concave turbines, we would assume their gassing factors are similar to those of the Phasejet. These gassing factors are likely about the same for upper and lower impellers, as there are almost no gas pockets. The gassing factor is essentially the density reduction factor.

Flooding is more complicated than it is for a Rushton, similar to the situation of the CD-6 impeller. This is shown in Figure 4.22, taken from Ref. [6]. Below an aeration number of 0.2, the maximum aeration number is roughly linear with respect to the Froude number. The slope increases as D/T increases. Above an aeration number of 0.2, the slope increases even more, and it increases by a larger factor as D/T increases. Since the BT-6 impeller has on occasion been applied at larger D/T ratios than those shown, it would be ideal if more data were presented. But we expect that its gas handling ability would continue to improve at larger D/T ratios.

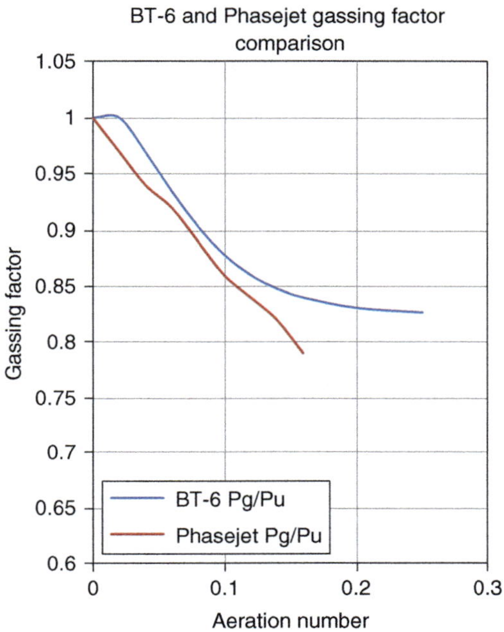


**Figure 4.21** BT-6 vs. Phasejet gassing factors.

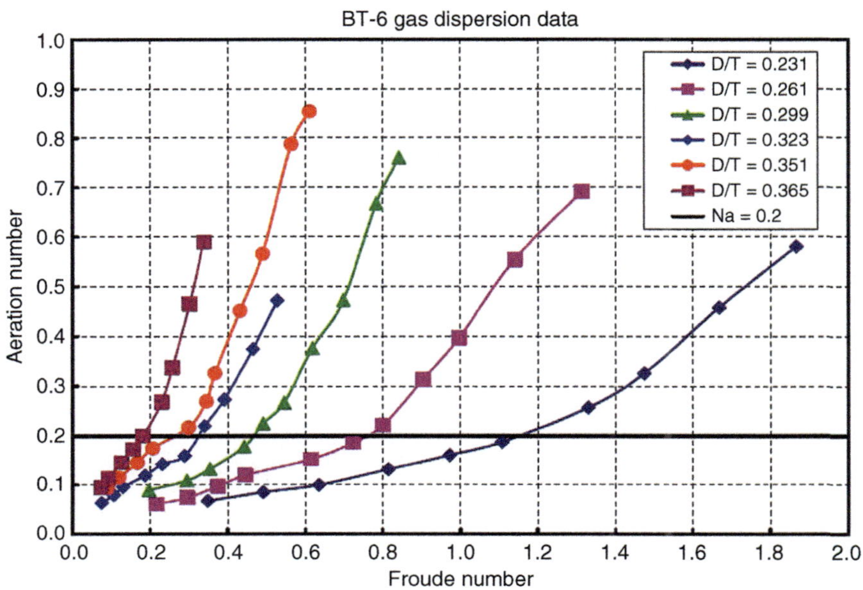


**Figure 4.22** BT-6 gas dispersion data. *Source:* Permission granted by Julian Fasano.

The behavior below an aeration number of 0.2 can be approximated by the following equation:

$$N_a(\max) = K_{fl} * N_{Fr}; \quad N_a < 0.2 \tag{4.14}$$

At an aeration number of 0.2, the slope increases, and there is a critical Froude number associated with that change. The behavior above an aeration number of 0.2 and above the critical Froude number can be approximated by the following:

$$N_a(\max) = 0.2 + K_{fl2}(N_{Fr} - N_{Frc}); \quad N_a \geq 0.2 \tag{4.15}$$

The constants are D/T dependent, as shown in Table 4.4. No correlation as a direct function of D/T has been developed by the author.

Regardless of these equations, it is not recommended to disperse gas below a Froude number of 0.035 for a BT-6 impeller.

Thus, the BT-6 impeller has the following advantages:

- Low sensitivity to viscosity; power number does not rise much until the Reynolds number drops below 600.
- Highest gassing factor of any impeller the author is aware of.
- Highest gas handling capacity of any impeller the author is aware of (about 5–5.5 times that of the Rushton at equal gassed power).

It has a few disadvantages as well:

- Heavier than Rushton due to its low power number and increased vertical load fluctuation. Possibly lighter than other deeply concave turbines, however, as its power number is a bit higher.
- In common with other deeply concave turbines, if subjected to an overload, the blades can open up, increasing overload and leading to catastrophic failure.

**Table 4.4** BT-6 flooding constants.

| D/T | $K_{fl}$ | $K_{fl2}$ | $N_{Frc}$ |
|---|---|---|---|
| 0.231 | 0.17 | 0.53 | 1.17 |
| 0.261 | 0.26 | 0.91 | 0.77 |
| 0.299 | 0.43 | 1.43 | 0.47 |
| 0.323 | 0.62 | 1.45 | 0.325 |
| 0.351 | 0.8 | 1.75 | 0.25 |
| 0.365 | 1.1 | 2.27 | 0.18 |

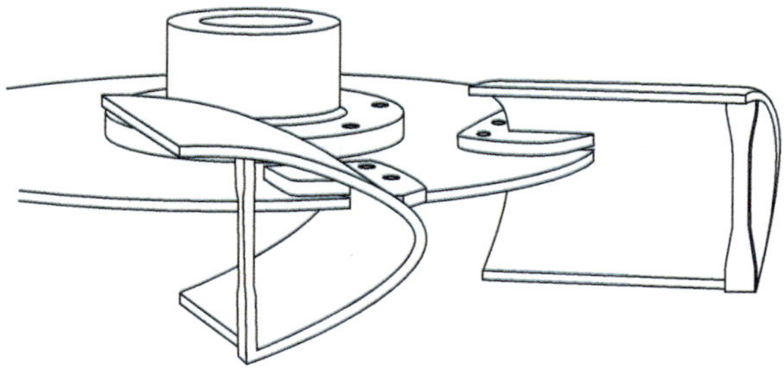

**Figure 4.23** Blade struts.

However, this can be prevented in the design phase by adding struts as shown in Figure 4.23. The strut design must avoid von Karman vortex shedding.

## Examples of Mixed Flow Impellers

Though one could imagine a number of ways to get to a mixed flow pattern, the main way is to have a nominally axial flow impeller but with a blade angle greater than the boundary separation angle. In turbulent flow, that is usually somewhere around 40°. Up until hydrofoil impellers came to be popular, the standard 45° pitched blade turbine was called an axial flow turbine. Now we recognize it as a mixed-flow impeller. It is depicted in Figure 4.24.

The version shown in Figure 4.24 is pretty much the industry standard. Its blade w/D is 0.2 and has a turbulent power number of 1.37 (based on the flat-to-flat diameter) over a range of D/T ratios from 0.25 to 0.5.

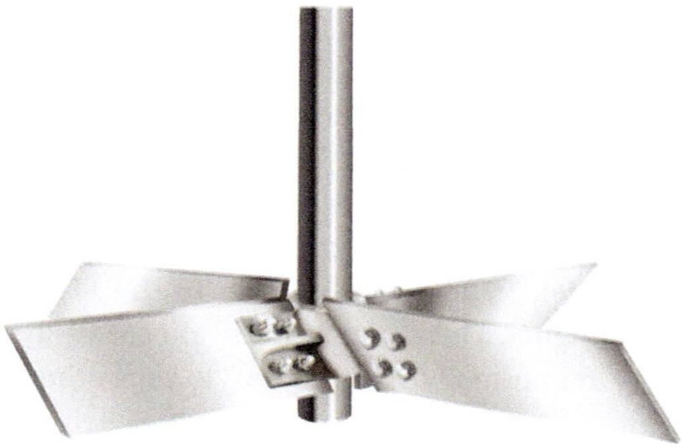

**Figure 4.24** Pitched-blade turbine. *Source:* NOV, Inc.

Pitched blade turbines can be considered to be of mixed-flow character with an angle ranging from 40 to 75°. Below 40, they should probably be considered axial flow, and above 75°, they are mostly radial.

Variants of the pitched blade turbine include numbers of blades from 2 to 8, and they can be built with a disc or open style.

The best use of a pitched blade turbine is as a top impeller when the gas in the headspace must be sucked in, as is commonly the case with hydrogenators. They are also useful for pulling in dry powders from the surface by means of a vortex. Sometimes the tank baffles are cut off below the surface to increase the vortex depth.

Though not much has been studied about the gassing factors of pitched blade turbines, it is believed they are similar to hydrofoils.

Advantages include:

- Cheap
- Versatile

Disadvantages include:

- Not very effective at dispersion.
- Require 2–3 times as much torque for equal liquid motion performance as low-solidity hydrofoils. This is true even in laminar flow, and especially true for pseudoplastic and Bingham plastic fluids.

## Examples of Chaos Impellers

A number of manufacturers produce such impellers. They are mainly categorized by having an inner section that pumps one direction (usually down) and an outer section that pumps the opposite direction (usually up). Examples of such an impeller include the Ekato Intermig, Figure 4.25a, and the ProQuip Doubly Pitched impeller, Figure 4.25b. Another example, which puts a new "twist" on the design, is the Chemineer JT-2, Figure 4.26. This impeller is claimed to greatly reduce blend time compared to competitive products.

Reference [7] lists the power number of the Intermig (two-blades) as 0.23 Private communications report it as ranging anywhere from 0.21 to 0.32. Most likely, the power number is D/T dependent. The power number of the JT-2 is unknown to the author, but is probably higher. These impellers are almost always applied using more than one, and they are normally staggered 90° to each other.

Such impellers have a variable flow pattern that is D/T dependent. When the transition from the inner section to the outer section is set at 70.7% of the tank diameter, the area available for up flow and down flow is equal. This produces the most axial flow pattern, though there is still a lot of fluctuation. If the impeller is

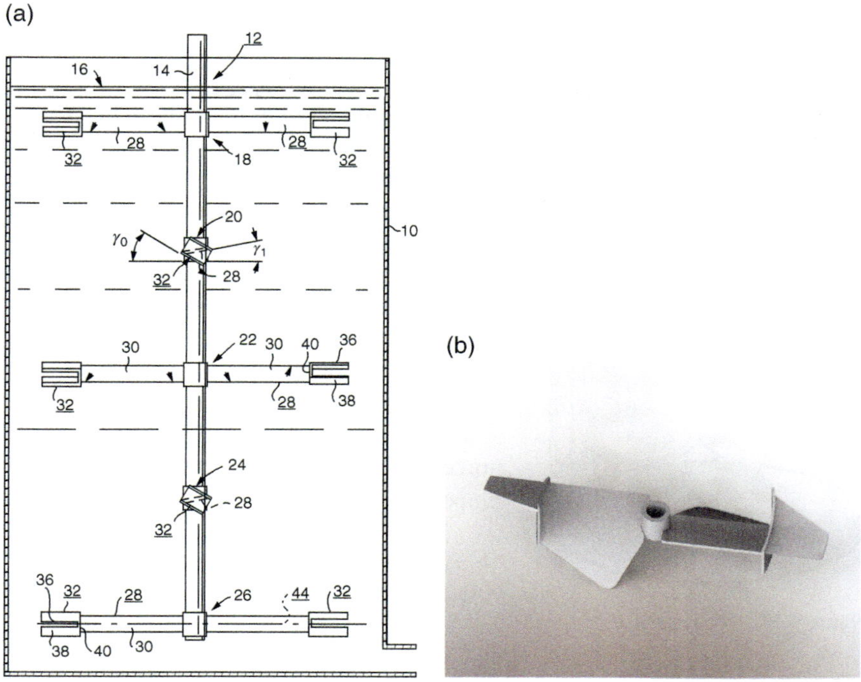


**Figure 4.25** (a) Ekato Intermig patent drawing. *Source:* U.S. Patent/Public domain. (b) ProQuip Doubly Pitched impeller. *Source:* Used with permission of PRO-QUIP.

applied at a larger diameter, more swirl (tangential flow) occurs. Smaller diameters create a flow pattern that is both more chaotic and more radial.

Though these impellers are less effective than most others for blending in either laminar or turbulent flow (so they are rarely used in the biofuel industry), they can excel in transition flow. Normal axial, radial, and pitched blade turbines can create torus-shaped dead zones which are very slow to blend, illustrated in Figure 4.27. These zones have a virtual surface that rotates around a circular core, and the entire torus slowly rotates around the shaft axis. There is usually one of these above and below each impeller when the Reynolds number is in the low hundreds. Under such conditions, the part of the tank volume outside the

**Figure 4.26** Chemineer JT-2. *Source:* NOV, Inc.

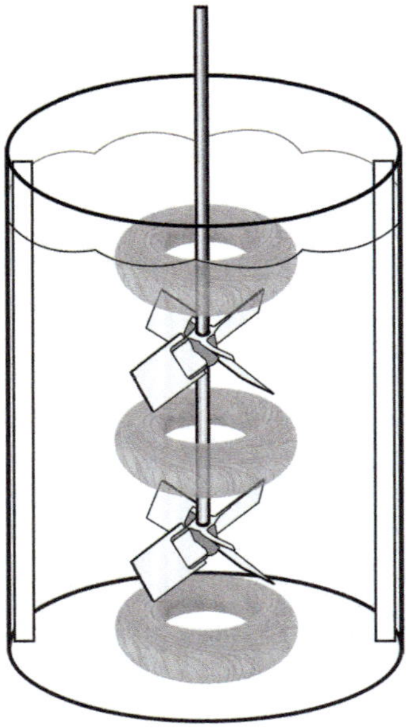

**Figure 4.27** Torus zones.

torus structure may blend in seconds or minutes, but the inside of the torus intermingles with the bulk fluid very little, so it may take hours to complete the blending. The cure to such phenomena is to introduce more irregularity to the flow pattern. Ways of doing that with conventional impellers include off-center mounting, angle mounting, asymmetric baffling, unequal impeller diameters, different impeller types on the same shaft, ramping the shaft speed up and down with time to change the Reynolds number and staggering the impellers, among other things. Some of these methods increase the hydraulic force imposed on the shaft. However, chaos impellers introduce the needed irregularity to avoid the formation of these torus structures, and can dramatically reduce blend time in the transition flow regime.

## Shear Rate, for Viscosity Determination

Conventionally, the average shear rate of an impeller in the volume swept by the impeller has been thought to be proportional to shaft speed, and mildly dependent on impeller type, using the Metzner–Otto concept [7]. This can be expressed as:

$$dv/dx = K_s * N \tag{4.16}$$

Table 4.5 gives commonly used values of $K_S$.

**Table 4.5** Metzner–Otto constants for common impellers.

| Impeller type | $K_s$ |
|---|---|
| High efficiency hydrofoil | 10 |
| 45° pitched blade | 11 |
| 90° straight blade | 11 |
| Rushton turbine | 11.5 |
| Ekato Intermig, D/T = 0.8 | 40 |

Reference [8] suggests that the value of $K_S$ may be quite a bit lower for axial flow impellers, as this work indicates that $K_S$ is proportional to $N_Q$, and may be as low as 3.5 in the turbulent regime, and even lower in laminar flow. Regardless of which set of values is used for $K_S$, the resulting calculated shear rates are quite low, especially in production-scale equipment, where the shaft speed is often less than 100 rpm.

## Helical Ribbon, Aka Helix

Though such impellers probably have no use in this industry, I am including them for completeness, and because some have used them in pilot studies to agitate lignocellulosic slurries. When I have observed such studies, the agitation was terrible, as I will describe later. But first I will describe the impeller type and what it is useful for.

A helix impeller has one or more helical ribbons extending vertically from the bottom of the vessel up to the maximum liquid level. A common embodiment is shown in Figure 4.28. This embodiment has two ribbons, or "flights," as they are commonly called. The flight width is typically 8–12% of the impeller diameter, and the pitch (distance between the same flight and a 360° rotation) is usually between 0.5 and 1.5 times the impeller diameter, with 1 being most common. Such impellers are applied at D/T ratios between 0.9 and 0.98, thus requiring the vessel to be accurately made, with very little deviation from roundness. This limits their use to smaller vessel sizes; up to perhaps 3–4 m diameter.

Helix impellers excel in viscous, laminar flow applications, with viscosities typically ranging from 50 000 cP up to perhaps 30 million cP. They typically blend the tank contents in about 75 revolutions when a dual flight, 1:1 pitch impeller is used, in a Newtonian fluid. They are mostly used at shaft speeds less than 10 rpm. Power requirements are low, but torque is high. Their main use is in bulk polymer reactors, such as polystyrene.

Conventional biofuel and renewable chemical applications typically have viscosities ranging from 1 to less than 1000 cP, so there is no need to use such an impeller in those industries.

The exception is lignocellulosic slurries, which can have low-shear viscosities of greater than 500 000 cP. Thus, it is tempting to use a helix impeller for this application. However, for a helix to work effectively, the wall viscosity must not be substantially lower than the bulk

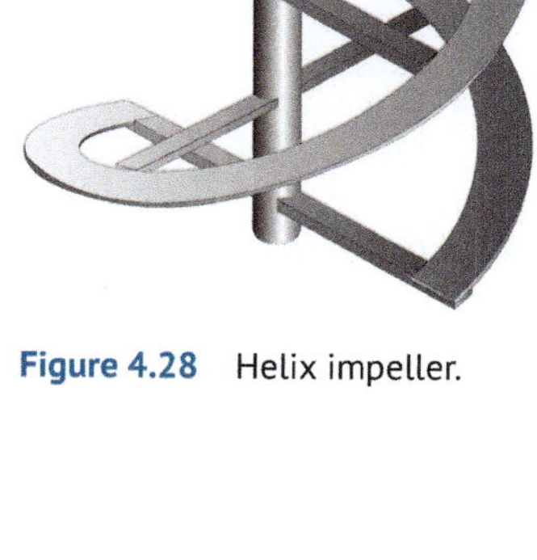

**Figure 4.28** Helix impeller.

viscosity. If it is, the fluid will "slip" at the wall, and the tank contents will rotate en masse with the impeller, thus providing movement but little or no mixing. When can this happen? If the fluid is strongly shear thinning, because the shear at the wall is orders of magnitude higher than the bulk shear, the apparent viscosity will be low at the wall. This can happen with some polymers and polymer emulsions. It can also happen if the tank contents are being heated; the higher temperature at the wall will lower the wall viscosity.

In the case of lignocellulosic and other high-solids fibrous slurries, the rheology is always shear thinning and usually has a yield stress. Every attempt I have personally witnessed to agitate such materials with a helix has resulted in mass rotation of the batch contents (like a wad of wet paper), which is a useless process result. Fortunately, simple turbines applied at the correct D/T ratio can mix such slurries very well, as will be described in more detail in Chapter 13.

## What Impeller Is Best for Most Biofuel Applications?

This is a complicated subject, and it depends on the application. However, for flow-controlled applications (which will be covered in Chapter 5, and which most biofuel applications fall under), there is a rational and objective basis, provided one can get appropriate quantitative data from the vendors.

Since power cost is directly related to power draw, and capital cost is closely related to torque (P/N), a rational basis is to see how much flow various impellers can produce when applied at equal power and equal torque, which also means equal shaft speed when power draw is fixed. Because different impeller styles have different power numbers, this means the impeller diameters will be unequal. This method of comparison makes more sense, in my opinion, than comparing at equal diameter and equal speed, or equal diameter, equal flow (which means different speeds and torques). The following method was derived by my former colleague, Dr. Julian B. Fasano.

From the definition of power number, one can see that:

$$D^5 \propto 1/N_P \tag{4.17}$$

From the definition of pumping number, one can derive that:

$$Q \propto N_Q D^3 \tag{4.18}$$

This can be substituted and rearranged to:

$$D^5 \propto \left(Q/N_Q\right)^{5/3} \tag{4.19}$$

Table 4.6 Impeller comparison.

| Impeller type | D/T | $N_P$ | $N_Q$ | $N_Q/(N_P)^{0.6}$ |
|---|---|---|---|---|
| Pitched blade | 0.3 | 1.37 | 0.8 | 0.662 |
| Narrow hydrofoil | 0.409 | 0.29 | 0.53 | 1.11 |
| Low angle profiled, double bend narrow hydrofoil | 0.425 | 0.24 | 0.57 | 1.34 |
| Wide hydrofoil | 0.316 | 1.05 | 0.72 | 0.7 |

Combining Eqs. (4.17) and (4.19) and eliminating $D^5$:

$$1/N_P \propto (Q/N_Q)^{5/3} \tag{4.20}$$

Re-arranging the above to express pumping (Q) directly:

$$Q \propto N_Q/N_P{}^{3/5} \quad \text{or} \quad Q \propto N_Q/N_P{}^{0.6} \tag{4.21}$$

Thus, the right hand of the proportionality indicates the relative flow produced at equal power and shaft speed, by inputting the pumping number and power number of the impeller in question.

So, how do various impellers on the market stack up? Though it is difficult to get good data on pumping and power numbers directly from vendors, especially for proprietary impellers, enough data have been published to make some generic comparisons. Table 4.6 compares four common impeller styles, starting with the generic pitched blade turbine, applied at a D/T ratio of 0.3. The data are based on turbulent flow. One can see that hydrofoils can pump much more than pitched blade turbines at equal power and shaft speed. The best can pump over twice as much. However, the best hydrofoils also have low power numbers, which means they must be applied at a larger D/T ratio to achieve the same results. Had we started out with a pitched blade turbine with a D/T ratio of 0.4, the comparison would have required hydrofoil impellers with a D/T ratio of more than 0.5, which can create poor flow patterns in low viscosity liquids. Still, the hydrofoils will generally be the preferred impeller design, as they will always save torque for equal pumping, and usually save power as well.

## Summary of Chapter

We have identified the major categories of impellers used in biofuels and renewable chemicals, and given examples of each. When possible, we have provided a limited amount of quantitative performance data for some example impellers. Such data may include power numbers, pumping numbers, gassing factors, and flooding characteristics. Most such data are held as private information by various manufacturers, so we must depend on them to do final calculations. The data presented here should be sufficient for ballpark estimation of impeller sizes.

We have presented a rational basis for choosing and comparing impellers for flow velocity controlled applications.

We have also presented some information on a helix impeller, and why it should not be used in this industry as we understand it today.

Impeller development is a continuing process. More specialty impellers will be on the market as time progresses. But the basic principles presented herein will apply to all future impellers.

## List of Symbols

| | |
|---|---|
| $D$ | Impeller diameter |
| $F_g$ | Alternate symbol for gas flow rate |
| $g$ | Acceleration due to gravity |
| $K_{fl}$ | Flooding proportionality constant |
| $K_{fl2}$ | Flooding proportionality constant above $N_a = 0.2$ for BT-6 |
| $K_s$ | Shear rate constant |
| $N$ | Shaft speed |
| $N_i$ | Alternate symbol for shaft speed |
| $N_a$ | Aeration number |
| $N_{Fr}$ | Froude number |
| $N_{Frc}$ | Froude number where slope changes for BT-6 flooding curve |
| $N_P$ | Power number |
| $P$ | Power draw |
| $P_g$ | Gassed power draw |
| $P_u$ | Ungassed power draw |
| $Q$ | Impeller pumping rate, Vol/time |
| $Q_g$ | Gas flow rate |
| $T$ | Tank diameter |
| $V$ | Liquid volume |
| $w$ | Blade width or height |
| $\rho$ | Liquid density |

## References


**1** Benz, G. (2021). *Agitator Design for Gas-Liquid Fermenters and Bioreactors*. Wiley InterScience.

**2** Middleton, J.C. and Smith, J.M. (2004). Gas-liquid mixing in turbulent systems. In: *Handbook of Industrial Mixing: Science and Practice* (ed. E. Paul, V. Atiemo-Obeng and S. Kresta), 585–638. Hoboken, NJ: Wiley.

**3** Myers, K.J., Janz, E.E., and Fasano, J.B. (2006). Gas dispersion capabilities of the CD-6 impeller. *Proceedings of the 12th European Conference on Mixing*, Bologna (27–30 June 2006).

**4** Vrabel, P., van der Lans, R.D., Luyben, K.C.A.M. et al. Mixing in large scale vessels stirred with multiple radial or radial and axial up-pumping impellers: modelling and measurements. *Chemical Engineering Science* 55: 5881–5896.

**5** Daub, A. (2015). Hydromechanical stress in aerated stirred tanks. PhD thesis, Rhein-Westfalia Technical Institute, 28 May 2015.

**6** Myers, K.J., Janz, E.E., and Fasano, J.B. (2006). Gas dispersion capabilities of the BT-6 impeller. *Annual AIChE Meeting, Multiphase Mixing I: Gas-Liquid Mixing, Paper 554b*, San Francisco, CA (16 November 2006).

**7** Metzner, A. and Otto, R. (1957). Agitation of non-Newtonian fluids. *AIChE Journal* 3: 3–10.

**8** Wu, J., Graham, J., and Mehidi, N. (2006). Estimation of agitator flow shear rate. *AIChE Journal* 52 (7): 2323–2332.

# 5

# Agitator Design for Liquid Motion

## Introduction

The vast majority of agitator applications in the biofuels industry may be classified as liquid motion. By that we mean that the agitator's main function is to make sure there is liquid motion throughout the tank. That is, there should be no "dead" spots or stratified layers. The required degree of agitation can depend on the kind of fluid being agitated, as well as any fluid property differences between fluids being blended. The principal property differences can be expressed as density difference ratio and viscosity ratio.

This chapter lays out the fundamentals of how to design for liquid motion, emphasizing turbulent flow, as all non-cellulosic biofuel and renewable chemicals applications the author is aware of have low enough viscosities that the tank will be in turbulent flow. Cellulosic materials are sufficiently special that Chapter 13 is devoted to them.

The main principles presented here were published by Chemineer in Ref. [1]. Here we have updated them for current impeller technology and some of the author's experience.

### Concept of Characteristic Fluid Velocity

It has been found through decades of empirical experience that the ability to create liquid motion and avoid dead spots in a tank is momentum driven. In practice, this means it relates to fluid velocity. But what fluid velocity should be used as a measure of overall agitation intensity? The velocity at the impeller tip is essentially equal to the impeller tip speed, and is much higher than the average velocity. The velocity on solid surfaces such as the tank wall or baffles is zero, by boundary layer theory. And the velocity near such surfaces is lower than the average throughout the tank. In the 1960s, staff at Chemineer, Inc. proposed the concept of, what they called at the time, a bulk fluid velocity, by analogy with pipe flow.

*Agitator Design Technology for Biofuels and Renewable Chemicals*, First Edition. Gregory T. Benz.
© 2022 John Wiley & Sons, Inc. Published 2022 by John Wiley & Sons, Inc.

They defined it as impeller discharge rate divided by the cross-sectional area of a tank having an equal diameter and liquid height (Z/T = 1), also known as a "square batch." Mathematically,

$$v_c = Q/A_{eq} \tag{5.1}$$

The implication of the term "bulk velocity" is that it represents the average velocity throughout the tank. However, since the impeller discharge does not pass through the tank, but, rather, recirculates back into the impeller, the average velocity is actually twice as high as the velocity calculated in Eq. (5.1). Therefore, a better term would be "characteristic velocity." That is the term we will use in this book.

To help visualize this concept, the actual velocity in the tank tends to match the calculated characteristic velocity approximately at the general location depicted in Figure 5.1.

## How to Calculate

Equation (5.1) is the basis for calculating characteristic velocity. There are two parts of the equation: Q, or pumping capacity, and $A_{eq}$, which is the equivalent area of a volume of liquid equivalent to a cylinder with an aspect ratio (Z/T) of 1. This is often referred to as a "square batch" geometry. Let us deal with each of these parts separately.

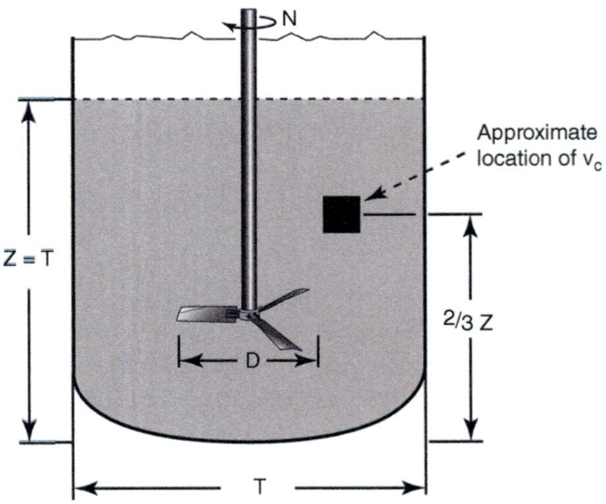


**Figure 5.1** Characteristic velocity visualization.

Q is calculated as follows:

$$Q = N_Q N D^3 \tag{5.2}$$

So where do we get the value of the pumping number, $N_Q$? The pumping number is a function of impeller type, D/T ratio, and Reynolds number, primarily. (There can be other geometric parameters which affect it, but we will focus on these, as they are the main ones.) Given that there are many dozens of impeller styles on the market today, and many are proprietary designs with unpublished performance data, it may be difficult to get pumping number data for some of them. However, broadly speaking, most impellers used for flow velocity-controlled applications (which encompasses most applications in the biofuels industry) fall into either low-solidity hydrofoils or standard 45° pitched blade turbines. Pitched blade turbines, of course, have been around a long time, and their performance characteristics are fairly well known, though not all references report the same values.

Though there are many hydrofoil impellers on the market, the most common ones in use have fairly similar characteristics. They typically have three blades and a turbulent power number between 0.28 and 0.32 at a D/T of 0.33 and a C/T of about 0.33. They typically have a blade angle at the hub between 25 and 35° and a chord angle at the tip of about 18–22°.

So, for the purposes of this book, we will use the typical performance characteristics of a pitched blade turbine and a generic hydrofoil. These characteristics are tabulated in considerable detail in Appendices 1 and 2. For this chapter, we will show a graphic view of power and pumping numbers to illustrate the concept, and will use the appendices for calculation purposes. The values in the appendices are intentionally conservative, especially for power numbers in laminar flow, so as to cover any dips in the power number curve in transition flow and to allow for some vendor variations. They are based on data from several sources, and may not agree exactly with some that have been published. But they should be within a 10% error band in most cases except in laminar flow, where they may be as much as 20% high. Note that both power number and pumping number depend not only on Reynolds number, but on D/T ratio. Many vendors and publications neglect this fact. Some additional impeller data can be found on the website listed as Ref. [2], as well as in the book listed as Ref. [3].

The pumping number for a generic hydrofoil is illustrated in Figure 5.2, for D/T ratios of 0.25, 0.35, and 0.45. Similarly, the pumping numbers for a typical pitched blade turbine are shown in Figure 5.3. Note the visible dependency on D/T ratio.

One may notice that the pumping numbers are lower for hydrofoils than for pitched blade turbines. This means that a hydrofoil must be larger in diameter than a pitched blade turbine, or rotated at a higher shaft speed, to pump the same amount. That may seem to be a negative for hydrofoils. However, when we look at

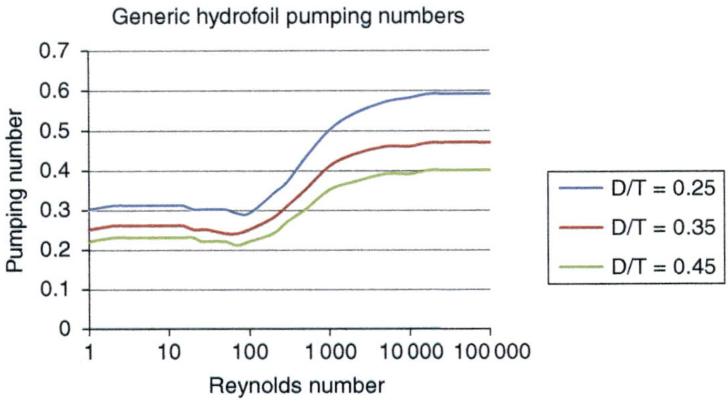


**Figure 5.2** Hydrofoil pumping numbers.

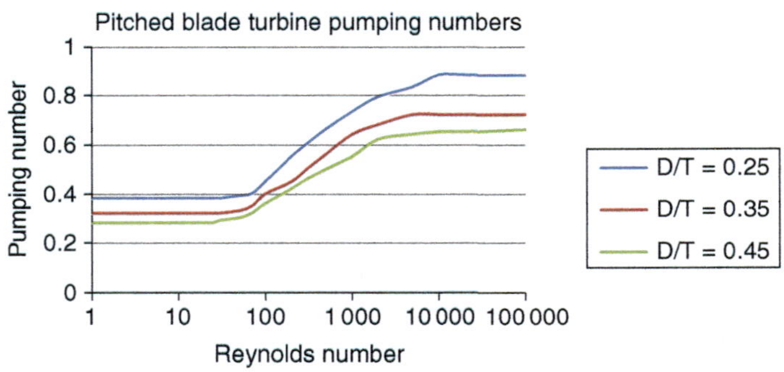


**Figure 5.3** Pitched blade turbine pumping numbers.

the power numbers, a different picture emerges. Figures 5.4 and 5.5 illustrate power numbers over a similar range.

On average, the turbulent power number of a pitched blade turbine is about 4.5 times as high as that of a hydrofoil. We will see that at equal pumping, a hydrofoil uses much less torque and usually much less power, depending on relative diameters.

Now we are finally ready to illustrate a pumping capacity calculation. Let us compare a hydrofoil with a D/T of 0.45 in a 10 m tank to a pitched blade turbine with a D/T of 0.35, operating in water, at a shaft speed of 25 rpm. From Appendix 1, our hydrofoil will have a pumping number of 0.4. From Appendix 2, our pitched blade turbine will have a pumping number of 0.72. So, now we can calculate pumping rate.

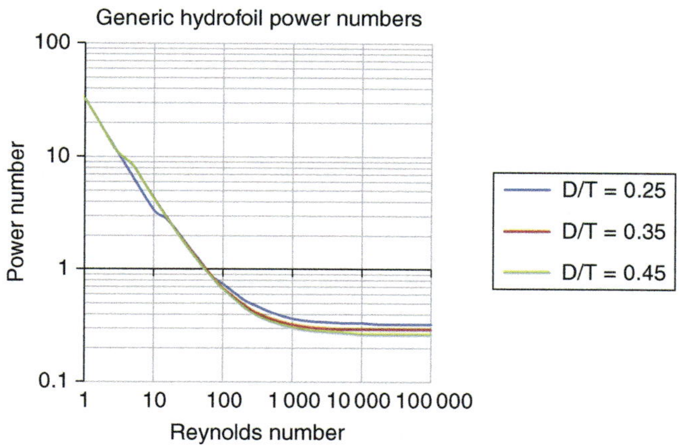


**Figure 5.4** Hydrofoil power numbers.

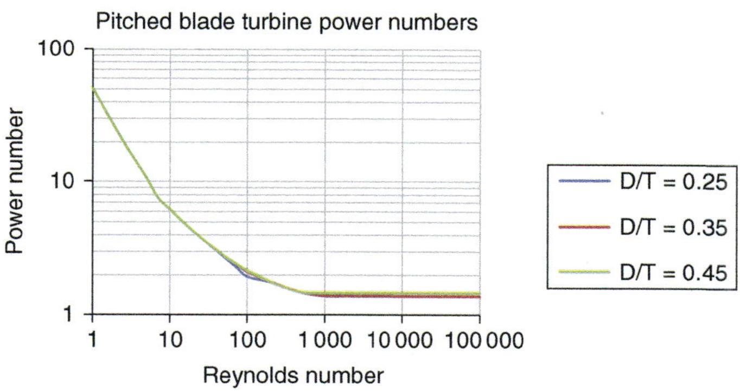


**Figure 5.5** Pitched blade turbine power numbers.

For the hydrofoil:

$$Q = N_Q N D^3 = 0.4 * 25/\text{minute} * \left(0.45 * 10\text{m}\right)^3 = 911\ \text{m}^3/\text{min} \tag{5.3}$$

For the pitched blade turbine:

$$Q = N_Q N D^3 = 0.72 * 25/\text{minute} * \left(0.35 * 10\ \text{m}\right)^3 = 772\ \text{m}^3/\text{min} \tag{5.4}$$

So, in this case, the larger hydrofoil impeller pumps more. But let's compare them at equal pumping. If we speed up the pitched blade turbine to (911/772)*25, or 29.5 rpm, the pumping will be the same. But pumping is not the only thing we must calculate. We also must compare power draw. Appendices 1 and 2 assign turbulent power numbers of 0.26 and 1.37, respectively, for the hydrofoil and pitched blade turbines. Using the definition of power number and SI units, we have the following results.

For the hydrofoil:

$$P = N_P \rho N^3 D^5 = 0.26 * \left(1\ 000\ \text{kg}/\text{m}^3\right) * \left(25/60\right)^3 * \left(4.5\ \text{m}\right)^5 \\ = 34\ 706\ \text{W} = 34.7\ \text{kW} \tag{5.5}$$

For the pitched blade turbine:

$$P = N_P \rho N^3 D^5 = 1.37 * \left(1000\ \text{kg}/\text{m}^3\right) * \left(29.5/60\right)^3 * \left(3.5\ \text{m}\right)^5 \\ = 85\ 521\ \text{W} = 85.5\ \text{kW} \tag{5.6}$$

Thus, in this case, the larger hydrofoil pumps the same amount as the smaller pitched blade turbine at 41% of the power and 48% of the torque. The general rule of thumb is in turbulent flow, hydrofoils require half of the torque of pitched blade turbines for velocity-controlled applications.

Part two of the calculation is to calculate the velocity, $v_c$. Using the same example, if the liquid level equals the tank diameter, then $T_{eq} = T = 10\,\text{m}$. Therefore:

$$v_c = Q/A_{eq} = Q/\left(\pi T^2/4\right) = 911\ \text{m}^3/\text{min}/\left(\pi * \left(10\ \text{m}\right)^2/4\right) = 11.6\ \text{m}/\text{min} \tag{5.7}$$

But what if the actual geometry is not a square batch? We must calculate as though it were. This may be done by defining $T_{eq}$:

$$T_{eq} = \left(4V/\pi\right)^{1/3} \tag{5.8}$$

A 10 m square batch would consist of a volume of $785.4\,\text{m}^3$. But let us suppose the same 10 m diameter tank had an actual volume of $850\,\text{m}^3$. Then, using Eq. (5.8), we would get $T_{eq} = 10.27\,\text{m}$. We would then re-do the previous calculations based on this diameter and on an equivalent area of $82.79\,\text{m}^2$, rather than the value of $78.54\,\text{m}^2$ based on a 10 m diameter. This would mean that $v_c$ would be about 5% less at equal pumping, which is only logical since the liquid volume is greater. We discuss this in more detail in the next section, "Effect of Aspect Ratio."

## Effect of Aspect Ratio

First of all, we must explain why all calculations are done on the basis of a square batch. Let us examine Figure 5.6, which depicts the exact same liquid volume contained in tanks of two different aspect ratios: 1 and 1.64. Although the volume is the same, the vessel with the higher aspect ratio has a smaller diameter, and therefore has less cross-sectional area. If we used actual tank cross-sectional area in our calculations, we would conclude that the same impeller and shaft speed would create a greater value for $v_c$ in the taller tank, even though the volume is the same.

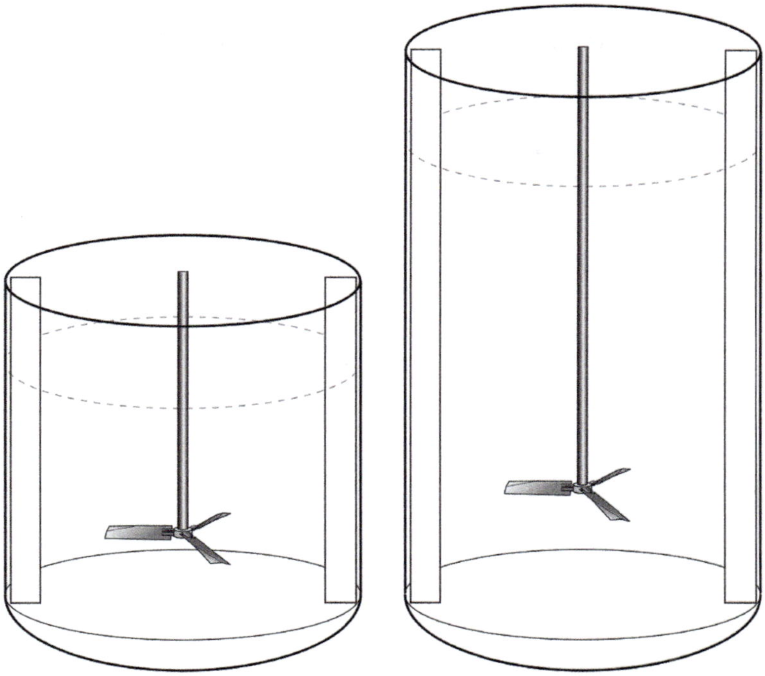


**Figure 5.6** Comparative aspect ratios.

This goes against reason, experience, and probably against conservation of momentum. What actually happens is that the velocity near the impeller is indeed greater, but the velocities in the upper regions of the tank are lower in the taller vessel, so the average balances out.

Though the calculations are done on the basis of a single impeller in a square batch geometry, aspect ratio, combined with viscosity (in the form of Reynolds number), determines how many impellers are needed, and their placement. Table 5.1 delineates recommended quantity and placement of impellers.

The aspect ratio, or Z/T value, for the first impeller represents the maximum that a single impeller can handle. Since flow patterns do not mesh completely for multiple impellers, the incremental aspect ratio that additional impellers can handle is less than that of the first impeller. For example, though a single hydrofoil operating in a low-viscosity liquid can handle up to a 1.4 aspect ratio, two impellers can handle only up to a Z/T of 2.5 (i.e. 1.4 + 1.1), three impellers up to 3.6, etc. If for some reason a lower impeller must be placed unusually close to the tank bottom (say, less than T/5 off bottom), the initial Z/T ratio allowed may be reduced somewhat. Other than money concerns, when in doubt, add another impeller.

**Table 5.1** Aspect ratio impeller guidelines.

| Impeller type | Reynolds number | Z/T, first impeller | Z/T incremental |
|---|---|---|---|
| Hydrofoil | <300 | 0.9 | 0.7 |
| Hydrofoil | 300–1000 | 1.1 | 0.9 |
| Hydrofoil | >1000 | 1.4 | 1.1 |
| Pitched blade | <100 | 0.7 | 0.6 |
| Pitched blade | 100–800 | 0.9 | 0.7 |
| Pitched blade | >800 | 1.2 | 1.0 |
| Radial | <50 | 0.5 | 0.4 |
| Radial | 50–300 | 0.6 | 0.5 |
| Radial | >300 | 0.8 | 0.7 |

## Special Case: Tickler Turbine

When a tank must be agitated down to the lowest possible liquid level, such as when draining the tank, it is common practice to place a lower open-style radial impeller as close as possible to the tank bottom. When an impeller is used this way, it is often called a "tickler" or "kicker" turbine. It should be sized so that it can properly agitate the remaining volume when the main impeller is no longer submerged in the liquid. In practice this usually means it must invest about one-third of the power of the main axial impeller located above it.

Though theoretically the radial impeller pumps less than an axial, and therefore the combination of an axial main impeller plus a radial tickler will pump less than if the same power were invested entirely in the axial impeller, I prefer to calculate the characteristic velocity as though all of the power were in the axial impeller. The reason is that the radial impeller actually improves the flow pattern on the bottom of the tank, avoiding dead spots. The end result is that the mixing effectiveness is not compromised by calculating this way, as evidenced by over 40 years of the author's experience. This can be understood better by referring to Figure 5.7.

In that figure, one can see that the tickler turbine creates a strong sweeping action on the bottom of the tank, avoiding a potential low-velocity zone directly under the main impeller. Contrast this to Figure 5.8, which has no tickler. If the D/T ratio is large enough, not only can a dead spot occur below the impeller, there can actually be a flow reversal, i.e. flow going up into the eye of a down-pumping impeller. In such a case, the tickler not only allows agitation at low liquid levels, it actually improves overall mixing by eliminating dead spots or flow reversals.

The subject of flow reversal is covered extensively in Ref. [4].

Figure 5.7 Axial impeller plus tickler.

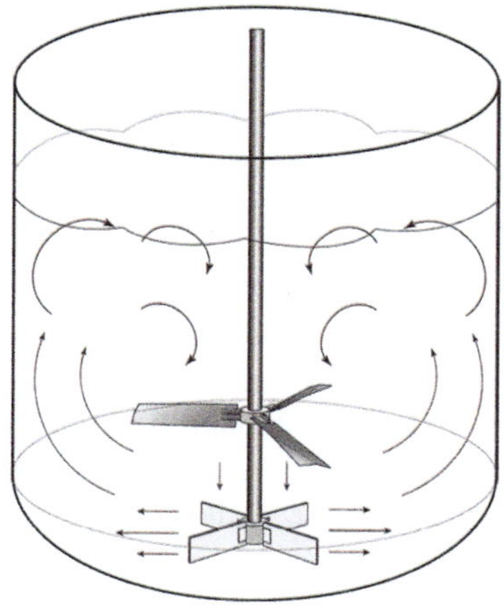

Figure 5.8 Axial flow without tickler.

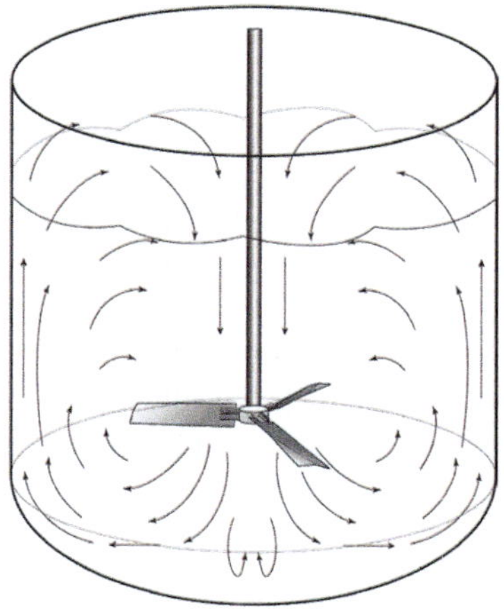

## Effect of Baffles

Proper performance of any impeller system will not be achieved if the correct baffle system is not used. In low-viscosity fluids (typically less than 2500 cP in a tank volume of less than 4 $m^3$, or 5000 cP in a tank volume of greater than 10 $m^3$), baffles are needed. The function of baffles is to prevent swirl and create the intended flow pattern, which is generally top-to-bottom in a flow-velocity controlled application. High viscosity acts the same as baffles, increasing drag at the wall, so baffles are not needed for high-viscosity liquids. As virtually all biofuel and renewable chemical applications are less viscous than the earlier guidelines, baffles will almost always be needed. The exception is in the early stages of reaction for lignocellulosic hydrolysis tanks and similar high-solids (generally above 6%) fibrous slurries. If baffles were used in these applications, they would create stagnant zones near the wall and around the baffles. More detail on this will be presented in Chapters 8 and 13.

So what number and size of baffles are to be used? In the United States, "standard" baffling consists of (4) baffles 90° apart, with a width of T/12, an offset of T/72 from the wall and T/36 off bottom. They normally extend to the maximum liquid level. The offset from the wall can be varied if there are solids present which may get lodged between the vessel and the tank wall.

In the EU, "standard" baffling is similar but is typically T/10 wide. This does not make a lot of difference. To compare the two, there is a dimensionless term called baffle number, $n_b$. It is defined as:

$$n_b = \text{number of baffles} * \text{width of baffles} / T \qquad (5.9)$$

When swirl is minimized, power draw is maximized for a given impeller system. The typical effect of baffle number on power draw is shown in Figure 5.9 for a D/T value of about 1/3.

One may readily calculate that the American standard baffling corresponds to a baffle number of 1/3, whereas the EU standard corresponds to 0.4. Figure 5.9 shows that there is little difference between the two.

Some have argued that when three-bladed impellers are used, three baffles should be used to avoid a beat frequency from passing blades. Unpublished studies I am aware of did not detect any such phenomena, with any combination of number of blades and number of baffles greater than two baffles. (Two or fewer baffles are hydraulically unstable.) Use of three baffles may save some money on the vessel cost, and it is perfectly acceptable regardless of the number of blades on

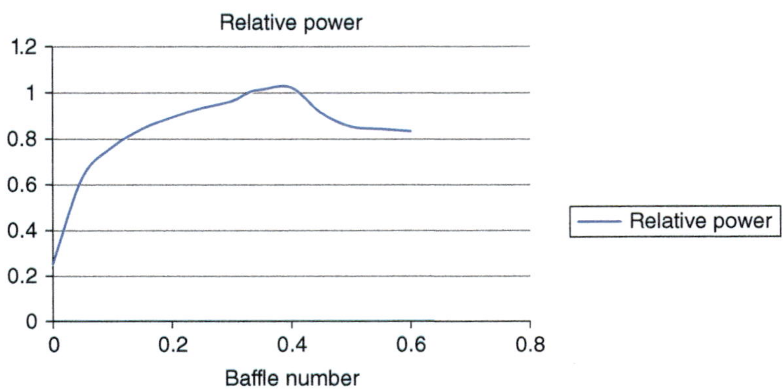


**Figure 5.9** Relative power vs. baffle number.

the impeller. However, the baffle width should be increased to T/9 ideally to maintain the same baffle number.

Some riggers and construction managers have noted that baffle supports are a very good place to attach rigging for installing and removing shafts and impellers. They note that it is much easier to rig from a four-baffle system than a three-baffle system. In the long run, the rigging labor savings may exceed any savings in the tank cost due to using three baffles.

## How to Relate Characteristic Velocity to Process Results

Up until now, in this chapter, we have focused on how to calculate characteristic velocity. However, this would be of no value if we could not relate it to process results. Fortunately, we can. There are two basic ways to do this. One is to relate the velocity to the ability to overcome fluid property differences (avoiding stratification or dead zones). The other is to relate to tangible field experience. We describe both ways in this book. But first, let's convert characteristic velocity into something simpler to use and which does not depend on more than one or two significant digits to be useful.

In the 1960s, Chemineer, Inc. (now a brand of NOV) devised a system of describing most agitator applications by means of a 1–10 scale, which they called ChemScale® level. Today, many have adopted this idea, but simply call it scale of agitation. That is the naming convention we will use in this book.

Originally, there was a scale of agitation for the three major classes of agitation: liquid motion, solids suspension, and gas dispersion. The latter two classifications have all but fallen out of use, as the 1–10 scale really does not work well for such applications. However, the scale is still used regularly for liquid motion applications, which includes most biofuel applications.

Though values of $v_c$ could range from 0 to very high values (up to the point where the liquid would be essentially ejected from the tank), Chemineer found, by experience, that about 90% of industrially significant applications in the liquid blending category have values of $v_c$ ranging from 6 to 60 ft./min, or 0.0305–0.305 m/s. They therefore assigned a 1–10 scale that is linear with characteristic velocity. In other words:

$$\text{Scale of agitation} = v_c\left(\text{ft.}/\text{min}\right)/6 = v_c\left(\text{m}/\text{s}\right)/0.0305 \qquad (5.10)$$

Though it is possible to calculate a scale of agitation less than 1 or greater than 10, applications at either extreme would be unusual, and would be treated as special cases. A scale of 1 is very mild; typically used for storage tanks where no active blending occurs, or very easy reactors and similar applications. A scale of 10 is normally used for very difficult blending or for cases where minor temperature gradients cannot be tolerated, such as in certain polymer reactors. As we shall see in later chapters, most biofuel applications fall in the 1–4 range, with 3 being most common.

Before getting into capability versus scale of agitation, there is a quick way to calculate scale based only on power draw, shaft speed, impeller type, and mass of fluid. This method is valid only for turbulent flow ($N_{Re} > 10\,000$), which is normally the case in biofuel applications. It is accurate within a couple of percent for common pitched blade turbines, and within 5–10% for common hydrofoils. The difference in accuracy is due to a different sensitivity to D/T ratio. The next equations will normally be slightly conservative for hydrofoils at D/T ratios less than 0.35.

For pitched blade turbines, we have:

$$\text{Scale of agitation} = 429\left(P/\left(M*N\right)\right)^{0.5}, \text{ where P is in kW, M is in kg and N is in rev/s} \tag{5.11}$$

Similarly, for common three-blade hydrofoils with a turbulent power number of about 0.3, we have:

$$\text{Scale of agitation} = 600\left(P/\left(M*N\right)\right)^{0.5}, \text{ where P is in kW, M is in kg and N is in rev/s} \tag{5.12}$$

Now we can relate scale of agitation to ability to overcome property differences and avoid stratification. The first measure of property differences is density difference ratio, defined as:

$$\text{DDR} = \left|\rho_a - \rho\right| / \rho \tag{5.13}$$

The greater this ratio is, the more agitation is needed to avoid stratification. Figure 5.10 gives a guide to how much agitation is needed as a function of DDR. For example, a scale level of 6 can handle a DDR of up to 0.6. This would be equivalent to adding a salt solution with a density of 1.6 g/cc to water. These guidelines are rough, and the scale of agitateon may need to be adjusted up or down according to how the additive is introduced. If the additive is denser than the bulk liquid, adding it to the surface or into the impeller swept area just above the impeller may reduce agitation requirements by one scale level. If the additive is less dense and added to the surface, the scale level may need to be increased. To the extent possible, the impeller should be submerged in the denser liquid. This allows it to transfer more momentum. So, if a tank starts out half full of a dense liquid and a less dense liquid is added, the impeller should be well covered by the dense liquid. Adding an upper impeller can help if a very low density liquid is added to the top, but if possible, low-density liquids should be added near the

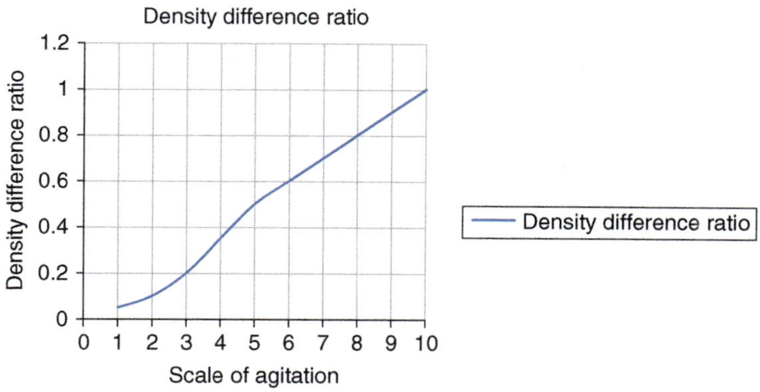


**Figure 5.10** DDR vs. scale of agitation.

bottom of the tank. Let gravity help out! (We have not studied agitation in low-g environments such as on the moon or Mars, but probably a lower scale of agitation would work there.)

The other property difference of importance is viscosity ratio. When liquids have different viscosities, they are more difficult to blend. This is particularly true of blending low-viscosity liquids into high-viscosity liquids. The impeller should be located in the viscous phase if the volume of said fluid allows, but if the viscous fluid is a minor additive, this will not be possible. Viscosity effect on blending difficulty can be expressed as:

$$\text{Viscosity ratio} = \mu_a / \mu_b \quad \text{or} \quad \mu_b / \mu_{a,}, \quad \text{whichever gives a value of 1 or more} \tag{5.14}$$

Figure 5.11 gives guidelines for scale of agitation required as a function of viscosity ratio. For example, if an additive is 300 times as viscous as the bulk fluid, we would normally need a scale of agitation of 3. If the additive can be injected into the impeller zone, the scale of agitation can be reduced if no other conditions require the higher scale level.

Those who have read Ref. [1] may note that the guidelines presented here differ from those in the reference. Specifically, Ref. [1] states that a scale level of 10 can handle a viscosity ratio of 100,000. The author's experience is that this is overly optimistic; such a viscosity ratio is very difficult to overcome. Thus, the guidelines we present in this book are adjusted downward.

Besides physical property differences, choosing the appropriate scale of agitation may be based on past experience, especially for slurries that have a sufficiently low settling rate that they act as liquids as long as enough agitation is applied. That experience may consist of what has worked in the past as

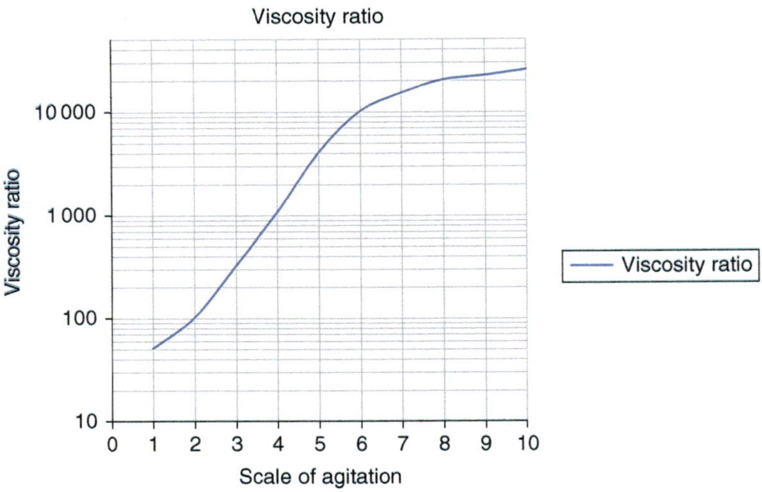


**Figure 5.11** Viscosity ratio vs. scale of agitation.

well as what has not worked. In fact, what has not worked can often be the more valuable information.

Biofuel and renewable chemical processes typically involve a number of such slurries, such as starch slurry, stillage, lignocellulosic residue, and others. The author has extensive experience with all of these. Specific application guidelines will be given in Chapters 7–10, and 13.

## Wide Range of Power Inputs Possible for Equal Results

The most important takeaway from this chapter is that there is not just one answer or power level that will meet the agitation requirements; there is a wide range of possibilities. D/T and impeller type are both involved. The general principle is that a small impeller must turn at a higher shaft speed than a larger one, and in so doing, it requires more power. This gets back to the basic principle that power = flow * head. A smaller impeller discharges at a higher velocity than a larger one pumping the same amount of fluid, and head is proportional to the square of velocity. So, the key to power savings is to use large impellers. However, this principle has limits. Years of research and experimentation, for example, have shown us that an installation with a D/T of greater than 1 will not be successful 😉. (This is to see if you are still awake!) On a more serious note, for low-viscosity fluids, D/T ratios larger than about 0.4 for pitched blade turbines and 0.5 for hydrofoils will often produce undesirable flow patterns, with too much radial flow and possibly flow reversal and/or dead zones at the tank bottom.

The wide spectrum of possible power inputs needed will be demonstrated in our example problems.

**Example 5.1** Problem 1 We wish to agitate a volume of 100 m$^3$ of liquid with a specific gravity of 1.15 and a viscosity of 50 cP, with a target scale of agitation of 3. Calculate speed and power required for both pitched blade turbines and typical hydrofoils, at D/T ratios of 0.25, 0.35, and 0.45.

**Answer**

We will go through the calculations for one case and then present the results in tabular format for all the requested cases.

Step 1: Calculate equivalent tank diameter using Eq. (5.8):

$$T_{eq} = \left(4V / \pi\right)^{1/3} = \left(4 * 100\text{m}^3 / \pi\right)^{1/3} = 5.031\text{m}$$

Step 2: Calculate required pumping capacity using Eq. (5.1), re-arranged and substituted with the definition of scale of agitation for velocity, and equivalent area:

$$\begin{aligned} Q &= v_c * A_{eq} = \text{scale of agitation} * 0.0305\ \text{m/s} * \pi T_{eq}^2 / 4 \\ &= 3 * 0.0305\ \text{m/s} * \pi * \left(5.031\ \text{m}\right)^2 / 4 = 1.82\ \text{m}^3/\text{s} \end{aligned}$$

Step 3: Calculate impeller diameter based on D/T ratio chosen, and equivalent tank diameter. We start with D/T = 0.25 and a pitched blade turbine:

$$D = 0.25 * 5.031\ \text{m} = 1.26\ \text{m}$$

Step 4: Look up pumping number from Appendix 2, assuming turbulent flow:

$$N_Q = 0.88$$

Step 5: Calculate required shaft speed, re-arranging Eq. (5.2):

$$N = Q / \left(N_q D^3\right) = \left(1.82\ \text{m}^3/\text{s}\right) / \left(0.88 * \left(1.26\ \text{m}\right)^3\right) = 1.034/\text{second} = 62\ \text{rpm}.$$

Step 5: Calculate Reynolds number to verify the assumption of turbulence

$$N_{Re} = D^2 N \rho / \mu = \left(1.26\ \text{m}\right)^2 * 1.034/\text{second} * 1150\ \text{kg/m}^3 / (50\ \text{cP} * \left(0.001\ \text{kg/m}-\text{s}-\text{cP}\right) = 37\ 756 \quad (5.15)$$

We have verified that it is turbulent, so turbulent pumping and power numbers may be used.

Step 6: look up power number.
From Appendix 2, the power number is 1.37.
Step 7: Calculate power, using the definition of power number:

$$N_P = P/\left(\rho N^3 D^5\right); P = N_P \rho N^3 D^5 = 1.37 * 1150\ \text{kg}/\text{m}^3 * \left(1.034/\text{second}\right)^3 * \left(1.26\ \text{m}\right)^5 = 5531\ \text{W} = 5.53\ \text{kW, or about 7.4 Hp.} \tag{5.16}$$

Now we have one data point for our solution. Table 5.2 presents the solution results for all cases requested. Rounding errors in the hand calculation have been eliminated.

A column showing relative torque has been added. In general, hydrofoils require about half the torque of pitched blade turbines, which means a smaller, less expensive gear drive may be used. This cost savings more than offsets the somewhat higher cost of the hydrofoil turbine. And the energy savings speaks for itself. Also, though results are shown for a D/T of 0.45, such a ratio may not work well for pitched blade turbines. A hydrofoil does not create as much radial flow at larger D/T ratios as a pitched blade turbine does.

This problem was solved from basic principles. It would be instructive to see how the results compare using the "quick" method of Eqs. (5.11) and (5.12). Table 5.3 shows the predicted scale of agitation using those equations for the same parameters presented in Table 5.2. Remember, the problem solutions in Table 5.2 were based on a scale of agitation of 3. Table 5.3 shows that Eqs. (5.11) and (5.12) result in predictions of the scale of agitation ranging from 2.93 to 3.03, or an error band of less than 3%. Since scale of agitation was never intended to be used with two significant digits for most applications, much less 3, Eqs. (5.11) and (5.12) are perfectly satisfactory methods of calculation for turbulent flow. For production scale equipment, this means they may be safely used for viscosities less than 100 cP.

**Table 5.2** Problem 1 summary of results.

| | Pitched blade turbine | | | | Hydrofoil | | | | |
|---|---|---|---|---|---|---|---|---|---|
| D/T | Pumping number | Power Number | Shaft speed (1/s) | Power (kW) | Pumping number | Power number | Shaft speed (1/s) | Power (kW) | Relative torque, hydrofoil/ pitched blade |
| 0.25 | 0.88 | 1.37 | 1.039 | 5.40 | 0.59 | 0.32 | 1.550 | 4.19 | 0.52 |
| 0.35 | 0.72 | 1.37 | 0.463 | 2.57 | 0.47 | 0.29 | 0.709 | 1.95 | 0.50 |
| 0.45 | 0.65 | 1.44 | 0.241 | 1.34 | 0.4 | 0.26 | 0.392 | 1.04 | 0.48 |

**Table 5.3** Comparison of quick turbulent calculation to rigorous method.

| D/T | Pitched blade | Hydrofoil |
|---|---|---|
| 0.25 | 2.93 | 2.95 |
| 0.35 | 3.02 | 2.98 |
| 0.45 | 3.03 | 2.93 |

The methodology for their use usually begins with choosing shaft speed. Most agitator gear drives have shaft speeds within ±4% of the AGMA (American Gear Manufacturers Association) standard speeds. These are, for production equipment, 5, 6, 7.5, 9, 11, 13.5, 16.5, 20, 25, 30, 37, 45, 56, 68, 84, 100, 125, 155, 190, 230, 280, and 350 rpm. The most common speeds in large biofuel plants will be between 20 and 84 rpm. The simplified procedure is thus:

1) Choose a shaft speed.
2) Based on either Eq. (5.11) or (5.12), depending on impeller type, calculate power required.
3) Round up to the nearest motor size.
4) Based on actual shaft speed (which may differ from the standard AGMA speed), size the impeller system to draw about 85% of motor power.

Problem 2 Use the same basic problem statement as Problem 1, but add the following additional constraints. The actual tank diameter is 3.5 m, due to shipping and plant layout constraints. A neutralization step consisting of adding 50% caustic, with a specific gravity of 1.5, is an additional requirement. What scale of agitation do we need? With hydrofoil impellers, how many do we need, and what might be several good power/speed/impeller diameter combinations to consider?

**Answer**

Step 1: The primary decision in agitator design for liquid motion is choosing the desired scale of agitation. Every other decision flows from this. For this problem, the determining factor is the addition of caustic. We use Eq. (5.13) to calculate density difference ratio (DDR):

$$\text{DDR} = \left(1.5 - 1.01\right)/1.0 = 0.5$$

Next we use Figure 5.10 to select the required scale of agitation. By inspection, we would choose a scale level of 5.

Step 2: Determine tank geometry. For simplicity, we will assume a cylindrical tank with a flat bottom. Most process tanks will have a dished bottom, which would make the liquid level slightly higher.

We have a liquid volume of $100\,m^3$, and our tank diameter is fixed at 3.5 m. We can thus solve for liquid level as follows:

$$Z = 4 * 100\ m^3 / \left(\pi * (3.5\ m)^2\right) = 10.39\ m.$$

Thus, we have Z/T = 10.39/3.5 = 2.97.

Step 3: Determine number of impellers. From Table 5.1, for a low-viscosity liquid, we see that the first impeller can handle up to a Z/T of 1.4, and additional impellers can handle 1.1 each. So, at a Z/T of 2.97, the first 1.4 times T requires one impeller. The remainder of 1.57 is greater than 1.1 but less than 2.2, so two additional impellers are required. Therefore, a total of three impellers is required.

Step 4: Determine power/speed impeller combinations that will satisfy the problem magnitudes. For this step, we will use the "quick" method of Eq. (5.12), and solve for the turbulent power number iteratively to size the impellers. We will go through this for one shaft speed, and then summarize the results in Table 5.4. We will start with 37 rpm.

Using Eq. (5.12), we have:

$$\text{Scale of agitation} = 600\left(P/(M * N)\right)^{0.5}; \quad P = (\text{scale})^2 * M * N / 360000$$

Assuming the additive is small, the density of the liquid is $1000\,kg/m^3$, and N is in 1/s or equals 37/60 = 0.6168/second. So, $P = 5^2 * 100000 * 0.6167/360000 =$ 4.283 kW (about 5.74 Hp).

Assuming a trial power number of 0.3, we would solve for diameter using the definition of power number, accounting for three impellers and expressing power in watts:

$N_P = P/\rho N^3 D^5$; $D = (P/(N_P \rho N^3))^{0.2} = ((4.283 * 1000/3)/(0.3 * 1000 * 0.6168^3))^{0.2} =$ 1.83 m. That corresponds to a D/T of 1.83/3.5 = 0.52. According to Appendix 1, the power number at such a D/T ratio is 0.26. So, we would iterate again and use a power number of 0.26, which would give us a D of 1.88 and a D/T of 0.54. That would raise the power number to 0.28. Interpolating with a power number of 0.27, we would get a D of 1.87 m, and a D/T of 0.53. Such a large D/T ratio is undesirable in low-viscosity liquids, as it will create more radial flow and possibly a flow reversal at the bottom of the tank (which could be ameliorated by use of a "tickler" turbine). So, we would continue our array at higher AGMA speeds. The results are presented in Table 5.4.

Although the lower speeds would be disqualified based on large D/T ratio, there is at least a 5 : 1 range of power draw for acceptable designs, based on a 2 : 1 range of impeller diameters. Had we looked at pitched blade turbines also, the range would have been even greater. It is possible that the higher speed ranges would be disqualified because of shaft design issues, especially critical speed,

**Table 5.4** Problem 2 results.

| AGMA speed (rpm) | Shaft power (kW) | Standard motor size (kW) | Impeller diameter (m) | D/T ratio | Comments | $N_P$ basis |
|---|---|---|---|---|---|---|
| 30 | 3.47 | 5.5 | 2.01 | 0.58 | D/T too large | 0.28 |
| 37 | 4.28 | 5.5 | 1.86 | 0.53 | D/T too large | 0.27 |
| 45 | 5.21 | 7.5 | 1.74 | 0.50 | Maximum D/T | 0.26 |
| 56 | 6.48 | 11 | 1.59 | 0.45 | | 0.26 |
| 68 | 7.87 | 11 | 1.47 | 0.42 | | 0.265 |
| 84 | 9.72 | 15 | 1.33 | 0.38 | | 0.28 |
| 100 | 11.58 | 15 | 1.24 | 0.35 | | 0.29 |
| 125 | 14.47 | 18.6 | 1.12 | 0.32 | | 0.3 |
| 155 | 17.94 | 22 | 1.02 | 0.29 | | 0.31 |
| 190 | 21.99 | 30 | 0.94 | 0.27 | | 0.32 |
| 230 | 26.62 | 37 | 0.87 | 0.25 | | 0.32 |

which will be addressed in Chapter 15. Final unit selection would depend on mechanical feasibility, capital cost, and present worth of operating cost (principally power cost). However, unless there were a compelling reason to use the higher power/speed combinations, the 45 or 56 rpm designs would almost certainly win in any economic evaluation that included the power cost.

## Geometries Other Than Vertical Cylindrical Vessels

As stated in the Preface to this book, our focus is on design for vertical cylindrical vessels. However, other geometries have been used. Such geometries often are prone to slow moving areas, and the overall scale of agitation must be increased to compensate.

One such geometry is square or rectangular tanks. Generally, the scale of agitation should be increased by about 20% for such tanks. If the length of the sidewall is greater than 1.5 times the tank width, more than one agitator will be required, in increments of 1.5 widths. Even with that increase in agitation, solids may settle in the corners. These kinds of tanks are generally not recommended for slurries, especially lignocellulosic biomass. Testing would be needed to verify the design.

Another such geometry is horizontal cylindrical. The author's experience indicates that the scale of agitation should be increased about 33%. For example, if one would design for a scale of agitation of 3 in a vertical cylindrical tank, the scale of agitation would be increased to 4 in a horizontal cylindrical tank. Again, this geometry is not recommended for lignocellulosic applications; testing would be needed to verify that it would work.

## Summary of Chapter

We have demonstrated how to calculate pumping capacity and characteristic liquid velocity, and how to relate these parameters to physical process results. We have shown that there are many possible designs for any given application, covering a range of speeds, power inputs, and impeller sizes and types. Economic analysis should be part of the final design choice.

## List of Symbols

| | |
|---|---|
| $A$ | Tank cross-sectional area |
| $A_{eq}$ | Equivalent tank cross-sectional area |
| $D$ | Impeller diameter |
| DDR | Density difference ratio |
| $g$ | Acceleration due to gravity |
| $M$ | Mass of fluid agitated |
| $n_b$ | Baffle number |
| $N$ | Shaft speed |
| $N_P$ | Power number |
| $N_Q$ | Pumping number |
| $N_{Re}$ | Reynolds number |
| $P$ | Power draw |
| $Q$ | Impeller pumping or discharge rate |
| $T$ | Tank diameter |
| $T_{eq}$ | Equivalent tank diameter |
| $V$ | Liquid volume |
| $v_c$ | Characteristic fluid velocity |
| $w$ | Blade width or height |
| $Z$ | Liquid height |

## Greek Symbols

| | |
|---|---|
| $\rho$ | Liquid density |
| $\rho_a$ | Density of additive (layer absent the impeller) |
| $\rho_b$ | Bulk density (impeller should be in denser phase) |

$\mu$ Liquid viscosity
$\mu_a$ Viscosity of additive phase
$\mu_b$ Bulk liquid viscosity

## References


1 Hicks, R., Morton, J., and Fenic, J. (1976). How to design agitators for desired process response. *Chemical Engineering Magazine* 22–30.
2 Post, T. and Post Mixing Optimizations and Solutions, LLC (February 2013). Website, https://postmixing.com/mixing%20forum/impellers/impellers.htm (accessed 23 November 2021).
3 Paul, E., Atiemo-Obeng, V., and Kresta, S. (ed.) (2004). *Handbook of Industrial Mixing*. Wiley InterScience.
4 Myers, K., Bakker, A., and Corpstein, R. (1996). The effect of flow reversal on solids suspension in agitated vessels. *Canadian Journal of Chemical Engineering* 74 (6): 1028–1033.

# 6

# Side-Entering Agitator Design

## Purpose of Chapter

This chapter will describe both the why and the how of employing side-entering agitators in biofuel/renewable chemical processes. It will compare their pros and cons with top entering and describe where they are most likely to be used.

## Why Choose Side Entering vs. Top Entering

In a word: cheap. Side-entering agitators have shorter shafts of smaller diameter than typical top-entering units. They also have only one impeller, and of a fairly small diameter. Gear-driven units usually have only one reduction stage, as compared to top-entering units, which typically have two or three reduction stages, making the side-entering gear drives less expensive. Some side-entering units have belt drives, which are less expensive still.

Mounting requirements for side-entering agitators are simple. They mount to a nozzle on the side of the tank. Since low-pressure tanks may be unable to support the load entirely from the nozzle, auxiliary support is usually needed. Two common methods are for tie rods and turnbuckles tied in to the tank side wall, and use of a pipeleg or auxiliary beam support under the agitator to the concrete pad. Figure 6.2 depicts a belt drive side-entering unit with an auxiliary support beam. Figure 6.3 shows a gear drive side-entering agitator, with a tie rod/turnbuckle assembly not shown, but which would be used. These figures are shown later under the section comparing belt and gear drives.

On top of lower capital costs for the agitator themselves, the vessel structure needed to support a top-entering agitator can be very expensive, particularly on large vessels. These vessels can often be 2000 $m^3$ capacity or more in biofuel applications such as anaerobic fermenters. Since most such large vessels are operated

*Agitator Design Technology for Biofuels and Renewable Chemicals*, First Edition. Gregory T. Benz.
© 2022 John Wiley & Sons, Inc. Published 2022 by John Wiley & Sons, Inc.

at near-atmospheric conditions, the vessel walls are quite thin. The top of the tank is often a 15° cone, which is not very strong. To accommodate top-entering agitators, various methods of reinforcement have been employed. The most common way, in this author's experience, is to extend large gussets or ribs from the agitator mounting nozzle to the sidewall of the tank, welding them to a support ring. Usually, there are six to eight of them, but there could be more. They are usually about 5% of the tank diameter in height. Effectively, the ribs, rather than the cone head, support the agitator loads. This method is illustrated in Figure 6.1.

Another common way is to build a beam structure spanning the tank diameter and tied into the sidewalls. (We do not have an illustration for this.) A third way seen by the author is to install vertical pillars inside the tank to assist in holding the roof up. These pillars also serve as tank baffles. They must be spaced far enough apart to clear the impeller diameter! (No illustration available for this method.)

The earlier measures may in some cases exceed the cost of the whole side-entering agitator! So why would anyone choose top entering over side entering? There are several good reasons:

- Process performance. In general, side-entering units do not suspend solids very well at all, and have long blend times. Even a fairly large side-entry unit is no match for the performance of a modestly sized top-entry unit (more on this later).
- Energy cost. Even with widely different performance results, a side-entry unit often requires 3–5 times the power of a top-entry unit (Ref. [1]).
- Maintenance cost. Side-entry units have a seal submerged in the process fluid, which can result in increased wear. Seal flushing will help, but would add significant dilution water over a fermentation cycle that lasts several days.

These differences are summarized in Table 6.1.

**Figure 6.1** Top-entering agitator on cone-top tank.

**Table 6.1** Side entering vs. top entering.

| | Side entering | Top entering |
|---|---|---|
| Capital cost | Low | High |
| Vessel capital cost | Low | High |
| Blending performance | Poor | Good |
| Solids suspension | Poor | Excellent |
| Contamination | Can be major | Low |
| Maintenance | High | Very low |
| Power cost | High | Low |

In general, side-entry units are used only for large tanks with very modest agitation requirements. Typical biofuel applications may include fermenters and biodiesel storage. Because of solids settling, infection may occur in the fermenters, reducing product yield. The author is aware of this happening in some installations. More is discussed under the heading "Flow Pattern Issues."

## Belt vs. Gear Drive

Side-entering agitators achieve speed reduction from motor speed to shaft speed in one of two ways: belt reduction or gear reduction. Typical images are presented in Figures 6.2 and 6.3.

Generally speaking, belt drive is less expensive than gear drive, and the maintenance is simpler. But belt drive is less efficient, and requires more frequent maintenance and adjustment. Although a properly tensioned V-belt drive can achieve above 95% efficiency, to maintain this level of efficiency would require re-tensioning almost daily. A more realistic planning value would be 92%, based on re-tensioning about every month or two. Table 6.2 compares some attributes of belt- and gear-drive units.

## Why More Power Is Needed from Side-Entering Agitators

There are two main reasons why much more power is needed when using side-entering agitators than when using top-entering units, even when a much lower process result is being achieved. These reasons fall broadly under the headings "D/T Effect" and "Flow Pattern Effects."

**Figure 6.2** Belt-drive side-entering agitator. *Source:* Courtesy Fusion Fluid.

**Figure 6.3** Gear-drive side-entering agitator. *Source:* Courtesy Fusion Fluid.

**Table 6.2** Belt vs. gear drive.

| | Belt | Gear |
|---|---|---|
| Capital cost | Low | Higher |
| Efficiency (typ) | 92% | 97% |
| Periodic maintenance | Adjust belt tension every 1–2 months | Change oil once/year |
| Replace belt or gear | ~2–3 years | ~15 years |
| Simplicity of belt or gear replacement | Easy | More difficult |

## D/T Effect

Recalling from Chapter 5, there are many combinations of power, speed, and impeller diameter that will produce the same process results. For flow velocity-driven applications, that generally means about the same impeller pumping capacity. Small impellers (low D/T ratio) operating at a high speed produce a higher discharge velocity, and hence, produce more head, than large impellers at lower shaft speeds for a given pumping capacity. This translates into more power required for small D/T designs.

Most top-entering agitators have a D/T ratio between 0.2 and 0.5. Almost all side-entering agitators have D/T ratios less than 0.1; most have less than 0.05. Some even have D/T ratios less than 0.02. If designed to match the pumping capacity of a top-entering agitator, a side-entering agitator could easily require more than 20 times the power of a top-entering unit.

D/T ratio also affects the flow pattern. While a small D/T ratio requires more power, the higher discharge velocity makes it more likely that the flow will reach the opposite wall of the tank before turning around and returning to the inlet side of the impeller. This is why we do not see side-entering agitators with relatively large impellers.

## Flow Pattern Effects

For flow velocity-controlled applications, top-entering agitators are normally equipped with axial flow impellers, which create a top-to-bottom flow pattern throughout the tank. With Newtonian fluids, there is little likelihood of any dead spots, as the flow velocities are significant throughout the tank.

Side-entering agitators, on the other hand, discharge directly across the tank. The flow pattern is quite complex, having two main rotating circuits in plan view and a complex asymmetric profile in the elevation view. We have attempted to illustrate this flow pattern in Figure 6.4. The illustration is for Z/T ratios up to 0.7, which is normally the maximum aspect ratio a side-entering agitator can effectively agitate.

When the Z/T ratio is greater than 0.7, the flow pattern splits. The tank becomes divided into vertical zones or "cells" in increments of Z/T = 0.7. The velocities in the bottom zone are reduced about 30% due to losing momentum to the next zone above it. In each upper zone, the velocities are about 50% as high as the one below it. So, successful agitation of tall vessels with a single side-entering agitator is all but impossible. Placing additional side-entering agitators along the sidewall at high elevations can improve mixing in the upper regions, but does little to create top-to-bottom mixing throughout the tank, and greatly complicates installation and maintenance.

Additionally, with a single agitator, the flow directions are more or less reversed in each zone. This is illustrated in Figure 6.5. The relatively ineffective flow pattern created by side-entering agitators contributes to the need for more power compared to a top-entering unit, even with an inferior process result.

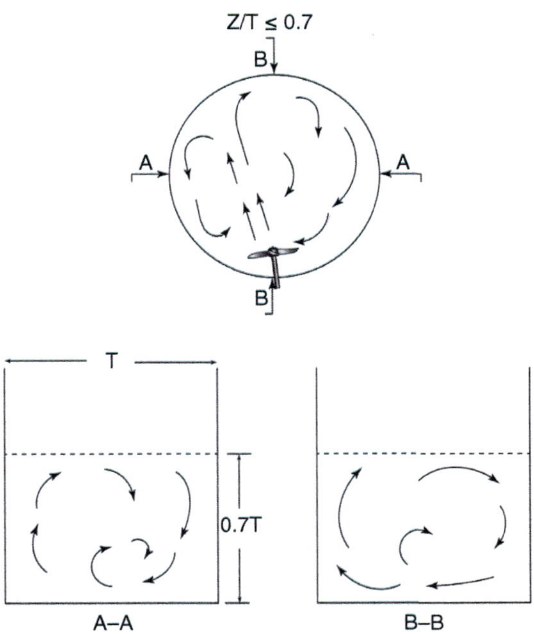


**Figure 6.4** Side-entering flow pattern, Z/T = 0.7.

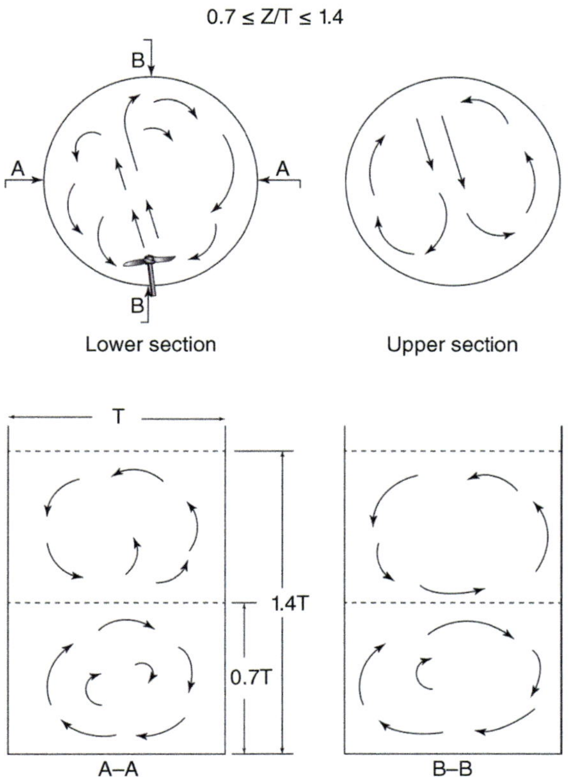


**Figure 6.5** Side-entering flow pattern, Z/T = 1.4.

## Why Equivalent Process Result to Top Entry Is Not Practical

As we have just seen, the tank geometry in which a side-entering agitator can be effective is quite limited compared to what is possible with top-entering units. The flow pattern is less effective for blending and solids suspension. And the power required for equal process result is many times as high. Therefore, side-entering agitators are not normally applied in a way as to equal the process results typical of top-entering agitators. They are normally used for very mild agitation where the result is not critical.

## Solids Settling Patterns

Though the jet flow from a side-entering agitator can be reasonably strong and go to the opposite tank wall, the flow must complete a circuit and return to the impeller inlet. In plan view, most of the return flow is along the sidewalls. The flow splits at the wall opposite the impeller discharge direction, and forms two return loops. In the middle of each loop, there is a low-velocity area. If solids have a tendency to settle, they will do so in those areas. The settling pattern is conceptually illustrated in Figure 6.6.

The author has seen such a pattern in a corn dry-mill-based ethanol fermenter. The solids formed permanent fillets, which became infected with undesirable septic organisms, reducing process yield. The tank had to be shut down for manual cleaning every several months.

Some might ask, Why not just use larger side-entering agitators? The answer is that they would need to be much larger—likely 3–5 times the typical power input that would be commonly selected. The petroleum industry often uses side-entering units with an oscillating swivel mount to sweep trace solids and water so that it can be discharged rather than build up on the tank bottom. This concept may require only 2 times the typical power. However, the author has not seen this approach applied in the biofuels industry, except perhaps in some biodiesel storage facilities, where the size requirements are very nominal.

The bottom line is that side-entering agitators are not the best choice for suspending solids.

**Figure 6.6** Solids settling pattern.

## Side-Entering Impellers

The original side-entering impeller was a classic marine impeller, with a power number of about 0.4. Such impellers, due to their fairly complex geometry, have traditionally been castings. But, due to the cost of casting molds, they have generally been available in relatively few sizes. Given that power is proportional to $D^5$, if one wants to be able to match desired power draw within 10%, the size increments should be about 2%. However, in practice, they are available in ½″ increments up to about 14″ (356 mm) diameter, and whole inch (25.4 mm) increments above that. This can result in much larger increments of power draw than 10%, which can lead to less process performance than intended.

For this reason, and also because marine propellers tend to be rather heavy, most agitator manufacturers today fabricate their side-entering impellers. Generally, they resemble high-solidity axial flow turbines. Most have three blades. Power number typically ranges from about 0.3 to –0.6. Some manufacturers have developed impellers with forward rake, camber, a retreating profile, and other features that are claimed to improve performance. In particular, some claim to create a more collimated, jet-like flow pattern which helps insure that the flow reaches the opposite wall of the tank before turning around. These claims could very well be true. However, the author does not have independent corroboration of these claims. Therefore, the sizing recommendations in the next section are based on the performance characteristics of marine propellers. If modern impellers outperform marine propellers, possibly smaller agitator sizing could be used. However, it is the author's opinion that, given the weak overall performance of side-entering agitators, it would be better to keep the sizing the same and get somewhat better performance rather than to downsize the equipment.

## Selection for Common Biofuel Applications

Most biofuel applications use top-entering agitators. However, there are two applications where side-entering agitators have sometimes been used. They are ethanol fermenters and biodiesel-storage tanks. The most common ethanol process is the dry grind process (typically corn, or maize, in the United States, but other grains or starch sources can be used also). In such a process, the whole grain is ground, but only the starch fraction is converted to sugar, which ferments to ethanol. The remainder of the solids is carried through until the end, where it is sold as distiller's grains. Though we will go over the essentials of this process in Chapter 7, we will not cover side-entering sizing there. Instead we will do so here. The intent of this sizing is to keep solids in suspension for at least the part of the fermentation where $CO_2$ is vigorously evolving, aiding in agitation. Solids will

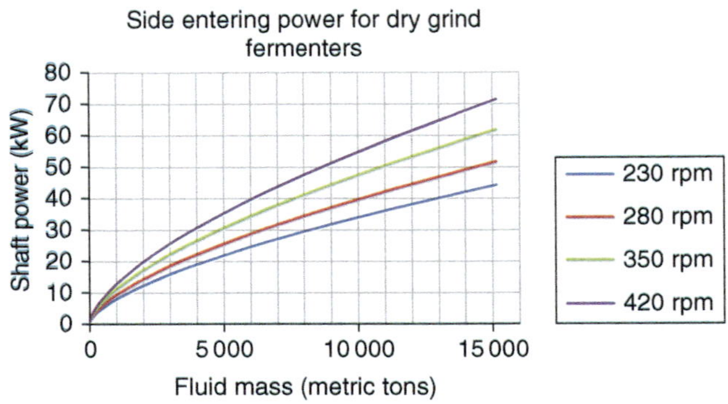


**Figure 6.7** Side-entering sizing for dry grind fermenters.

usually settle out after a couple of days with this sizing. The required power input depends on shaft speed, and is somewhat more complicated than torque alone. Figure 6.7 plots the required shaft power in kW vs. fluid mass in metric tons, at four of the most common speeds used for side-entering agitators: 230, 280, 350, and 420 rpm. Power required increases as shaft speed increases, though impeller diameter (not shown) decreases.

Some ethanol plants are based on a wet mill process, or a sugarcane process. In such plants, ideally there are no suspended solids other than yeast cells, which are virtually neutrally buoyant. Any other solids present are usually in trace quantities. The end result is that much less agitation is required. Figure 6.8 plots the requirements for wet mill fermenters. Such sizing can also be used for biodiesel storage, which may have trace amounts of water and some solids.

As an additional perspective, the edible oil industry has traditionally used a rule of thumb of investing 1 hp/100 000 lb or 1 kW/61 MT. Shaft speed typically has not been stated in this rule, but presumably it would be at 350 rpm, as this is a very common speed. Compared to Figure 6.8, this rule of thumb matches fairly well for fluid masses of 100 tons or less, and is conservative for larger quantities.

The reason why side-entering agitators do not scale on a torque/volume basis is because they do not create equal superficial velocity in different scales. They are closer to creating equal bottom sweeping velocity, letting the velocity in the upper parts of the tank decrease with scale.

## Example Problem

Suppose we have an ethanol fermenter with an operating capacity of 2 000 000 l. Further suppose that it has a specific gravity of 1.1, bringing the fluid mass to

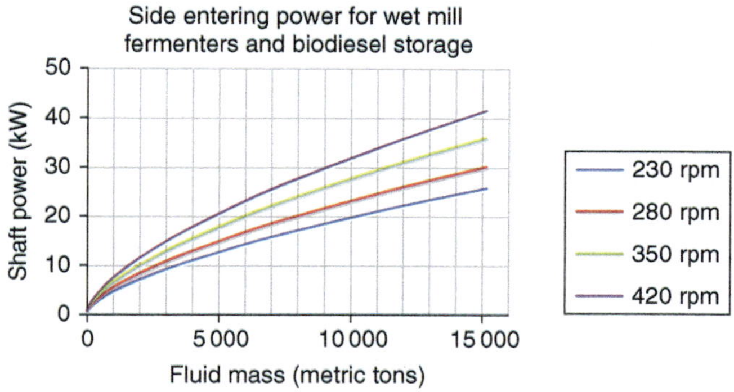


**Figure 6.8** Side-entering sizing for wet mill fermenters and biodiesel storage.

2200 metric tons. Let us also suppose that we will choose a tank so that the aspect ratio is 0.7, so that the flow pattern does not split for the side-entering agitator. Thus, we would have T = 15.4 m. Z would be about 10.75 m. According to Figure 6.7, we would need a shaft power of approximately 10 kW at 350 rpm or about 7 kW at 230 rpm. Choosing 350 rpm, the nearest standard motor size would be 11.2 kW. For reference, if our impeller has a power number of 0.4, its diameter would be about 0.64 m, resulting in a D/T ratio of 0.041. With a pumping number of about 0.6, this would result in a pumping capacity of about 55 $m^3$/min, and a tank turnover time of about 36 minutes.

## Comparison to Top Entry Design for the Same Problem

As noted in Chapter 5, there are many solutions for a given problem statement. And, as we shall see in Chapter 7, several different scales of agitation have been used in dry grind ethanol fermentation, according to the process response required. For the purposes of this problem, we will do a "pre-cap," as opposed to a "recap." To achieve full solids suspension at all stages of fermentation, a scale of 3 is required. In the early stages of fermentation, when $CO_2$ is vigorously evolving, a scale of 2.5 has been found to keep solids in suspension, but will allow solids to settle when $CO_2$ evolution diminishes. A scale level of 1.5–2 will allow settling even with gas evolution, requiring periodic shutdown for tank clean-out. So, for this problem, we will look at the impact of scale levels of 2, 2.5, and 3 on the design, at several shaft speeds for each scale of agitation. The results will be presented in tabular format. However, we will go through the calculations for one case to demonstrate how they are done.

For speed of calculation, we will use the simplified method of Eq. (5.12):

$$\text{Scale of agitation} = 600\left(P/(M*N)\right)^{0.5}, \text{ where P is in kW,} \\ \text{M is in kg, and N is in rev/s} \tag{5.12}$$

Re-arranging the above to solve for shaft power, we get P = (scale of agitation/600)$^2$ * M * N.

For our calculation, we will choose a shaft speed of 13.5 rpm, or 0.225/second. Our mass of fluid is 2.2E6 kg. We will design for a scale of agitation of 3. Thus, at 13.5 rpm, the required P = (3/600)$^2$ * 2.2E6 * 0.225 = 12.38 kW. Assuming a power number of 0.3, our impeller diameter may be calculated by re-arranging the power number definition and using SI units:

$$D = \left(P/\left(\rho N_P N^3\right)\right)^{0.2} = \left(12\ 380/\left(1100 * 0.3 * 0.225^3\right)\right)^{0.2} = 5.053 \text{ m.} \tag{6.1}$$

This gives a D/T ratio of 0.33. For simplicity, we will not re-check the power number from Appendix 1, though it will vary somewhat with D/T, requiring an iterative process for accurate sizing. Pumping number varies more significantly with D/T.

We can calculate pumping capacity by re-arranging the definition of pumping number. From Appendix 1, we find that a typical hydrofoil has a pumping number of 0.5 at a D/T of 0.33. Therefore, our pumping capacity is:

$$Q = N_Q N D^3 = 0.5 * 13.5 \text{ rpm} * \left(5.053 \text{ m}\right)^3 = 871 \text{ m}^3/\text{min} \\ \left(\text{rounded to a whole number}\right) \tag{6.2}$$

Though not a design variable per se, it is sometimes useful to compare turnover time when looking at top-entering vs. side-entering agitators. A rule of thumb is that side-entering agitators blend the tank in about 1–1.5 turnovers. Reference [2] has detailed fairly complex ways of calculating blend time for side-entering agitators. Suffice it to say that the probable blend time for our side-entering selection, exclusive of the settled solids or any additions with physical property differences, is around 45 minutes to 1 hour. For top-entering units, the blend time does not track closely with turnover time, but is often between 1 and 3 turnovers. With a pumping rate of 871 m$^3$/min in a 2000 m$^3$ volume, our turnover time is about 2.3 minutes, versus the 36 minutes for the side-entering unit we cited.

Top-entering blend time for a single hydrofoil impeller in turbulent flow may be estimated by using Eq. (6.3):

$$16.4 = \theta_{99} * N * \left(D/T\right)^{1.7} \left(Z/T\right)^{-0.5} \tag{6.3}$$

Table 6.3 lists a wide array of possible designs as a function of scale of agitation and shaft speed. Blend times based on Eq. (6.3) are included for reference.

**Table 6.3** Top-entering comparison problem. Basis: fluid mass = 2200 MT.

| Shaft speed (rpm) | Scale of agitation | Power needed (kW) | Impeller diameter (m) | D/T | Pumping number | Pumping capacity ($m^3$/min) | Turnover time (min) | $\theta_{99}$ blend time (min) |
|---|---|---|---|---|---|---|---|---|
| 6 | 2 | 2.44 | 5.94 | 0.39 | 0.43 | 541 | 3.69 | 11.5 |
| 6 | 2.5 | 3.82 | 6.50 | 0.42 | 0.42 | 691 | 2.89 | 9.9 |
| 6 | 3 | 5.50 | 6.99 | 0.45 | 0.40 | 819 | 2.44 | 8.8 |
| 7.5 | 2 | 3.06 | 5.43 | 0.35 | 0.47 | 566 | 3.53 | 10.7 |
| 7.5 | 2.5 | 4.77 | 5.94 | 0.39 | 0.43 | 677 | 2.96 | 9.2 |
| 7.5 | 3 | 6.88 | 6.39 | 0.42 | 0.42 | 822 | 2.43 | 8.2 |
| 9 | 2 | 3.67 | 5.05 | 0.33 | 0.50 | 580 | 3.45 | 10.1 |
| 9 | 2.5 | 5.73 | 5.52 | 0.36 | 0.46 | 698 | 2.87 | 8.7 |
| 9 | 3 | 8.25 | 5.94 | 0.39 | 0.43 | 812 | 2.46 | 7.7 |
| 11 | 2 | 4.48 | 4.66 | 0.30 | 0.53 | 591 | 3.38 | 9.5 |
| 11 | 2.5 | 7.00 | 5.10 | 0.33 | 0.50 | 729 | 2.74 | 8.2 |
| 11 | 3 | 10.08 | 5.48 | 0.36 | 0.45 | 816 | 2.45 | 7.2 |
| 13.5 | 2 | 5.50 | 4.30 | 0.28 | 0.55 | 589 | 3.40 | 8.9 |
| 13.5 | 2.5 | 8.59 | 4.70 | 0.31 | 0.52 | 727 | 2.75 | 7.7 |
| 13.5 | 3 | 12.38 | 5.05 | 0.33 | 0.50 | 871 | 2.30 | 6.8 |
| 16.5 | 2 | 6.72 | 3.96 | 0.26 | 0.58 | 596 | 3.35 | 8.4 |
| 16.5 | 2.5 | 10.50 | 4.33 | 0.28 | 0.55 | 739 | 2.71 | 7.2 |
| 16.5 | 3 | 15.13 | 4.66 | 0.30 | 0.53 | 886 | 2.26 | 6.3 |

One can readily see that the predicted blend times for all top-entering selections are significantly shorter than that for our side-entering selection.

As some may be surprised to see, some of the top-entering designs could require even more power than the side-entering selection we chose previously. However, from a performance standpoint, the side-entering agitator provides bottom sweeping performance roughly equivalent to a top-entering scale of agitation of perhaps 1–1.5. *It does not pump as much as a top-entering agitator would at even a scale of agitation of 1*. However, its flow is directed across the tank bottom in such a way to produce a higher sweeping velocity than a top-entering agitator would at similar pumping capacity. But note that a top-entering scale of 2 would provide more flow velocity at the tank bottom than the side-entering selection, achieving comparable or slightly better solids suspension performance as well as much better overall tank blending. So, when comparing the top-entry selections to the side-entry selections, all of the designs in Table 6.3 will produce superior results. Some selections save considerable energy. At 6 rpm, with a scale of 2, we would need only ¼ of the power of the side-entry unit, and would still have about 10 times the turnover rate. The author's recommendation would be the 9 rpm design with a scale of agitation of 3, with an 11.2 kW motor, to keep the D/T ratio at or below 0.4. (It is OK to go slightly larger, but some solids may accumulate directly under the impeller unless it is placed close to the tank bottom.) Although this design uses the same motor size as our side-entering selection, it would keep the tank bottom clean at all times, and, in another preview of the next chapter, it may even increase ethanol production rate, yield, and maximum ethanol titer.

## Summary of Chapter

We have demonstrated the why and how of using side-entering agitators in their most common applications in the renewable fuels and chemicals industries. We have also pointed out that their performance is quite different from top-entering units, resulting in possible compromise on process results and higher energy costs for comparable results.

## List of Symbols

| | |
|---|---|
| D | Impeller diameter |
| M | Mass of fluid agitated |
| N | Shaft speed |
| $N_P$ | Power number |

$N_Q$ Pumping number
P Shaft power
Q Impeller pumping capacity
T Tank diameter
Z Liquid height

## Greek Symbols

$\rho$ Fluid density
$\theta_{99}$ Blend time to 99% attenuation of concentration differences

## References

**1** Kehn, R. (2010). Comparing top entry versus side entry agitator performance in low viscosity blending. *Presented at Mixing XXII Conference (North American Mixing Forum)*, Victoria, BC, Canada (June 2010).

**2** Giacomelli, J., Boyer, B., Johnson, S., and Grenville, R. (2018). Blending in oil storage tanks with side-entering agitators. *Presented at 16th European Conference on Mixing*, Toulouse, France (9–12 September 2018).

# 7

# Application Guide for Starch and Sugar-Based Ethanol Processes

## Purpose of Chapter

This chapter will provide application guidelines for starch and sugar-based biofuel processes. Although the processes described are for producing ethanol, the agitation guidelines would be the same for similar processes producing butanol, other biofuels, and various renewable chemicals that use anaerobic processes, as long as similar fluids are being processed. Aerobic processes are covered in detail in Ref. [1].

Likewise, though corn-based and sugarcane-based processes are the prime examples in this chapter, similar applications are found in processes using other grains, starch, and sugar sources, so the agitator application guidelines apply to these processes as well, with notes added in the text for specific cases where agitation requirements may be different.

Though process details differ for these different sources of starches and sugars, the common thread connecting them all is the fact that the yeast ultimately ferments the monosaccharide, glucose, a C6 sugar, even if the process feed is sucrose, a C12 sugar. This is because sucrose is readily hydrolyzed to C6 sugars in a weak acid environment. This means all of these processes operate the fermenters under similar conditions of temperature, pH, and type of yeast used. The main process differences are in the steps leading up to fermentation.

Basically, sugars are the starting point for a great many fuels and renewable chemicals. There are too many possible chemicals and reaction routes to cover all processes here. We are using ethanol production as a proxy for agitation purposes for other possible processes, by analogy to similar applications.

*Agitator Design Technology for Biofuels and Renewable Chemicals*, First Edition. Gregory T. Benz.
© 2022 John Wiley & Sons, Inc. Published 2022 by John Wiley & Sons, Inc.

## Sources of Starch and Sugar

There are many sources of starch and sugar available around the world. In addition to their use as a feedstock for biofuels, they have many other uses. Most of these uses are for food additives, though there is a considerable use of starches for industrial purposes such as paper sizing and coatings, adhesive manufacturing (e.g. dextrin), and, of course, starch can be converted to various sugars both for sweetening and as a feedstock for both aerobic and anaerobic fermentation. Aerobic fermentation is covered extensively in Ref. [1].

Because starches and sugars have considerable utility as foodstuffs, to some extent there is an inherent conflict between using them for food vs. fuel and other industrial uses. Aside from any issues pertaining to the need for human nutrition, there is a practical economic issue. Starches and sugars sold for human consumption tend to be of higher purity and command a higher price than those used for non-food industrial purposes. Therefore, it only makes economic sense to convert starches and sugars into non-food products if there is a surplus available beyond what is needed for food. This makes it an incremental use. However, in most nations that produce such commodities, especially those with a strong economy, advanced farming and manufacturing technologies, there is a very large surplus of starches and sugars produced. In fact, some say the surplus is so large as to more than fulfill the global food supply needs including developing nations. This is why we see so much non-food industrial use of these commodities.

Sources of sugar in nature include a few well-known plants and a few surprising ones (Ref. [2]). Some we are all fairly familiar with are sugarcane and other cane species, sugar beets and sorghum, aka sorgo. Less obvious are various trees, such as various species of maple (especially those collectively called "Sugar Maple") and various species of palm trees or bushes. Most likely, very little tree-derived sugars are used for ethanol or other biofuel production. By far, the most common source of sugar (not including sugar made from starch) used for ethanol production is sugar cane, and Brazil is the leading country manufacturing ethanol this way. But sugar beets and sorghum have also been used. Though process details differ, the agitation requirements are similar.

There are also many sources of starch in nature. Broadly speaking, there are two main plant types that can produce starch: grain crops, e.g. maize (called corn in the United States), wheat, milo, rice, oats; and root or tuber crops, principally potato and cassava (also called tapioca). Sorghum can also produce some starch. Starch from these sources differs in several ways, such as particle size, particle shape, and waxiness, among others. However, these differences have little effect on agitation requirements, except for particle size. Table 7.1 lists particle size ranges for several common starches.

**Table 7.1** Starch particle size range.

| Starch source | Size range (μm) |
|---|---|
| Maize (corn) | 3–26 |
| Wheat | 2–45 |
| Rice | 2–10 |
| Milo | 3–26 |
| Potato | 3–110 |
| Cassava (Tapioca) | 4–35 |

One can readily see that most starches have a similar particle size range. The two outliers are rice, which has smaller than average particles, and potato, which has larger than average. Past experience indicates that potato starch requires more agitation than the others to stay in suspension. We will give specific guidelines under the appropriate spreadsheet section.

## Generic Process Flowsheet: Wet Mill

Wet mill processes are those which separate the grain, to the extent possible, into its main physical constituents. It is called wet milling because the grain is soaked or steeped in liquid, softening it and making it easier to mechanically separate into various streams. The stream which is used in ethanol production is the starch stream.

Although in principle there could be unique wet mill processes for a number of grains, the principle grain which is wet milled is maize (corn). Therefore, we will use corn wet milling as a generic flow sheet for this process.

Figure 7.1 is a greatly simplified block diagram of such a process. Though it does not show all the separation and recycle steps, the key areas where agitators are used are included. Corn has four main components which can be separated: fiber, gluten (or corn protein), starch, and the germ or endosperm.

Though we will focus on the starch stream, as this is the only stream which goes into ethanol production in most cases (the fiber can go into a cellulosic ethanol process), we will briefly go into a couple of applications upstream of that point.

There is no agitation involved in the steeping process. Therefore, we will start downstream of that step. After steeping, the corn undergoes several stages of milling, each followed by a separation step. Each fraction is a slurry, with agitation requirements as described next. The applications from starch onward will be detailed after the generic flow sheets are discussed.

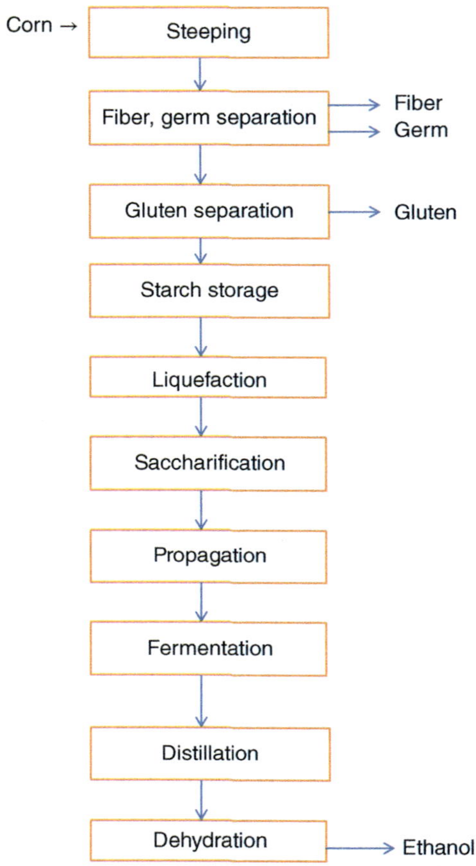


**Figure 7.1** Wet mill process.

## Fiber Slurry

As separated from the rest of the kernel, the fiber slurry is rather dilute: typically 3% by weight or less. For such slurries, the recommended scale of agitation is 3. If an unusual tank geometry is used, such as horizontal cylindrical, the scale should be increased to 4. The specific gravity is about 1.03 and the viscosity is water-like.

Sometimes, the fiber slurry is concentrated downstream of the separation step. It can become non-Newtonian, though the apparent viscosity is still low (say, less than 30 cP). At 6–8% solids, the specific gravity is about 1.05 and the recommended scale of agitation is 5, increased to 6–7 in a horizontal cylindrical tank.

## Germ Slurry

The germ, or endosperm, contains almost all of the oils in the kernel. As such, it actually will float in water. Thus, the slurry-specific gravity will be a bit less than 1. A scale of agitation of 4 is normally sufficient.

## Gluten Slurry

Most gluten slurries have a specific gravity between 1.1 and 1.15 and a design viscosity of about 30 cP. Gluten is easily kept in suspension with a nominal scale of agitation of 3.

## Starch Storage

After separation and purification, the starch slurry is stored to feed the rest of the process. It may be modified into various industrial starches, or it may be converted to various sugar streams to,make corn syrup, ethanol, etc. For this book, we are only concerned with converting it to the simplest sugar monomer (monosaccharide), dextrose or D-glucose, as this is what feeds the ethanol process.

In rare instances, a small ethanol plant may be built using dry starch shipped in. This adds a starch makedown requirement upstream of the storage tank. We will discuss agitation requirements after the flowsheet summaries.

## Liquefaction

Starch is basically a condensation polymer of sugar units as monomers. To produce the monomers (sugars), it must be hydrolyzed by catalytic addition of water. The first step of hydrolysis is called liquefaction. In this step, enough hydrolysis occurs to convert the starch into soluble sugars. However, they are still composed of more than one monomer unit.

The liquefaction step can be either acid-catalyzed or enzyme-catalyzed. The latter is more common these days, as it produces fewer by-products. Most liquefaction processes are continuous flow.

## Saccharification

After liquefaction, the resultant solution must be further hydrolyzed down to simple sugars or monosaccharides. Hence, we have the name "saccharification." This process is rather slow and is carried out with the aid of enzymes. Though this sounds similar to liquefaction, the temperature, pH, and enzymes used are all different. Thus, it is a distinct process step.

Because it is slow, most saccharification processes are carried out in sequential batch tanks, called sacc tanks for short. However, some facilities use a continuous flow process. In some facilities, especially cellulosic, the fermentation step is carried out concurrently with saccharification. This process is called SSF, short for simultaneous saccharification and fermentation.

## Propagation

This step grows, or propagates, the yeast cells (typically saccharomyces cerevisiae). The propagation is ideally an aerobic process for maximum growth rate. In older facilities this was a separate vessel optimized for production of yeast. In many new plants that use continuous fermentation, propagation is carried out in the first stage of fermentation, by sparging a small amount of air into the tank. This must be done with care, as too much air can result in production of acetic acid and other undesirable products. That is why it is not done in the beverage industry. The taste of the final product in that industry is important. However, for fuel ethanol, taste per se does not matter.

## Fermentation

This is where the yeast cells or other organisms convert the sugar into ethanol. In a wet mill process, there are virtually no other suspended solids besides the yeast. This affects agitation requirements, as we will see later.

Fermentation can be either batch or continuous. If continuous, it is usually tanks in series, as pipe loops or agitated columns are not practical for the large volumes required. Most commonly, there are four tanks in series in a continuous process. The first tank may also serve as a propagator.

$CO_2$ evolves vigorously in the first two stages, less in the third stage, and almost not at all in the fourth stage. The final "product" from the fermenter is an aqueous solution, generally less than 10% by weight ethanol. This may change as new yeast strains are developed.

## Distillation

This is the primary step for separating ethanol from the high-water content solution it is produced in. The maximum concentration of ethanol in the distillate is 95%, due to that being the azeotropic concentration with water. Co-distillation with gasoline can produce a three-way azeotrope with lower water content.

## Dehydration

Since the distillate still has some water in it, it is not yet suitable for fuel use. It must be dehydrated to almost 100% water-free. There are several ways of doing this.

One way is by chemical absorption. Another is my use of osmotic membranes. Possibly the most common way is by pressure-swing absorption. None of these involve agitation, except possibly for feed or discharge tanks, where the agitation requirements are minimal. After absorption, the commercial grade ethanol is ready for sale.

## Generic Process Flow Sheet: Dry Grind

Because wet milling allows a fairly complete separation of the major grain components, it can produce the greatest total product value per unit of feedstock. It can produce many different products as well. Examples include fiber feed, gluten feed, a wide variety of corn syrups (including high fructose syrup), standard starch, a wide variety of modified starches, corn germ, corn oil, hydrogenated corn oil, ethanol, butanol, bioplastics, and other products. But this flexibility comes with a high capital cost and is not necessarily needed when the objective is to produce ethanol. In fact, if all ethanol were produced in wet mill plants, there would be a surplus of many of the co-products, driving the market price down and reducing the return on investment for building a wet mill.

As a result, many or even most ethanol facilities are built using a dry mill process, which has much less capital expense. As the name implies, the grain is milled in a dry state before being fed into the process. In the older plants, this meant that the rest of the grain was carried through the process in slurry form along with the starch while the starch was being converted to sugars and then fermented. That is, the grain was not separated into its constituent parts.

In most new plants, there is a degree of separation attained by a process known as dry fractionation. Though not as complete a separation as attainable in a wet mill, it does permit developing some higher protein and/or oil-bearing fractions which command enough of a premium in the marketplace to justify the incremental cost of the fractionation equipment. Figure 7.2 gives a simplified overview of the dry grind process, which we will describe next.

### Milling

When grain is shipped to the plant, the first step is storage. But the first actual processing step is milling. As the term "dry grind" implies, the grain is ground in a dry state. In principle, all of the components of the grain are commingled. However, only the starch is converted to sugars and ethanol. In older plants, the rest is carried through to become part of a feed product known as distiller's grains. More recent plants will tend to add a dry fractionation step, as per the next block on the flow sheet.

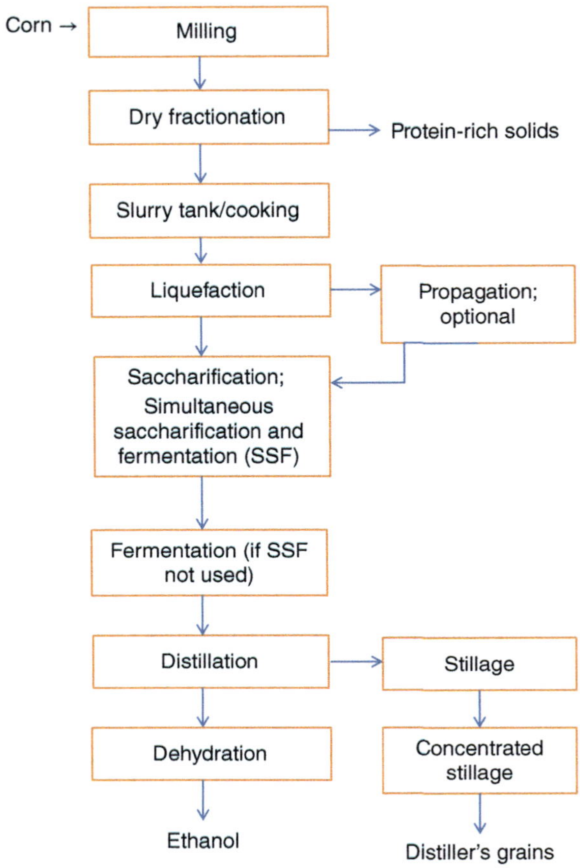


**Figure 7.2** Dry grind process.

## Dry Fractionation

This is a process improvement that separates the component parts of the kernel to some degree. It is briefly described in a North Dakota State University blog, Ethanol Production – Dry Fractionation – Energy (ndsu.edu), accessed by this author last on 11/28/2021. To quote from the linked article: "Dry fractionation involves removal of the germ, fiber, and endosperm from the corn kernel (Murthy, et al., 2006)." Essentially, the corn kernel is separated into component parts. The germ portion of the corn kernel is high in corn oil, which is a potential feedstock for biodiesel. Removal of this oil and fiber in the kernel leaves a higher percentage of starch for ethanol production. This reduces the volume of material entering the plant which increases throughput and overall efficiency.

With less material entering the plant and a higher proportion of the kernel (starch) being converted to biofuel, fractionation reduces the volume of distillers dried grains (DDGS) produced. The lower quantity of DDGS produced does contain higher protein levels though." "This benefits producers of non-ruminant animals by increasing digestibility. Ethanol-production equipment manufacturer ICM calls these higher quality DDGS "Hi-Pro DDGS" because of the increased protein content per ton of dry matter.

Another benefit of dry fractionation is that it can greatly increase biofuel plant production efficiencies. DDGS production decreases under fractionation because other parts of the corn kernel are utilized more efficiently. However, higher quality (high-protein) DDGS can lead to increased value as hog and poultry feed which increases the value of the co-product to animal feed manufacturers. While use of DDGS for domestic animal feed and exports have both steadily increased, continued growth of the ethanol industry produces larger amounts of DDGS which poses a marketing challenge for the ethanol industry. Dry fractionation, which leads to lower volumes of DDGS produced, helps ethanol producers maintain stable DDGS prices."

As an agitator designer, this author is not familiar with the actual mechanics of how that system works.

## Slurry Tank/Cooking

These are actually two different process steps combined into one block on the spreadsheet. Up until this point, all grain handling has been in the dry state. The rest of the processing to make ethanol is in a liquid state, so the ground grain must be made into a slurry. Essentially, the grain is added to water with fairly vigorous agitation to pull the grain into the water and break up any soft agglomerates which form. Though it may be tempting to use hot water for this step due to it being followed by a cooking step, hot water can cause the starch to gelatinize or paste, creating "fisheyes" or starch balls, which are hard to break up. This author recommends making the starch slurry at temperatures below the gelatinization temperature of the particular starch species used.

The cooking step is usually carried out at a temperature of 100 °C or more, and may be done in an agitated tank, but is more commonly done in a jet cooker, where steam is injected and the slurry travels through a pipe reactor with a retention time of about 20 minutes, give or take. The effluent is partially liquefied, but has only been thermally treated at this point.

## Liquefaction

This is basically the same process step as done in a dry mill, with the same enzymes. However, the slurry also has grain solids instead of just starch and enzymes. It can be batch but is usually continuous. All of the starch should be solubilized by the time the slurry exits this step.

### Propagation (Optional)

As in a wet mill process, the purpose of this operation is to propagate yeast cells. Older facilities may have a dedicated tank for this purpose. Many newer facilities allow the yeast to propagate in the first state of fermentation, same as in a wet mill. Small amounts of air are added in the first stage to facilitate this.

### Saccharification

As in wet milling, this step converts the complex sugars to monosaccharides. However, since yeast can grow and ferment when enough simple sugar is present, propagation, saccharification, and fermentation are often combined in a continuous flow process. This is called SSF, or simultaneous saccharification and fermentation.

### Fermentation

Sometimes, this is a separate process step, with its own vessel. More often, it is just the last stage in a continuous flow, tanks-in-series approach wherein the SSF scheme described earlier essentially is finished with the saccharification but the fermentation needs more time. The finished fermentation broth is often transferred to a surge vessel known as a Beerwell.

### Distillation

The primary separation of water from the ethanol "beer" is by distillation, producing an azeotrope of 95% ethanol, 5% water. Since there are solids present in the still, these have to be dealt with also, in the stillage tanks. The solids loading may be reduced by centrifugation or filtration prior to distillation.

### Dehydration

The water can be basically eliminated by means of chemical absorption, pressure swing absorption, semi-permeable membranes, and perhaps a triple azeotrope distillation, producing the product: anhydrous ethanol.

### Stillage

After the distillation removes ethanol from the broth, the unconverted grain solids remain as still bottoms. The solids are high in protein and other nutrients.

These still bottoms are typically transferred into a temporary holding tank and are referred to as whole stillage or heavy stillage.

### Light Stillage

The whole stillage may have much of the solids removed by screening, centrifugation, or other methods. The centrate or filtrate may then be called light stillage. It can be concentrated by evaporation into a kind of syrup, though little or no sugar should be present. It is still a slurry at this point and is stored in an agitated tank. The next step is to dewater the stillage, making a product called WDG, or wet distiller's grains (WDGS adds solubles to the term), and it can also be dried to below 10% moisture for stability, and then it is called dried distiller's grains (DDG or DDGS). This is a valuable co-product.

## Generic Process Flow Sheet: Sugar-Based Processes

Unlike starch-based processes, sugar-based processes do not need to convert feedstock to sugar. The sugar is already present in the plant. Therefore, there is no liquefaction or saccharification step. The process, therefore, is designed to remove as much of the non-sugar plant matter as possible before fermentation. The method used depends on the plant matter used as feedstock.

Though it is possible to extract sugar from various trees, the author is unaware of any ethanol plant using such a feedstock. (That would be a very unfortunate waste of maple syrup, for instance!) Thus, we will focus on sugarcane, sorghum, and sugar beets. All of these involve shredding or crushing to free up the sugar, along with possible washing steps. As a stand-in for all three kinds of feedstock, we will describe a simplified form of the sugarcane process, as outlined in Figure 7.3.

### Milling, Cleaning, Extraction

Like any other natural crop, there can be a fair amount of dirt in sugar cane. Therefore, the harvested crop is washed to remove as much of the dirt as possible before further processing. From there, the stalks are often cut to shorter lengths, sliced longitudinally to free up internal area, and even shredded to further improve access to the sugar. The resulting pulp is then crushed to extract a sugar solution, called juice. Water or steam may be added and may flow counter currently to aid in extraction of the sugar, enhancing yield but resulting in a more dilute juice. These steps do not involve fluid agitation.

The "products" of these steps are the juice and bagasse, which is basically everything left behind after extraction of sugar.

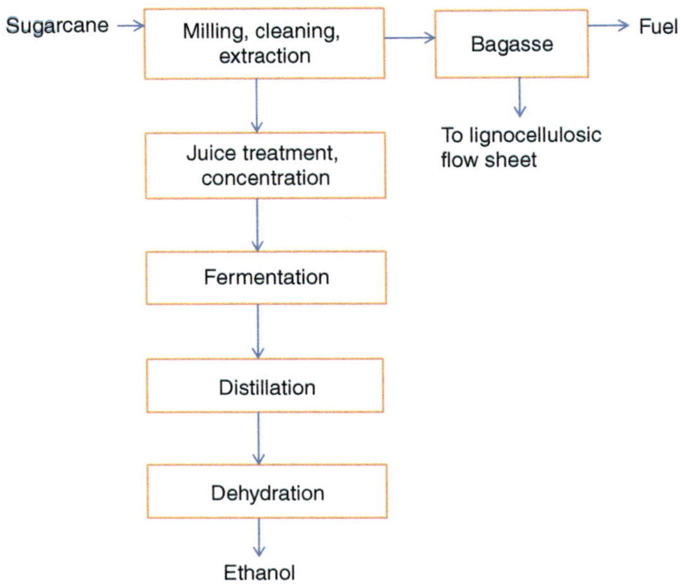


**Figure 7.3** Sugarcane process.

## Bagasse

Bagasse may be dried and used for fuel, either within the process to provide energy or as a commercial product. It may also be sent on as a feedstock to a lignocellulosic ethanol process, described in Chapter 8.

## Juice Treatment, Concentration

The juice is fairly dilute. It may be filtered to remove residual dirt and bagasse particles. For food grade uses, the pH is adjusted to about 7 to retard hydrolysis of sucrose into glucose. For fuel ethanol purposes, the pH may be adjusted to about 4.2, as that appears optimal for growth as well as facilitating hydrolysis of sucrose (Ref. [3]). It may also be concentrated to reduce the required fermenter volume. For food use, it will typically be evaporated using multiple effect evaporators up to a concentration of at least 65 wt%. For ethanol production, it need not be as concentrated; probably between 20 and 40% sugar is practical.

There may be agitation in various tanks for pH adjustment and holding tanks between the extraction step, the evaporation step, and the fermentation step.

### Fermentation

Once again, this could be batch or continuous. Since the process feed to the fermenter is almost solids-free, little agitation is needed to maintain suspension. However, there are other purposes for agitation besides maintaining suspension, as we will see later.

When $CO_2$ is vigorously evolving at the early stages of fermentation, no additional agitation is needed to maintain suspension. In fact, alcoholic beverages are almost always fermented with no mechanical agitation. But fuel and chemical ethanol almost always is produced using mechanical agitation, as this influences rate and yield. We will discuss this more later, in the application section on fermentation.

There may be a beer well between the fermenter and the distillation step.

### Distillation

Ethanol is thermally separated from water by distillation, producing a 95% ethanol azeotrope. The 5% water must be removed in the dehydration step to produce anhydrous ethanol as a product.

### Dehydration

As in the starch-based processes, water can be mostly eliminated by means of chemical absorption, pressure swing absorption, semi-permeable membranes, and perhaps a triple azeotrope distillation, producing the product: anhydrous ethanol.

## Agitator Application Guides

In the previous sections, we crudely described the processes. Any mistakes in the process description are mine alone, and can be attributed to the fact that I have never actually designed or operated an ethanol plant. However, the agitation requirements and principles detailed in this section should be quite accurate, based on successful design for many installations as well as troubleshooting and analyzing where agitation was insufficient.

Many of these applications are found in more than one process type, but are basically similar. Differences important to agitator design will be pointed out.

### Starch and Other Slurry Makedown

This process step is not found in wet mill processes, but may be found in processes using powdered starch as a feedstock, as well as in all dry grind processes. It is not found in sugar-based processes.

The general design approach is to meter the starch or ground grain in at a reasonable rate, agitating the bottom portion of the tank using a scale of agitation of 4–5 and the upper part of the tank with a scale of 6–8. The bottom part of the tank should use a hydrofoil to save energy. The upper part should use a pitched blade turbine located about one impeller diameter or somewhat less below the surface. Tank baffles should be cut off about D/2 below the surface to aid in forming a vortex to pull down solids. Thus, we are essentially doing two agitation problems: makedown and slurry mixing. However, the shaft speed must match between the top and bottom design calculations. No one makes a two-speed agitator shaft (smile).

This approach will result in about 75–80% of the total power being in the upper impeller, in a two-impeller design.

If a specific rate of incorporation (mass/tank cross-sectional area-time) is needed, lab testing will be required. It empirically scales up on a P/V basis, as described in Chapter 11.

Final slurry concentration varies, but it can be up to 50% in some processes.

Because the freshly prepared slurry may often be fed into a starch cooker or liquefaction unit, it is tempting to make the slurry using hot water. But if the hot water is above the gelatinization temperature of the starch (which ranges from about 50 to 85 °C, depending on the starch), fisheyes or starch balls will form, with a gelatinized layer on the surface and dry starch or grain on the inside. The shear forces created by the makedown agitator will not break these up unless the unit runs for days. The result will be poor conversion downstream and possible transport problems in the pipes.

A better approach would be to thoroughly wet out the solids using water well below the gelatinization temperature, then add steam downstream of this tank to raise the temperature to the cooking temperature desired. That will prevent the fisheyes from forming with dry starch inside.

## Starch Ball Buster Agitator

No, it is not called that on any flow sheet I have seen. In fact, it is usually not planned for up front. Where it has been used is in unfortunate cases where starch was added to very hot water, resulting in starch balls. This is generally not anticipated by the process designer. Then a high-shear agitator must be added to break up the fisheyes. This can be a unit mounted eccentrically in the same tank used for makedown, inside a pipeline or in a separate vessel. If an open-style impeller is used, its tip speed needs to be about 15–25 m/s to break up the fisheyes. Rotor-stator devices may also be used, especially as inline devices.

## Starch and Other Slurry Hold Tanks

Starch or grain slurries may be held at various points in the process. For example, they may be held after makedown, after separation from germ and gluten in a wet mill, etc. In general, most slurries of vegetable matter can be maintained in suspension using a nominal scale of agitation of 3. "Nominal" means ±½ scale unit. So, a nominal scale of 3 means a scale ranging from 2.51 to 3.50. This works well for dry grind slurries that are finely ground. However, for coarse slurries or for starch, the rules are s bit more stringent. Field experience in large corn starch tanks has resulted in observable settling using a scale of agitation of 2.6, but none at 2.8. Therefore, the minimum scale of agitation for corn starch is 2.8. This has been successfully applied in tanks as large as 5000 $m^3$. So, the procedure is to select based on a scale of 3 but then do detailed calculations to make sure it is at least 2.8.

Another exception is potato starch. Due to its larger grain size, field experience has indicated some settling at a scale level of 3. Recommended practice is to design for a nominal scale of 4.

Physical properties of starch slurries are tabulated in Table 7.2. Grain slurries used in dry grind processes will have slightly lower densities.

## Starch/Grain Cooker/Liquefaction

The purpose of this step is to convert the starch slurry to a solution of complex sugars. This can be done using several reaction schemes. It can be done on a batch or continuous basis. It can be done thermally by raising the temperature well above 100 °C. It can be done by enzymes under controlled conditions of temperature and pH. It can also be done by acid hydrolysis.

When it is done thermally, a common configuration is the jet cooker. This basically consists of a tubular reactor through which the slurry is pumped, with

**Table 7.2** Starch slurry properties.

| % by weight starch | Density (g/cc) | Design viscosity (cP) |
|---|---|---|
| 10 | 1.04 | 5 |
| 20 | 1.08 | 10 |
| 30 | 1.13 | 20 |
| 40 | 1.18 | 30 |
| 50 | 1.23 | 50 |

steam injected under pressure to raise the slurry to the desired temperature. Retention time is a few minutes. There is no mechanical agitation involved.

After the cooker, the pressure is let down by letting the slurry go into a vacuum chamber, set at an absolute pressure designed to get the temperature down to that required for the next stage. In this next stage, further conversion is done by means of acid or enzymes. This is often done using agitated columns (Figure 7.4) or pipe loops (Figure 7.5) that have inserts such as Static Mixer® units, triangular plates normal to the pipe axis, or other devices that break up the tendency to form a laminar flow parabolic velocity distribution, which would cause a deviation from the desire to approximate plug flow. Though the viscosity is not very high in the loops, the velocity is low (typically ~0.3 m/s), so such devices are needed to disrupt formation of a laminar flow velocity profile. Static Mixer elements may be slightly more effective than other devices, and probably create less pressure drop. However, they are more expensive.

Specific reaction kinetics depends on the system employed and is beyond the scope of this book. But, if the designer can specify a residence time distribution needed, the following principles can be employed to predict the results and guide agitator design. Reference [4] goes into this subject in considerable detail.

For an agitated column, as per Figure 7.4, the design approach is to use the minimum agitation needed to keep solids in suspension in each stage and to make sure each stage acts as an idealized backmixed stage. That level of agitation

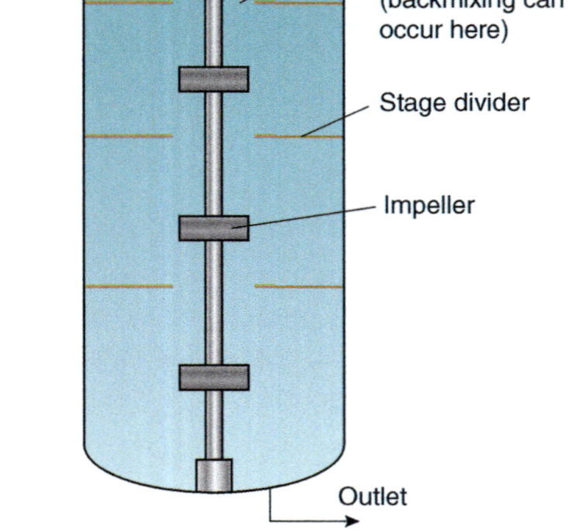


**Figure 7.4** Agitated column. *Source:* Permission of CEP.

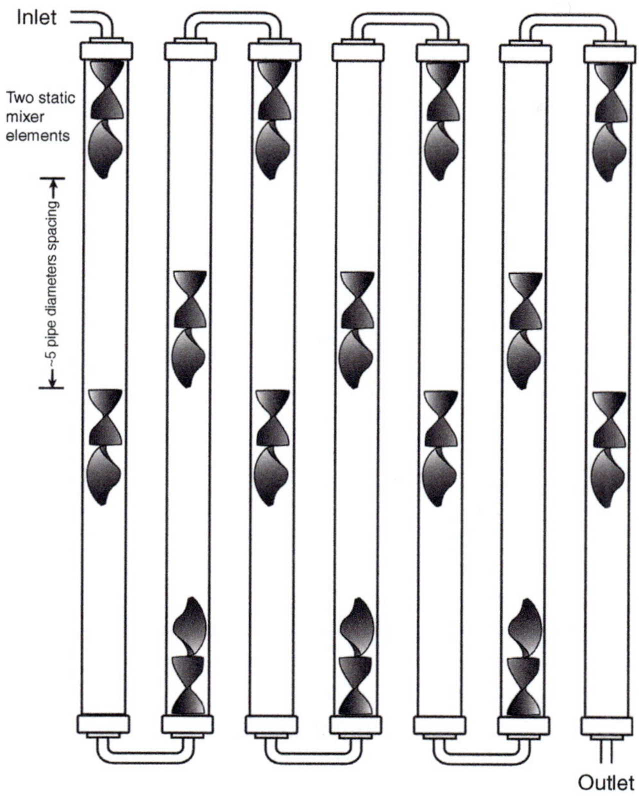


**Figure 7.5** Pipe loops.

depends on the process. If starch cooking or acid conversion is done in the column, the first couple of stages are designed for a scale of agitation of 8–10 and a viscosity of about 10 000 cP. Later stages have a lower viscosity and need a lower degree of agitation—generally a scale of 3–4. If the process is strictly enzyme based, the viscosity is not so high and there is no danger of pasting, so a scale of 3–4 is satisfactory throughout. Too much agitation leads to lower stage efficiency. This can be better understood by considering a staged column to be equivalent to tanks in series with backmixing between stages, as visualized by Figure 7.6.

Because an opening exists between each stage of a multistage compartmented tank, it is possible for random turbulence created by the agitator to cause interstage flow in addition to the flow produced by process throughput. Generally, this interstage, or backmix, flow is expressed as a fraction of the process feed flowrate:

$$\beta = F_B / F \tag{7.1}$$

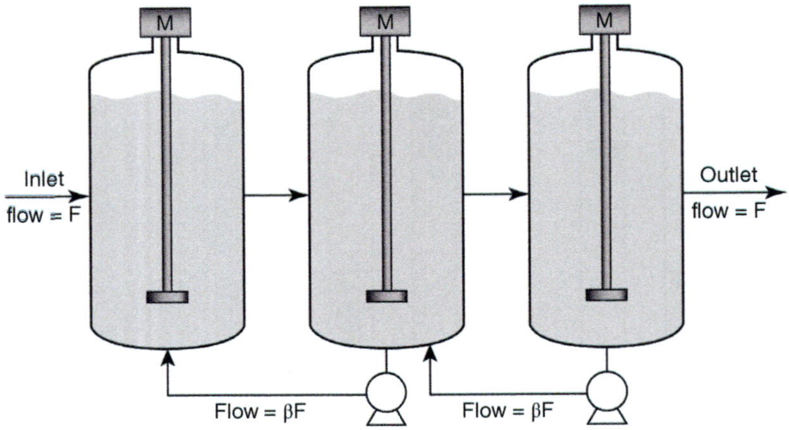


**Figure 7.6** Staged tank model.

where β is the backmix ratio, $F_B$ is the interstage backmix flowrate, and F is the net feed flowrate. β is a function of agitation parameters, process flow, and construction details. Modeling such a system, for example, in terms of residence-time distribution (RTD) and reaction conversion, typically treats each stage as perfectly mixed with recycle streams between stages.

To minimize the number of physical stages needed, the value of β must be kept as low as possible. In practice, this is done by minimizing the open area between stages and, where practical, including a standpipe (i.e. a short vertical length of pipe projecting above or below the stage divider).

When setting clearances between stationary vessel parts and rotating agitator parts, it is essential to account for the additive tolerances between parts (called "stack-up") as well as shaft and vessel head deflections under load. For example, consider a 10-m-tall vessel with the following tolerances: vessel eccentricity of ±3 mm, agitator mounting structure eccentricity of ±1 mm, shaft straightness of 0.25 mm/m (or 2.5 mm over the 10-m length), shaft deflection under load of 40 mm, nozzle angularity tolerance of 0.25° (44 mm over 10 m), and nozzle deflection under load of 0.1° (17.5 mm over 10 m). The potential misalignment of the shaft relative to the theoretical tank centerline could be as much as 108 mm, the sum, or stack-up, of the individual terms. The diametric tolerance stack-up is double the radial stack up, or 216 mm. Thus, for this example, the interstage opening should be at least 250 mm larger than the shaft diameter to provide a reasonable safety cushion in the design.

Because this subject is rather specialized, the reader is encouraged to read the Appendix "Multistage Reactor Design" for a detailed description of how to calculate stage efficiency, residence time distribution, etc. Even more information is available in Ref. [4], especially with regard to construction details. The quick executive summary is that production-scale columns generally can be built with greater than 90% stage efficiency. Small columns (those too small for a person to work inside) tend to be less efficient; about 75% stage efficiency is reasonable.

Pipe loops constructed with inserts to break up laminar flow as shown in Figure 7.5 are very close to plug flow. A reasonable design basis would be to apply 5% more retention time than the kinetic model would require for true plug flow.

## Saccharification

This is where the process of converting starch to simple sugars (monosaccharides) is completed. It is done by using different enzymes from those used for liquefaction, and it takes a considerable amount of retention time.

This step can be either batch or continuous. When carried out as a batch process, there are generally multiple tanks, i.e. a tank farm approach. This allows downstream equipment to be fed continuously by switching tanks sequentially.

In a wet mill process, there are almost no solids. Those that are present include yeast cells, small amounts of floating oils, some enzyme solids, and trace amounts of dirt. The batch "sacc tanks" require very little agitation. A scale level of 1–2 is typical. And the tanks do not need continuous agitation. Often, they only run about 5 minutes per hour except when filling and draining. Though a scale of 1–2 may not provide fully uniform solids suspension, when the tank is draining, the scale of agitation increases to 3–4 due to the lower impeller agitating a smaller volume. This is sufficient to re-suspend any solids which have settled and to incorporate any floaters into the discharge stream.

In a dry mill process, substantial amounts of suspended solids are present. The agitator must run continuously and should provide a nominal scale of agitation of 3.

When saccharification is carried out on a continuous flow basis, it is most commonly done as tanks in series, due to the large retention times and volumes involved. Only in pilot scale operations would an agitated column be practical. Since solids and floaters cannot be allowed to accumulate, a scale of agitation of 3 is recommended for both wet mill and dry mill processes.

An alternative to tanks in series is to compartmentalize a large tank both vertically and horizontally. This is illustrated in Figure 7.7.

The example configuration as shown has 12 actual stages and can probably deliver about 10–11 theoretical stages. It requires four agitator assemblies. Because of the irregular stage geometry, more agitation is needed to maintain full motion in the corners. A scale level of 3 is recommended for a wet mill process, and a scale of 4–5 is recommended for a dry mill process.

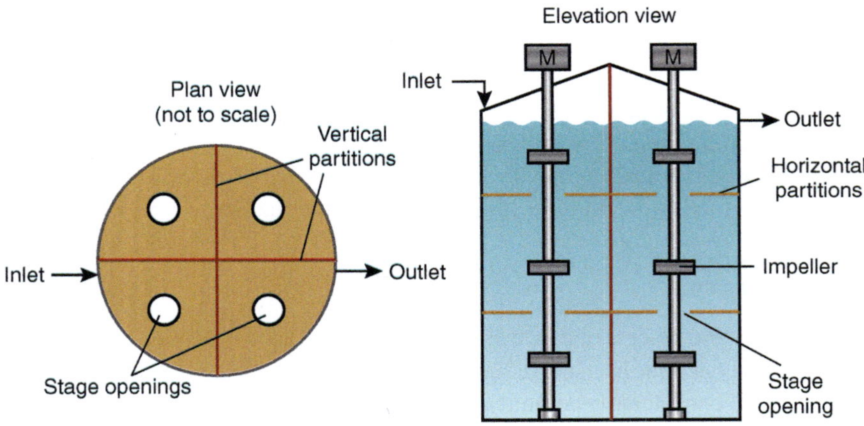


**Figure 7.7** Large compartmentalized tank. *Source:* Permission of CEP.

## Propagation

As mentioned previously, most new plants perform propagation in the first stage of fermentation. Some older plants have separate propagation units. If new organisms are developed, they may be better served with dedicated propagators. These are basically aerobic fermenters, and ideally they should be designed using the principles detailed in Ref. [1]. Lacking the level of information required to design that way, most engineers use rules of thumb. The typical approach is to apply that agitator with a P/V level of 2W/l and an airflow rate of 0.5VVM. A lower radial flow turbine plus high solidity up pumping axial flow impellers in the upper positions are used. Power should be distributed in the ratio of 7 for the lower radial impeller and 3 in each of the upper axial impellers.

Most such propagators are not large vessels, so optimization may not be deemed worthwhile.

## Fermentation

This application is not as straightforward as some might expect. If the purpose of agitation is to keep solids in suspension, then the recommendations are simple. However, agitation affects more than just solids suspension in an anaerobic fermenter. And it is interesting to note that beverage ethanol fermenters are typically not agitated, whereas fuel ethanol fermenters "always" are. It has been demonstrated by early experience that mechanically agitated fermenters are more productive than unagitated fermenters. However, the processes producing flavor are not enhanced to the same degree by agitation as ethanol production is, so alcoholic beverage products produced with agitation have plenty of alcohol but not much taste.

So, let us first cover minimum agitation requirements to keep solids in suspension. In a wet mill process or a sugar-based process, there are only trace solids other than yeast. Gas evolution alone is sufficient to keep solids in suspension in the early stages of the process. When the gas evolution stops, as little a scale of agitation of 1 is sufficient to keep solids in suspension. So, for a wet mill process, a batch fermenter should have a scale of agitation of 1–2. For a continuous flow process composed of tanks in series, about the first half of the tanks may get by with no agitation, and the second half should have a scale of 1–2. For example, in a four-tank system, at least the last 2 should be agitated.

A dry mill process has a substantial solids loading in the fermenter. Though some have argued that agitation is not needed while gas is vigorously evolving, actual field experience indicates that this is not true. Many installations have had to periodically shut down the first stage of fermentation to manually shovel out the tank. This has happened even with mechanical agitation, though not when the proper scale of agitation is applied. Table 7.3 outlines the process result as a function of continuous flow stage, based on a four-stage system, for several

**Table 7.3** Dry mill fermenter agitation results.

| Process stage | Scale of agitation | Process result (typical) |
|---|---|---|
| 1 | <1 | Massive settling |
| 1 | 1.5 | Settling; periodic cleanout required |
| 1 | 2 | Full suspension |
| 2 | <1 | Massive settling |
| 2 | 1.5 | Settling; frequent cleanout required |
| 2 | 2 | Minor settling |
| 2 | 2.5 | Full suspension |
| 3 | <1 | No suspension; inoperable |
| 3 | 1.5 | Massive settling |
| 3 | 2 | Settling; frequent cleanout required |
| 3 | 2.5 | Settling; periodic cleanout required |
| 3 | 3 | Full suspension |
| 4 | <1 | No suspension; inoperable |
| 4 | 1.5 | Massive settling |
| 4 | 2 | Massive settling |
| 4 | 2.5 | Settling; periodic cleanout required |
| 4 | 3 | Full suspension |

degrees of agitation. Because we are looking at minimum levels of agitation, we are reporting agitation in fractional units rather than rounding to whole numbers.

Examining the table, one can see that, unless one likes to shut down and clean out tanks periodically, certain minimum levels of agitation are needed. To minimize capital and power cost, one should plan for a minimum scale of agitation of 2 in the first stage, 2.5 in the second, and 3 in the third and fourth stages. However, since a fermentation can fail and gas can stop evolving, this author recommends using at least a scale of 3 in all stages.

The guidelines in Table 7.3 are based on minimum agitation to avoid operating problems. But, as mentioned in the first paragraph of this section, fermenters are more complex than that. Many years ago, a consultant to the ethanol industry told me that there was quite a bit of variation in rate of production, yield, and ethanol content in fermenters across the country, even using the same process, same yeast strain, etc. The only difference he could see was in the degree of agitation.

As it turns out, at least one academic researcher studied this. In Ref. [5], Professor Galindo and others studied the effect of agitation on a simple saccharomyces cerevisiae and sucrose fermentation, with no extraneous solids in suspension, in three scales of operation: 2, 14, and 130 l.

In Figure 7.8, I have re-plotted Galindo's data in an effort to quickly summarize his results. Rate of production, yield, and maximum % ethanol concentration produced are plotted versus agitation power/volume (P/V). Since the study was done on several tank sizes and all of the data seem to resolve as a function of P/V, that is how I chose to plot it. However, it is not clear why P/V should be a scaling parameter. Normally, P/V is associated with maintaining equal interfacial area/volume when two phases are present and one is being dispersed into the other,

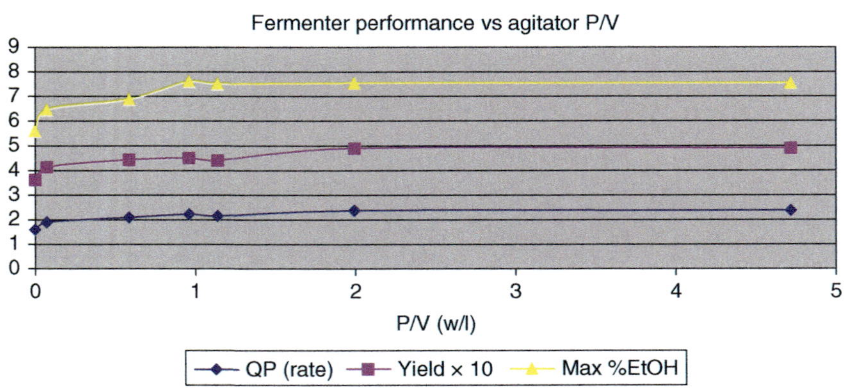


**Figure 7.8** Effect of agitation on fermenter performance.

such as liquid/liquid or gas/liquid contacting. Evolving gas does not fall under this category. More research needs to be done to uncover the mechanism(s) of the effects of agitation on such fermentations, so that we may be more certain how to scale-up the results.

Nonetheless, upon examining the data, we see that agitation has a profound effect going from no agitation (P/V = 0) to very high agitation (P/V = 5 W/l). Rate of production, yield, and maximum % ethanol concentration all increase, but the results level out around a P/V of 1.8–2.0. Specifically, rate increased 49% and both yield and % ethanol increased 36% compared to the unagitated case. Compared to modest agitation, the increases were about 15% on rate, 20% on yield, and 10% on concentration. These are significant increases, meriting further study. However, a P/V level of 1.8 W/l is far higher than any ethanol fermenter agitator operating in the industry today, based on the author's knowledge of the industry. (Please feel free to contact me if I am wrong on that point.) Most production installations have a P/V of less than 0.02 W/l. So, there is potential room for improvement without going all the way to 1.8 W/l.

My usual advice to clients has been to see if they have any money left over in their budget after everything but the fermenter agitators is accounted for, and then, if there is any surplus of funds, use it to buy bigger fermenter agitators. Alas, I have yet to hear of a case where there was any surplus in the overall budget!

## Beerwell

The beer well is a surge vessel between the fermenter and the distillation equipment. In a wet mill or a sugar-based process, it will have very low solids content, mostly yeast. It may not even have that if the yeast is first centrifuged out for re-use. Some wet mill beer wells have no agitation. For better uniformity to the still, agitation is recommended, with a scale level of 1–2 unless short retention time makes it blend time limited.

In a dry mill process, the high solids loading is still present. A scale of agitation of 3 is recommended.

## Stillage

This term applies primarily to the dry grind process.

After leaving distillation, the still bottoms, called stillage in this industry, still have the solids they entered the still with. The ethanol has been removed by the still. Generally, there are two stages of stillage. The heavy stillage is as it comes from the fermenter. It is fairly low in solids (<30%). These solids are mostly removed by screening, filtration, or centrifugation, and the centrate, called light stillage, is lower in solids, and they are finer. This stillage is often put through another concentration step (evaporation, for example) to increase the solids content, in preparation for final concentration and drying for sale as distillers' grains.

From an agitation viewpoint, the two kinds of stillage require almost the same degree of agitation: a scale level of 3. The main difference is that the fluid density is higher in heavy stillage, which may require more power for a given tank volume. However, some plants use a lower scale of agitation than 3 for light stillage.

### Miscellaneous Applications (Hold Tanks, pH Adjust)

Generally, most miscellaneous tanks can be handled with a scale level of 3 if any grain solids are present. If no solids are present, even less agitation may be needed. The principles in Chapter 5 may be applied. pH adjustment tanks should normally have at least a scale of agitation of 3 to avoid control system lag time problems.

## Combination Processes

Some plants are built with combination processes, such as adding cellulosic ethanol to the mix. For example, a corn-based dry mill plant may add sugar produced by hydrolysis of corn fiber to the fermenter feed. A cane sugar ethanol plant may hydrolyze the bagasse to sugars and add these to the fermenter feed. The agitator may be designed based on whatever process step is discussed in the flowsheet discussions herein.

One important thing to bear in mind is that starch and sugar-based processes are relatively mature, so their agitation requirements do not change much. The biggest change in this author's career has been the change from pitched blade turbines to hydrofoils, which has saved the industry considerable capital and operating costs. Cellulosic processes are both newer and much more complex from an agitation viewpoint. Much experimental work is needed for each different feedstock and process scheme. That is why, in this book, we have devoted not only a whole chapter to the cellulosic process (Chapter 8) but also two more chapters (12 and 13) devoted to the topics of developing the necessary data for design, especially for hydrolysis reactors.

### Application Summary Chart

For the sake of compactness, most of the previous guidelines have been summarized in Table 7.4 as a quick reference.

**Table 7.4** Application summary chart.

| Process type | Application | Nominal scale of agitation | Minimum scale of agitation | Comments |
|---|---|---|---|---|
| Wet mill | Starch makedown | 6–8 | | Avoid temperature >65 °C |
| Dry mill | Grain makedown | 6–8 | | Avoid temperature >65 °C |
| Wet mill | Starch ball buster | N/A | | Use if fisheyes form. Tip speed 15–25 m/s |
| Dry mill | Starch ball buster | N/A | | Use if fisheyes form. Tip speed 15–25 m/s |
| Wet mill | Starch slurry hold | 3 | 2.8 | Double check with detailed calculations |
| Dry mill | Grain slurry hold | 3 | | |
| Wet, dry mill | Cooking/liquefaction | N/A | | Specialized procedure. See text discussion |
| Wet mill | Batch saccharification | 2 | 1 | Intermittent operation |
| Wet mill | Continuous saccharification | 3 | 2 | |
| Dry mill | Batch saccharification | 3 | 2.5 | |
| Dry mill | Continuous saccharification | 3 | 2.8 | |
| Any | Propagation | N/A | | If separate step, use aerobic fermenter design approach |
| Wet mill, sugar | Batch fermentation | 2 | 1 | Higher degree of agitation may improve rate, yield and ethanol content |
| Wet mill, sugar | Continuous fermentation stages 1–2 | 1 | | Higher degree of agitation may improve rate, yield and ethanol content |

(*Continued*)

**Table 7.4** (Continued)

| Process type | Application | Nominal scale of agitation | Minimum scale of agitation | Comments |
|---|---|---|---|---|
| Wet mill, sugar | Continuous fermentation stages 3+ | 2 | | Higher degree of agitation may improve rate, yield and ethanol content |
| Dry mill | Batch fermentation | 3 | | Higher degree of agitation may improve rate, yield and ethanol content |
| Dry mill | Continuous fermentation stages 1–2 | 3 | 2 | Higher degree of agitation may improve rate, yield and ethanol content |
| Dry mill | Continuous fermentation stages 3+ | 3 | 2.5 | Higher degree of agitation may improve rate, yield and ethanol content |
| Wet mill, sugar | Beer well | 1 | | |
| Dry mill | Beer well | 3 | | |
| Dry mill | Stillage | 3 | | Light stillage may use scale of 2 in some facilities |

## Example Problems

**Example 7.1** Starch Storage
Let us design a starch slurry holding tank for a 30 wt% slurry, with a capacity of 6000 $m^3$. The vessel will be 18.5 m diameter with a sloped bottom. The liquid level above the tank centerline will thus be 22.25 m, giving us a Z/T ratio of 1.2. We will only need one main impeller for this geometry. From Table 7.2, we see that 30% starch slurry has a specific gravity of 1.13, or a density of 1130 kg/$m^3$. Thus, we have a mass of 6.78E6 kg to be agitated. Per Table 7.4, the nominal scale of agitation required is 3, and we must not have less than 2.8.

For such a large tank, we know that the shaft speed will be low. Let us choose the AGMA (American Gear Manufacturer's Association) speed of 16.5 rpm or 0.275/seconds for this example. For our initial selection, we will use Eq. (5.12) to select the power input:

$$\text{Scale of agitation} = 3 = 600\left(P/(M*N)\right)^{0.5} = 600\left(P/(6.78E6*0.275)\right)^{0.5}$$

Solving for P, we get P = 46.6 kW. The nearest standard motor size is 55 kW, which is at 85% loading. Now, to verify that this design will produce at least a scale of agitation of 2.8, we must calculate the impeller size and the equivalent tank diameter. We use Eq. (5.8) for the latter:

$$T_{EQ} = (4V/\pi)^{1/3};\ \text{thus, } T_{EQ} = 19.69\ \text{m.}$$

To size the impeller, we will start by assuming a power number of 0.3:

$$P = 46.6\,\text{kW} = 46\,600\,\text{W} = N_P \rho N^3 D^5 = 0.3*1130*(0.275)^3 D^5;\ D = 5.81\,\text{m.}$$

This gives us a D/T ratio of 0.31. From Appendix 1, we estimate the actual power number, based on turbulent flow, to be pretty close to 0.3, so our initial estimate holds. Now we calculate pumping, based on $D/T_{EQ}$, which is 0.295. Again, from Appendix 1, our pumping number is about 0.53–0.54. Using 0.53, we get a pumping rate of 0.53 * 0.275/seconds * (5.81 m)$^3$ = 28.6 $m^3$/s. Based on $T_{EQ}$, our equivalent cross-sectional area is 304.5 $m^2$. Therefore, our characteristic fluid velocity, or Q/A, equals 28.6/304.5 = 0.094 m/s. Using Eq. (5.10), we get:

$$\text{Scale of agitation} = v_c(\text{ft./min})/6 = v_c(\text{m/s})/0.0305 = 0.094/0.0305 = 3.08$$

Therefore, our design checks out. Special note: if agitation is required at a low liquid level, it is permissible to reduce the power of the main impeller by 25% and invest that power in a lower radial "Tickler" turbine with no loss of performance.

For a more complete design, we would repeat the process above for several other shaft speeds, especially those lower than 16.5 rpm, until the D/T ratio was up to a maximum of 0.45, and then estimate capital and present value of operating costs to get the most economical design.

As a footnote, such agitation levels have actually been successfully applied in starch storage tanks this big.

**Example 7.2** Dry Grind Slurry Makedown
Let us suppose we have a 60 $m^3$ tank for this purpose, with tank dimensions of 3.7 m diameter and 6.5 m straight side, with a working liquid level of 6 m. We will make down the slurry to 30% solids, with a specific gravity of 1.11. For this tank geometry and this application, we would use two impellers: an upper pitched blade turbine for incorporating the dry solids, and a lower hydrofoil, to promote overall mixing and suspension. To facilitate design, we will divide the tank into two zones conceptually. The lower zone would be designed for a scale of agitation of about 4. The upper zone would be designed for a scale of 6–8. We will use a scale of 7 for this example.

Since both impellers must rotate at the same shaft speed, we will need to choose one to start out with. Based on experience, we will start with an AGMA speed of 84 rpm or 1.4/seconds. For the lower 30 $m^3$ section, we can use Eq. (5.12), noting that the fluid mass for the lower half of the vessel = 30 * 1 110 = 33 300 kg:

$$\text{Scale of agitation} = 4 = 600\left(P/(M*N)\right)^{0.5} = 600\left(P/\left(33\ 300*1.4\right)\right)^{0.5}$$

Solving for P, we get P = 2.07 kW

For the upper impeller, we use Eq. (5.11):

$$\text{Scale of agitation} = 7 = 429\left(P/(M*N)\right)^{0.5} = 429\left(P/\left(33\ 300*1.4\right)\right)^{0.5}$$

Solving for P, we get P = 12.41 kW. Thus, our total power = 2.07 + 12.41 = 14.48 kW. The nearest standard motor size is 18.6 kW.

How large are our impellers? The lower one may be solved by starting with a power number of 0.3:

$$P = 2070\,W = 0.3*1110*1.4^3*D^5$$

Solving for D, we get D = 1.18 m, a D/T value of 0.32. Based on Appendix 1, our power number estimate looks good.

For the upper impeller, estimating the power number at 1.37:

$$P = 12\ 410\,W = 1.37*1110*1.4^3*D^5$$

Therefore, D = 1.24 m, a D/T value of 0.34. From Appendix 2, our power number estimate is confirmed. This would be a viable design. We would not want to go much lower in shaft speed or much larger in impeller size, lest we reduce the vortexing action. Likely, 68 rpm would still be OK, but maybe not 56 or 45 rpm. Vortex action can be enhanced by cutting tank baffles below the surface. For this problem, I would locate the upper impeller about 1 m below the surface and cut the baffles off at 0.5 m below the surface.

## Summary of Chapter

We have given an overview of common starch and sugar ethanol processes, recognizing that there may be several variations of such processes in use. However, the agitation guidelines should work in any of these variants as long as the agitator is called upon to perform similar functions.

We have provided specific guidance on how to do the process design of agitators for each major process step where agitators are employed.

## Multistage Reactor Design

For the system modeling concept presented in Figure 7.6 to be strictly valid, each stage must be perfectly mixed. That is, the composition throughout the stage is equal to the outlet concentration of that stage. That requires a blend time of zero. In practice, this is not possible. It is the author's opinion that a reasonable approach for design purposes is to require that the blend time to 99% uniformity in each stage be equal to or less than the mean retention time in each stage. A number of papers, such as Ref. [6], have been written on the topic of agitated blend time. For turbulent conditions only, the following relationships may be used with reasonable accuracy.

For marine propellers as used in the mixing industry:

$$t_{99} * N = (Z / T)^{0.5} * 12.4 / (D / T)^{1.73}$$

where $t_{99}$ is the blend time to 99% attenuation, N is the shaft speed, Z is liquid height, T is tank diameter, and D is impeller diameter.

For radial turbines (Rushton and similar):

$$t_{99} * N = (Z / T)^{0.5} * 4.8 / (D / T)^{2.3}$$

For four-bladed, 45° pitched blade turbines:

$$t_{99} * N = (Z/T)^{0.5} * 6.3 / (D/T)^{2.3}$$

Additionally, if there are two phases present, agitation must be sufficient to promote near uniformity of dispersion, or there may be buildup of one of the phases preferentially in the column. This can be a serious problem in slurry reactors.

## Calculating the Degree of Backmixing

The backmixing ratio as defined in Eq. (7.1) can also be defined in terms of velocity ratios, which will prove more convenient, as the backmix velocity, rather than backmix flow, is determined by impeller characteristics:

$\beta = V_B/V_F$, where $V_B$ is the backmix velocity and $V_F$ is the feed velocity, defined by
$V_F = F/A$, where A is the interstage open area.

The actual backmix velocity is a function of stage construction, impeller characteristics and impeller Reynolds number, $N_{RE}$, as well as the process feed velocity. Increasing the feed velocity actually decreases the backmix velocity, ultimately becoming zero at a high enough flow (Ref. [7]).

The actual backmix velocity is calculated by first calculating the backmix velocity at zero feed flow and turbulent conditions, and then multiplying it by shutoff factors (the last three terms in the next equation) for process flow, viscosity, and standpipe (only if one is used). The next equations are based on a central opening no larger than 50% of the tank diameter:

$$V_B = V_{B0} * C_F * C_\mu * C_T$$

The individual terms may be calculated thusly:

$V_{B0} = C_I * ND^2T^{-0.5}Z^{-0.5}$, where $C_I$ is a factor depending on impeller type
$C_F = k^{(VF/V_{B0})}$ for $V_F/V_{B0} < 6$; use 0 if $\geq 6$. k has an approximate value of 0.591 for Rushton turbines and 0.178 for axial hydrofoils.
$C_\mu = (N_{RE}/100\,000)^{0.177}$ for Reynolds numbers between 2000 and 100,000.
$C_T = 10^{(-0.2128 * L/DH)}$ for $L/D_H < 5$

$C_I$ has a value between 0.1 and 0.17 for most common impellers. It is about 0.142 for a traditional Rushton turbine, and less for hydrofoils. The earlier expressions are the author's curve fits to data presented in Ref. [7]. Do not extrapolate beyond the limits shown.

## Effect of Backmixing on Residence Time Distribution, Stage Efficiency

One way to characterize the performance of a reactor scheme is by its residence time distribution. There are several ways to represent residence time distribution. A typical way is to look at the exit age distribution, which is defined such that

$$\int_0^\infty E(t)dt = 1$$

The fraction of the fluid that spends a given duration, t, inside the reactor is given by the value of $E(t)dt$.

Often, results are expressed in dimensionless time, $\Theta$, or $t/\tau$.

For constant density systems, the following apply.

For a single ideal mixed reactor:

$$E(t) = 1/\tau * e^{-t/\tau} \quad \text{or} \quad E(\Theta) = e^{\Theta}$$

For ideally mixed tanks in series with zero backmix:

$$E(\Theta) = \left(n^n / (n-1)!\right) * \Theta^{(n-1)} EXP(-n\Theta)$$

The equations used to determine the exit age distribution for tanks in series with backmixing are very bulky and complex. The reader is referred to Ref. [8] for the calculation methodology. But in this chapter, we wish to illustrate the effect of backmixing on typical E curves. Figure 7.9 compares five ideal stages to a five-stage system with a backmix ratio of 0.2.

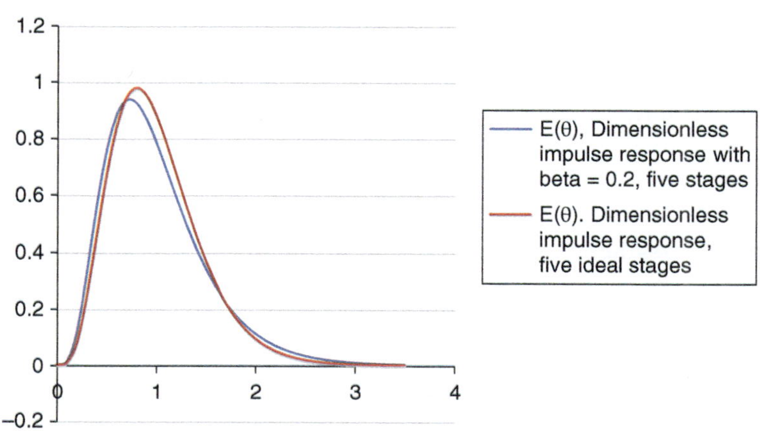


**Figure 7.9** Five-stage RTD curve.

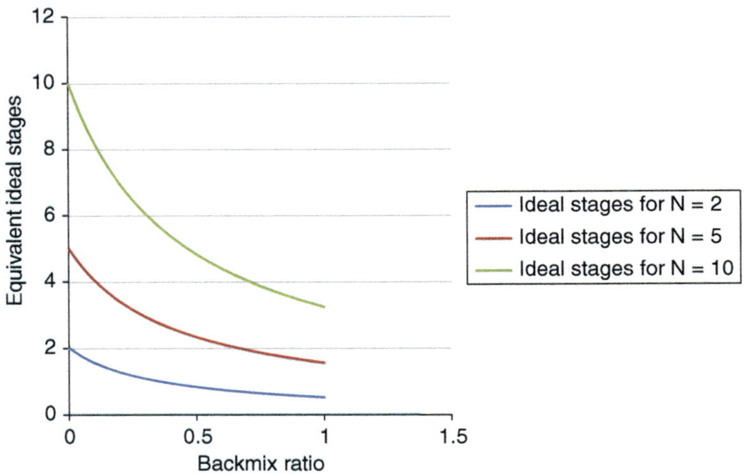


**Figure 7.10** Ideal stages versus beta.

Another way to evaluate performance is to compare on the basis of equivalent ideally mixed stages. For convenience, we define a variable, γ, to reduce the size of the standard deviation equation:

$$\gamma = \beta / (1 + \beta)$$

The standard deviation of the dimensionless E-curve or exit age distribution may be expressed as:

$$S = ((1 + \beta) / n^2) * ((n + \beta) * (1 + \gamma) + \beta * (1 + 2 * \gamma^n) - 4 * (1 + \beta) * \gamma - (1 + \beta) * \gamma^2)$$

The number of equivalent ideal stages is simply the reciprocal of the standard deviation:

$$n_{ideal} = 1 / S$$

Figure 7.10 visualizes the effect of backmix ratio on the number of ideal stages, for 2.5 and 10 actual stages.

## List of Symbols

| | |
|---|---|
| A | Interstage opening; tank cross-sectional area (depends on use) |
| $A_m$ | Poles of E-curve solution |
| $C_F$ | Shutoff factor for process flow (dimensionless) |

| | |
|---|---|
| $C_I$ | Impeller factor (dimensionless) |
| $C_T$ | Shutoff factor for standpipe (dimensionless) |
| $C_\mu$ | Shutoff factor for viscosity (dimensionless) |
| D | Impeller diameter |
| $D(\theta)$ | Denominator of transfer function |
| $D'(\theta)$ | Derivative of above |
| $D_H$ | Hole diameter in stage or stand pipe |
| $E(t)$ | Exit age distribution function |
| $E(\theta)$ | Dimensionless exit age distribution |
| F | Process flow rate |
| $F_B$ | Interstage backmixing flow rate |
| $G_n$ | Laplace transfer function of tanks in series with interstage backmixing |
| $g_n$ | Solution function in E-curve; equals $E(\Theta)$ |
| k | Shutoff factor constant (dimensionless) |
| L | Stand pipe height or length |
| m | Subscript for zeroes of transfer function |
| N | Shaft rotational speed |
| $N_{RE}$ | Impeller Reynolds number (dimensionless), $D^2N\rho/\mu$ |
| n | Number of stages |
| $n_{ideal}$ | Equivalent number of ideal stages |
| P | Agitator power |
| S | Standard deviation (dimensionless) |
| $s_m$ | Poles of E-curve solution |
| T | Vessel diameter |
| $T_{EQ}$ | Equivalent tank diameter |
| t | Blend time or time |
| $t_{99}$ | Blend time to 99% attenuation of a concentration disturbance |
| V | Reactor or stage volume |
| VVM | Volume of gas per volume of liquid per minute |
| $V_B$ | Backmix mean velocity |
| $V_{B0}$ | Backmix velocity at zero process flow, turbulent conditions |
| $V_F$ | Feed velocity through stage opening |
| X | Reaction conversion (dimensionless) |
| Z | Stage height/compartment or liquid level in a single tank |

## Greek Letters

| | |
|---|---|
| $\beta$ | Backmix ratio (dimensionless) |
| $\gamma$ | Backmix variable, $\beta/(1+\beta)$ (dimensionless) |
| $\rho$ | Liquid density |

## References


1 Benz, G. (2021). *Agitator Design for Gas-Liquid Fermenters and Bioreactors*. Wiley. ISBN: 978119650492.

2 Cropsreview. Sugar crops, sugar alcohols and alternative sweeteners. cropsreview.com (accessed 18 December 2020).

3 Buzas, S., Dallmann, K., and Szajani, B. (1989). Influence of pH on the growth and ethanol production of free and immobilized *Saccharomyces cerevisiae* cells. *Biotechnology and Bioengineering* 34: 882–884.

4 Benz, G. (2014). Designing multistage agitated reactors. *Chemical Engineering Progress* 110: 30–36b.

5 Galindo, E., Salvador, M., and Romain, B. (1992). Effect of mechanical agitation on alcoholic fermentation. *Paper presented at the AIChE 1992 Annual Meeting*, Miami, FL (2–6, November 1992).

6 Fenic, J. and Fondy, P. (1966). Application of similarity analysis to blending of miscible liquids. *Presented at AIChE Meeting*, Atlantic City, NJ (18–21 September 1966), Paper 4d.

7 Fasano, J., Penney, W., and Xu, B. (1993). Design and scale-up of compartmented, staged process equipment with emphasis on interstage backmixing. *Presented at 14th Bi-Annual Engineering Foundation Mixing Conference*, Santa Barbara, CA (20–24 June 1993).

8 Roemer, M. and Durbin, L. (1967). Transient response and moments analysis of backflow cell model for flow systems with longitudinal mixing. *IE&C Fundamentals* 6 (1): 120–129.

# 8

# Application Guide

Cellulosic Ethanol and Renewable Chemicals

## Introduction

As mentioned in Chapter 7, sugars are the building blocks for most renewable fuels and chemicals (except biodiesel and other oil-based processes, which are discussed in Chapter 9). The most common sources of sugars are either direct sugar-producing plants or starch producing plants, as starch can be easily hydrolyzed to C6 sugars (those having six carbons in the molecule). Both of these sources are mostly found in plants used as food.

But there is another plant-based potential source of sugar that is indigestible by humans, and thus cannot be used directly as food: lignocellulosic materials. By that we mean the matrix of various kinds of cellulose bonded together by lignin, a kind of biopolymer glue. The molecular structure and empirical formula of cellulose is related to starch, but has different spatial orientation of its atoms. This means that, though it can be hydrolyzed to sugars, it is far more difficult to do so. It requires different enzymes, pH and temperature. Moreover, the cellular structure and the presence of lignin make it difficult for the enzymes to even get to the cellulose, requiring pretreatment steps to disrupt the cellular structure. This resistance, or recalcitrance, to breakdown is what makes plants strong and what makes them difficult to process. So far, this makes processing costs for sugar made from cellulose more expensive than food-based routes. However, the feedstock is usually much cheaper.

Another factor is that cellulosic hydrolysis does not produce just C6 sugars; it also produces C5 sugars. These are not as readily fermented by common yeasts, so the overall productivity may be less, unless C5 and C6 sugars are separated and fermented by separate organisms. Some companies simply discard the C5 sugars.

Though such considerations affect economics, they do not necessarily affect agitation requirements, so we need not discuss fermentation biology strategies.

*Agitator Design Technology for Biofuels and Renewable Chemicals*, First Edition. Gregory T. Benz.
© 2022 John Wiley & Sons, Inc. Published 2022 by John Wiley & Sons, Inc.

## Sources of Feedstock

The world is covered in plants, and all of them have cellulose as part of their make-up. Therefore, the possible choices of feedstock are almost unlimited. However, such factors as ease of growth, ease of harvesting, difficulty in processing and economic value if used for other purposes all affect the choice. Broadly speaking, most sources fall into four main categories: by-products from food crop processing, waste products from other processes, purpose-grown crops that have no food value or major commercial value, and specialty cellulose products. We will give examples of each.

### Food-Processing Byproducts

These are among the most common sources for biofuels and renewable chemicals. Some examples include corn stover (Figure 8.1a), sugar beet pulp (Figure 8.1b), sugar cane bagasse (Figure 8.1c), Cassava fiber (Figure 8.1d), corn fiber (no image), fiber from various species of sorghum (no image), rice hulls and stalks

**Figure 8.1** (a) Corn stover. *Source:* Courtesy NREL. (b) Sugar beet pulp. *Source:* Image credit to Valérie Heuzé (AFZ). (c) Sugar cane bagasse. *Source:* Courtesy NREL. (d) Cassava fiber. *Source:* By author.

(no image), and spent algae (after oil processing; no image). It can also include straw from grain harvesting (no image). It can include peels and other waste from fruit processing facilities.

## Waste Products from Other Industries

Most of these are wood and paper related. For example, when a forest is logged, small branches and other debris can be left on site, or they can be burned as fuel, or they can be chipped and used for many purposes, including composite wood, biofuels and chemicals, etc. Wood chips are illustrated in Figure 8.2a. Sawdust can also be a feedstock.

Paper mills create a great deal of waste. The most useful is the fines from paper processing, as these have already been reduced to short fibers, rendering them more accessible to enzyme treatment (no image).

Related to the above is municipal solid waste. This can be in the form of waste newspapers and other paper products (Figure 8.2b) or mixed solid waste which must be separated to sort the lignocellulosic fraction (plus food waste, baby diapers, etc.) from metals, plastics, and glass (no image). It can also include sludge and paper waste from municipal waste treatment plants (ugh!), though this seems rare at the moment.

(a) (b)

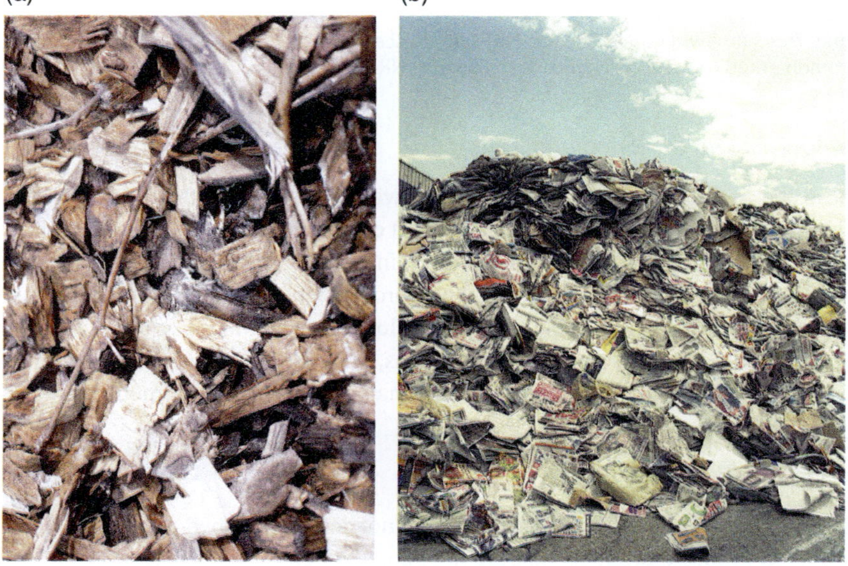

**Figure 8.2** (a) Wood chips. *Source:* NREL. (b) MSW. *Source:* NREL.

**Figure 8.3** (a) Switchgrass. *Source:* NREL. (b) Miscanthus. *Source:* Miya. M/Wikimedia commons/CC BY SA 3.0. (c) Napier grass. *Source:* NREL.

## Purpose-Grown or Nonedible Crops

These can include various grasses such as switchgrass (Figure 8.3a), Miscanthus (Figure 8.3b), or Napier grass (Figure 8.3c). It can include invasive species such as Kudzu (no image). It can also include certain species of trees that grow quickly and have little commercial value for wood production, such as certain species of poplar (no image). In principle, it could include such mundane sources as lawn clippings and leaves from residential areas, but the logistics and economics involved to collect such waste seem to be a major hurdle at the moment.

## Specialty Cellulose Products

This is kind of a catchall category for sources not described above. Mainly, it would be refined cellulose products that are not good enough for the fabric industry. An example is cotton linters. These are the short fibers that stick to the seeds after ginning. They can be combined with other short fibers that are mostly cellulose for bioprocessing. However, there are other commercial uses besides biofuels. The linters themselves are sold for batting and pillow fill material. They are also sold as "chemical cotton," as they are normally over 98% cellulose, and can be chemically converted to water-soluble polymers that act as gums and gels (such as CMC, aka carboxymethylcellulose, and others). These are used as drilling mud thickeners, thickeners for toothpaste, ice cream, salad dressing and other food products. These products are far more valuable than biofuel, so chemical cotton is unlikely to be used for biofuel. However, these products are still renewable chemicals, as they are made from a renewable feedstock. An emerging technology is the recycling of used clothing. The fibers are separated into the synthetics such as polyesters, polyolefins, nylons, etc., and the natural fibers, such as cotton, flax, and wool. The synthetics, which are thermoplastics, can be melted and re-spun into fibers. The cellulosic fibers are too weakened to be used in clothing, but the cellulosic fibers such as cotton can be fed into the same kinds of processes as other cellulose sources. I am not sure what can be done with animal fibers, such as wool. Perhaps they can be used as fuel for the process, or as a feedstock for pyrolysis.

There are too many individual processes to cover in this book. However, the agitation principles are the same for all cellulosic materials, so the procedures in Chapters 12 and 13 are fully applicable.

## Visual Comparison of Feedstocks

Though all potential feedstocks share some chemical and structural similarities, they can behave in quantitatively different ways in the process. A visual examination of some samples can help to show this. Figure 8.4 compares samples,

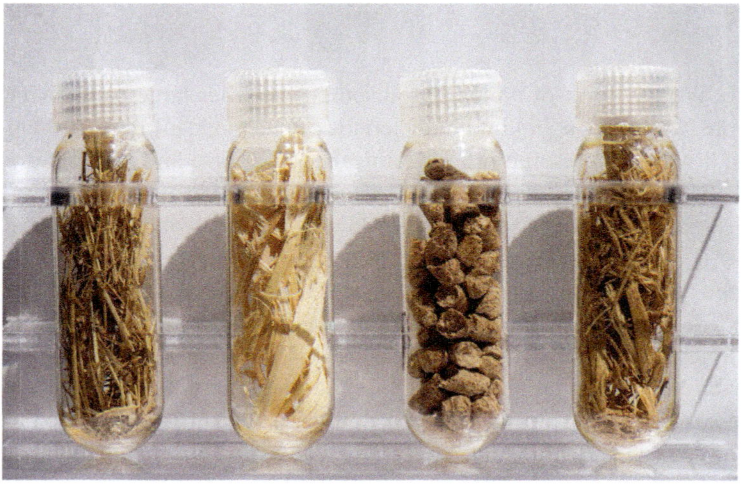

**Figure 8.4** Biomass comparison. *Source:* NREL.

from left to right, of switchgrass, corn stover, oak pellets, and bagasse. Even when milled, due to differences in fibrous structure, they behave differently in slurry form, affecting agitation and pumping. These differences necessitate a lot of "wet" lab testing for agitator design, which will be discussed in Chapters 12 and 13.

## Generic Cellulosic Sugar/Ethanol Production Block Process Diagram

In this section we will go over the major processing steps in a generic lignocellulosic ethanol process. A major portion of the process is devoted to converting cellulose to C5 and C6 sugars. After that point, the process can be used to make ethanol or many other products. Many of the other products are made by aerobic processes, which will not be covered here. These would be covered mostly in Ref. [1]. Therefore, we are using the ethanol process as an archetype of anaerobic lignocellulosic processes.

There can be many possible variations of process steps, so our block diagram will not cover all possibilities. We hope at least to cover enough agitator applications so that it will be possible to transfer the design philosophy to new processes as they are developed.

We will first go over the block diagram (Figure 8.5), and then go into the agitator applications in more detail where possible. All high-solids slurry handling will be covered in Chapters 12 and 13.

### Milling

In most cases, the raw form of biomass feed sources is either large particles or stringy material, which has poor flow characteristics when made into a slurry, and poor accessibility to enzymes and other chemicals. Microalgae and sawdust are examples of exceptions to this observation. Therefore, prior to further processing, most feedstocks are milled and screened to get down to a manageable particle size and shape.

What happens if this step is skipped? One of my clients had the idea that if the raw biomass feed could be slurried and processed without the milling step, this might save money in the long run, even if the hydrolysis step took a lot longer. The particular material in question was switchgrass, and it came into the facility in strands up to 200 mm long, with an average length of around 50 mm. We tried to agitate it in my test facility with a 0.3 m tank diameter and an appropriate impeller system. Figure 8.6 shows what happened.

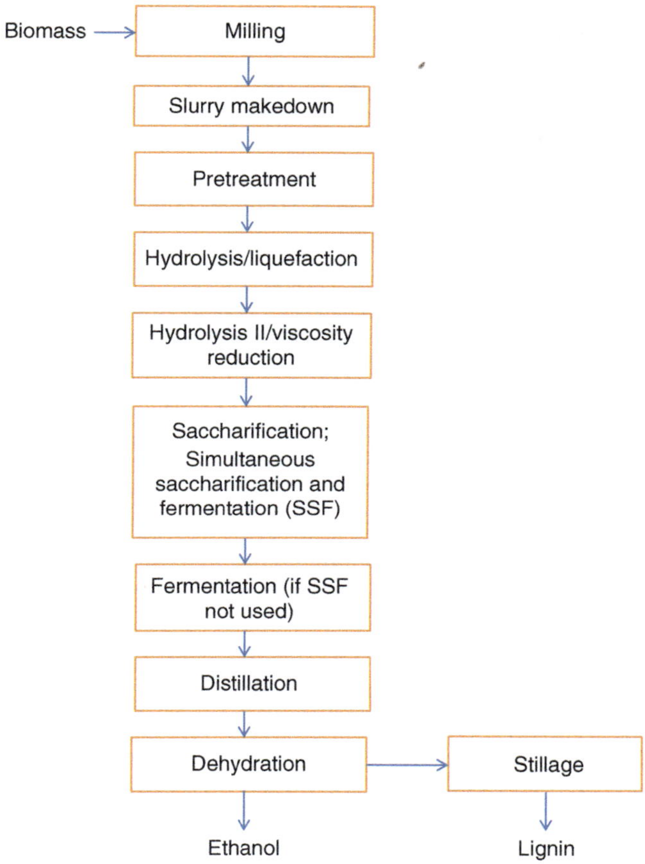


**Figure 8.5** Generic lignocellulosic ethanol process.

**Figure 8.6** Raw switchgrass on impeller.

Basically, the long strands wrapped themselves around the shaft and impeller. (Yes, there is an impeller inside that mess!) It was impossible to agitate this material under such conditions. It might be possible that on equipment large enough such that the impeller dimensions were much larger than the strand length, the strands would not have wrapped around the shaft or gotten hung up on the impeller. It would probably take at least a 2 m impeller diameter to have any chance of that working. That is a rather large scale in which to carry out a feasibility study that might not work anyway. So, it is this author's opinion that biomass should be milled before any other process step. In general, smaller particles make slurries that are easier to agitate. Less than 1 mm is ideal, but this author has successfully tested and agitated solids with up to about 20 mm length.

## Slurry Makedown

After milling, biomass solids are typically added to a tank where they are incorporated into water to form a slurry. This can be batch or continuous. The maximum % dry solids content that can be handled by a turbine agitator and also be pumpable depends on the nature of the biomass, but will generally be in the range of 12–15%. Above a certain concentration, the "slurry" will no longer be a liquid, but will behave as a damp solid. Then it can only be mixed by a solids mixer and conveyed by a conveyor, not pumped.

There are three criteria that determine the maximum % solids that a turbine agitator, or other fluid mixer, can handle:

1) The interior of the cells must be saturated with water.
2) The external, or "free" water, must completely fill the volume between particles.
3) 1–2% additional water content is needed to allow the slurry to flow as a liquid.

These criteria still apply after pretreatment and hydrolysis. But since those steps destroy the cellular structure, the internal water becomes external water, allowing higher concentrations to be handled.

Most of the time, the biomass is not saturated with water internally when it is added. In such a case, it will gradually absorb the moisture, becoming more difficult to agitate. This is akin to what happens when water is added to oatmeal. Therefore, additional retention time is needed to assure saturation before the next processing step. Sometimes this is accomplished by using a holding tank after the makedown tank. This is not shown on the block diagram.

Not all solids require this. Bagasse, for example, may be directly fed to such a process after being crushed and extracted for sugar, as it would be fully saturated with water at that point.

## Pretreatment

The purpose of pretreatment is to break down the cell walls as much as possible to allow access of enzymes and other chemicals to the lignocellulosic structure. There are a number of methods by which this is done. A few are listed next.

- Mechanical grinding to a small particle size
- Steam explosion (heating to a high temperature followed by a sudden drop in pressure, resulting in water inside the cells flashing to steam and rupturing the cells)
- Acid treatment
- Alkali treatment

Sometimes more than one pretreatment method is used. Most pretreatment processes are continuous flow, but some may be batch. Some use an agitated reactor, and some use a pipe reactor.

Generally, the effluent from the pretreatment reactor is screened before the hydrolysis step in order to remove some water and increase % solids. The pH will also need to be adjusted for the enzymes used in the hydrolysis reactor.

## Hydrolysis/Liquefaction

This step breaks down the cellulosic matrix and begins the process of converting cellulose to sugar. It can be batch, fed-batch, or continuous. There are considerable advantages to using fed-batch or continuous systems, as these allow higher % solids to be introduced; the agitator does not see the high solids feed, but the mixed contents of the tank, which will have a fair amount of hydrolyzed material. We will discuss more on this when we get to agitator concepts.

This step can be arranged as a tank farm, and liquefaction, saccharification and fermentation can in principle all be combined, if a combination of enzymes and organisms can be found that all use the same temperature and pH. So, the next step is not always used.

## Hydrolysis II/Viscosity Reduction

During hydrolysis, an enormous change in rheology occurs, and it does so rather quickly. At the maximum % solids feasible to agitate with a turbine agitator, the material in the tank resembles wet peat moss. Depending on how good the enzymes are and the dosing used, the viscosity will drop an order of magnitude or more in less than an hour. After several hours, it gets almost as thin as it will ever be, and resembles a mud or starch slurry. Yet the total hydrolysis cycle could take 60–90 hours to complete the conversion to simple sugars. So, rather than do it all in one kind of tank, it may be more economical to have separate tanks devoted to the final viscosity reduction, so that smaller agitation systems can be used. This probably saves capital and operating costs.

## Saccharification: Simultaneous Saccharification and Fermentation (SSF)

Rather than let all of the conversion take place in the previously listed tanks, it is not unusual to alter the enzyme system to more fully hydrolyze the sugars. This step is called saccharification. In most modern plants, fermentation is also conducted here, so that total system retention time and volume are reduced.

## Fermentation

If the plant does not employ SSF, it will have separate fermenters. While this would be unusual in today's ethanol plants, it may be used if other products are being made, where the conditions required might be different from those needed for saccharification.

## Distillation

This step concentrates the ethanol to 95%, an azeotrope with water. This will need to be denatured and dehydrated for fuel use. The denaturing and dehydration steps can be skipped for hand sanitizer and beverage ethanol production.

## Dehydration

The same steps are followed here as summarized under the sugar and starch based processes described in Chapter 7. Pressure swing absorption or other technologies are employed. Fuel-grade anhydrous ethanol is the product.

## Stillage

The main solids component of stillage in a cellulosic process is lignin. While not edible, this is a potentially valuable co-product. According to Ref. [2], the high heating value of lignin ranges from 22.2 to 28.5 MJ/kg, depending on feedstock type. This is often enough to provide all of the heating needs for the plant, with some left over. The excess lignin can be sold as fuel. There are also processes under development that can depolymerize lignin to make other products, or simply process it further for other uses.

The lignin can be separated and dried using conventional techniques.

## Application Guidelines

In this section, we go over agitation principles applicable to the process steps that employ mechanical agitation. The applications involving high concentrations of fibrous solids will be covered in more detail in Chapters 12 and 13.

### Makedown/Slurry Tank

In this tank, biomass solids are incorporated into a water-based slurry prior to pretreatment. The agitation requirements vary greatly according to what kinds of solids are being added, what concentration is desired, and the rate of addition. Some users prefer a more dilute slurry prior to the pretreatment system for ease of pumping. Others prefer to have as concentrated a slurry as possible. Dilute slurries (less than 5% dry substance) are best handled with a baffled vessel, as shown in Figure 8.7. Concentrated slurries (typically up to a maximum of 12–16% solids when not pretreated) require an unbaffled tank as shown in Figure 8.8, with other details as described in Chapter 13. Note that both of these photos show rather large particles, as they have not yet been pretreated.

Dilute slurries can work with a D/T ratio as little as 0.3, and a minimum scale of agitation of 3, unless more is needed based on incorporation rate desired. One impeller can handle a Z/T ratio of up to 1.2. Concentrated slurries usually require a D/T ratio of at least 0.5, and 0.6–0.7 is fairly typical. Two impellers per square batch are a typical design. Hydrofoils will do the job with one half to one third of the torque of pitched blade turbines. The concentrated slurries must be designed using the procedures of Chapters 12 and 13 to assure minimum motion.

Figure 8.7 Dilute un-pretreated woody biomass slurry in a baffled tank.

When a certain incorporation rate is needed, that may require more than minimal motion. The desired incorporation rate in the full scale equipment should be expressed as mass of solids/cross-sectional area of the tank per unit time, or F/At. When scaling down to the lab, this means the required time is proportional to the tank diameter, as mass/volume is the same but mass/area goes up. So, if the production tank is 2 m diameter and the lab tank is 0.3 m diameter, and we want a 30 minute incorporation time in the production tank, we need to get it into the lab tank in

**Figure 8.8** Concentrated unpretreated woody biomass slurry in an unbaffled tank.

4.5 minutes. When we find the shaft speed that will accomplish this, we empirically scale the speed up using a scaling exponent of 2/3. This procedure is described in Chapter 11.

Often, it takes some time for the biomass to fully absorb water, and the slurry will thicken up while this happens, because free water is being absorbed into the cells, reducing the available free water. It is optional to use a second tank or tank farm to complete this step. Only minimum motion is required for the second tank, which might be thought of as an aging step. Design must be based on lab testing plus scale-up, as per Chapters 12 and 13.

As a historical note, the earliest mixing equipment used in cellulosic bioprocessing was various forms of solids mixers such as horizontal ribbon blenders, horizontal paddle or plow blenders, and even rotating dryer assemblies. These can handle very high % solids; even 100%. But they are expensive, have scalability issues, and are overkill once the slurry becomes liquefied.

## Pretreatment

There are a number of ways of handling pretreatment. If continuous flow, chemicals, and/or steam can be added via a static mixer, and then fed into a pipe reactor, similar to that illustrated in Figure 7.5. One could also use a staged column as shown in Figure 7.6. Since the retention time is short, it is vital that the feedstock is vigorously agitated in the latter case to assure that chemicals or heat fully penetrate into the cellular structure and rupture most of it. The fibrous nature has not been altered much at this point, but the goal is to break the cells so that the internal water is released as extracellular or "free" water, enabling higher solids concentration in the hydrolysis step. The added shear may also help break down the cells, especially if they have been chemically softened.

If pretreatment is done on a batch basis, it is usually by chemical means, and may involve a tank farm. Agitator design should be done by lab test and scale-up, as per the procedures in Chapters 12 and 13. However, agitation should be much more than minimal. Some judgment is needed, but a shaft speed of 30–50% more than needed for minimal motion is suggested by the author. This will assure good movement throughout the vessel.

## Hydrolysis Reactor Schemes: Batch, Fed Batch, Continuous

There are a number of ways to conduct the hydrolysis reaction. The simplest is a batch reactor, illustrated in Figure 8.9. In such a reactor, the tank would be filled with the pretreated biomass, and then enzymes would be added. The reaction would continue in this reactor until completion of saccharification, unless transferred to other tanks as described under "Hydrolysis II." Possibly, fermentation could be done concurrently (SSF). As hydrolysis progresses, the viscosity would drop by at least two orders of magnitude, going from a material that resembles damp peat moss to a slurry similar to starch or drilling mud.

To use a simple turbine agitator, the solids content would be limited to perhaps 14–18%, depending on the feedstock. Using a solids mixer, such as a rotating drum with fins, a horizontal ribbon blender or a horizontal paddle or plow mixer, would allow the reaction to be carried out at any solids content that can be fed into the reactor. However, since the reaction product is a relatively thin liquid, even at 30–40% solids in the feed material, a solids mixer would just slosh around the product near the end, and such mixers are not available in large enough volumes for the biofuel industry. So, we recommend turbine agitators, sized as per Chapters 12 and 13.

A major disadvantage of the batch process is that the agitator must be capable of handling the maximum % solids fed in, even though the product drastically thins out with time. This means the required agitator is grossly oversized for 80–90% of the batch cycle time.

Another major disadvantage if turbines are used in a batch reactor is that the feed must be at least nominally liquid, meaning that the % solids it can handle is limited. Other reaction schemes do not suffer from this limitation.

Fed-batch or semi-batch reactors, as depicted in Figure 8.10, are generally a better choice. These start out with a heel of liquid, which can be water, water with enzymes, or partially hydrolyzed biomass, as long as the heel is relatively fluid. Pretreated biomass, which can be 20–30% solids or even more, is then added to the vessel at such a rate that the rheology of the tank is still nominally fluid, similar to 8–16% pretreated biomass. The agitator must be sized for the thickest material that will actually be in the tank, but not for the biomass fed into it. In essence, we are giving the enzymes time to do their job of thinning the biomass to make the

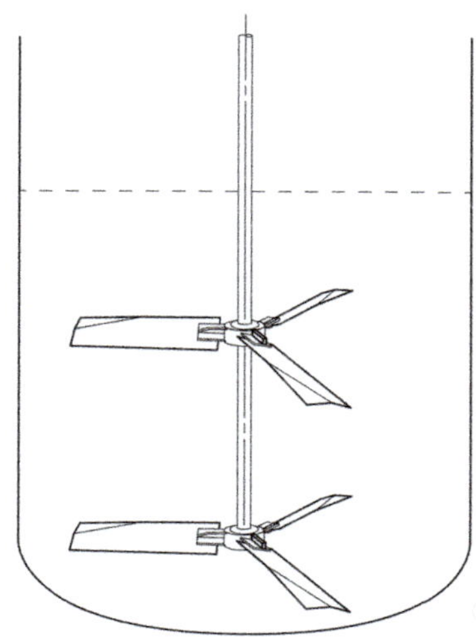

**Figure 8.9** Batch reactor.

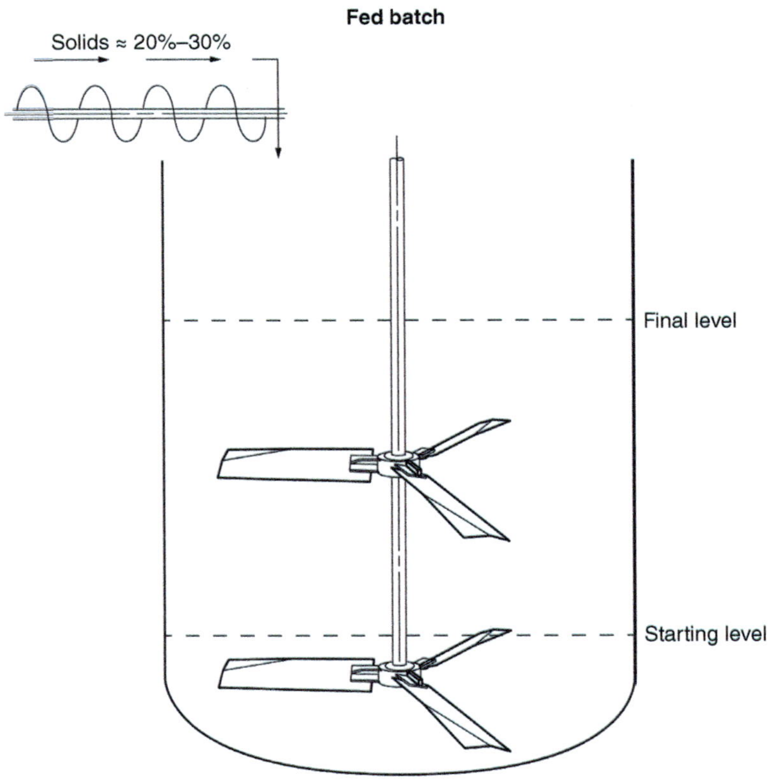


**Figure 8.10** Fed-batch process.

agitation job easier. However, the viscosity at 1/s may still be several hundred thousand cP. Depending on the feedstock, pretreatment and enzyme type and dosing, it may take 1–12 hours to fill the tank. After that, the reaction can continue to completion in the tank, or the slurry can be transferred to other tanks (Hydrolysis II) to finish the reaction, and this may be combined with fermentation (SSF).

This approach allows much less water to be carried through the process than a batch process requires, so the overall tank size is smaller, and concentrating the final product uses much less energy.

Continuous flow systems can be arranged a number of different ways, and may combine some batch process elements. Figure 8.11 depicts one possible arrangement.

In this system, the biomass feed is at a high solids concentration, but is fed into a small backmixed reactor. The retention time, along with enzyme dosing, is chosen such that the rheological properties within the tank resemble pretreated

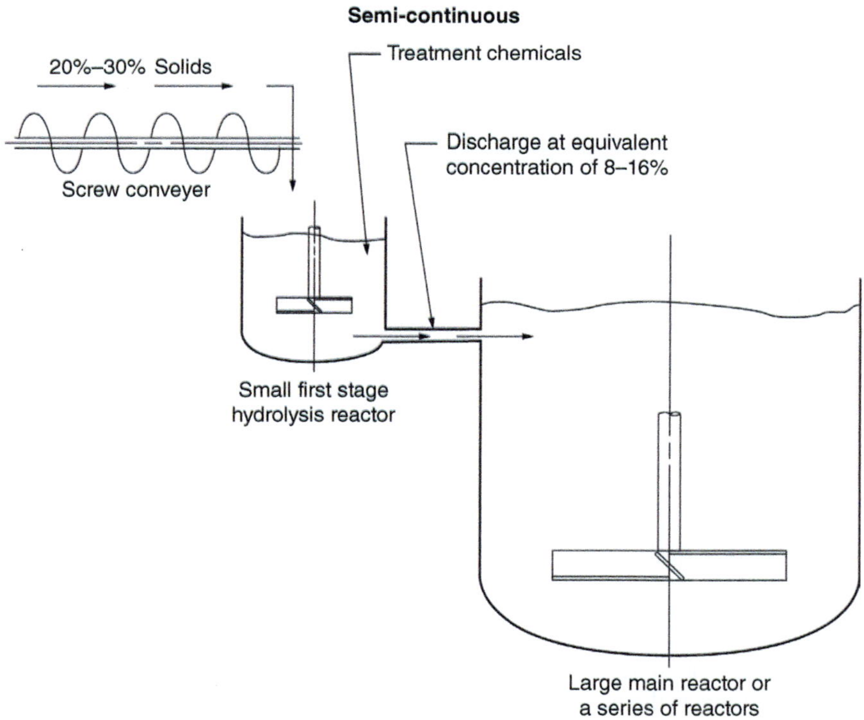


**Figure 8.11** Semi-continuous process.

feedstock at 8–16% solids. This may take anywhere from 30 minutes to several hours. In other words, the tank contents must be nominally fluid, though they can be quite thick. (Think of cooked oatmeal, for instance). Viscosity at 1/s shear rate may be several hundred thousand cP.

Often, the enzymes are added to the solids in a continuous flow conveying mixer, such as that depicted in Figure 8.12. This would take the place of the screw conveyor in Figure 8.11.

The discharge from this tank can be fed to tanks in series, which eventually become what we describe as Hydrolysis II. Or the larger tanks can be set up as a tank farm to either complete the hydrolysis or to feed to another series of tanks to complete the hydrolysis (Hydrolysis II). Basically, the required volume gets bigger as the product gets thinner, as the initial thinning out happens quite rapidly but the reaction completion may take close to 100 hours.

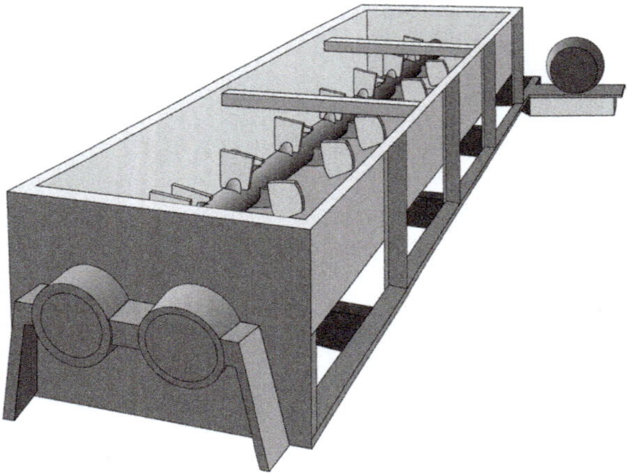

Figure 8.12 Continuous paddle mixer.

## Mixing System Rationale

Up until this point in the process, most of the materials we have been handling have been either damp solids or highly viscous, non-Newtonian liquids, closer to being pastes (at least before significant hydrolysis has occurred) than free-flowing liquids. As long as the material in question is actually a liquid, rather than damp solids (this means all spaces between particles are filled with liquid, plus 1–2% more free liquid to allow the whole mass to flow as a liquid), we have recommended the use of turbine agitators. Though we will get into more detail on sizing procedures in Chapters 12 and 13, we pause here to explain the rationale behind using turbine agitators.

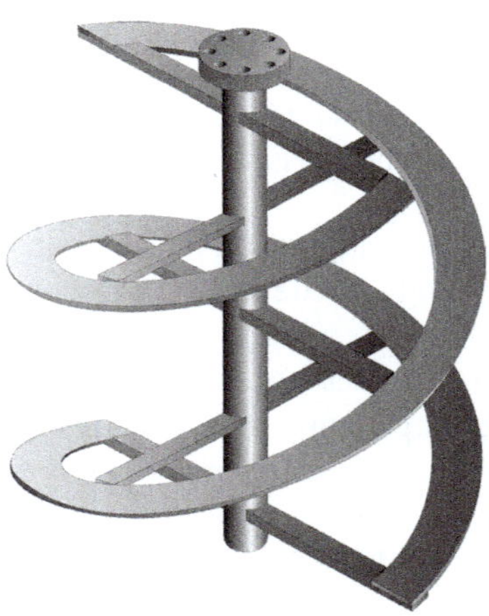

Figure 8.13 Helix impeller.

Those who are familiar with mixing of viscous liquids, as in bulk polymerization reactors, for example, are used to seeing this accomplished primarily with helical ribbon impellers, or, simply, a helix (Figure 8.13). These impellers are close-clearance devices, normally applied at D/T ratios of 0.9–0.98. They do an excellent job of blending viscous Newtonian and moderately non-Newtonian liquids at high viscosities. The author has seen them used successfully in liquids ranging from 50 000 to 20 000 000 cP. Since concentrated biomass slurries have viscosities at 1/s of perhaps 50000 to a bit less than 1000000cP, it would seem that they would be an ideal application for helix impellers. But, as noted in Ref. [3], helix impellers are not a good choice for lignocellulosic slurries for several reasons, repeated here:

1) Simply put, they don't work! Though helices work very well in Newtonian and moderately shear-thinning liquids if the wall viscosity is much lower than the bulk viscosity, the fluid will slip at the wall and tend to rotate en masse with the impeller. Lignocellulosic slurries generally have a yield stress as well as shear thinning behavior, which exacerbates the tendency to slip at the wall. The result is mass rotation and no mixing. The author has witnessed this on a number of occasions, with materials as diverse as repulped newspaper waste and cotton linters.
2) Hydrofoils and other turbine agitators are inexpensive. Even if both a helix and a turbine would work, the capital cost of a turbine agitator is much less than one equipped with a helix impeller.
3) Turbine agitators are scalable. Due to their close clearance design, helix units are limited in most cases to shop-fabricated vessels. Turbine impellers have no such limitations. Turbine impellers have been built with more than 10 m impeller diameter, and have been applied to vessels as large as 30000 $m^3$ capacity.

In a similar vein, many might ask, Why not use pitched blade turbines rather than hydrofoils? Traditionally, it has been assumed that hydrofoils were best for turbulent flow, pitched blade turbines were best for transition flow and that neither was very good in laminar flow. As Refs. [4, 5] show, these assumptions are false. Properly applied, turbine agitators can be used in laminar flow. And hydrofoils get the job done with one third to one half of the torque required from a pitched blade turbine. In Ref. [4], a simulation was done to show the comparative power and torque throughout a batch hydrolysis process, comparing pitched blade turbines to common hydrofoils. This was based on tests conducted in a mixed biomass slurry at different concentrations. Rheology and required shaft speed for full motion were measured. The results are presented in Table 8.1 and Figure 8.14. Also noteworthy is that the hydrofoils retained a more axial flow pattern that the pitched blade turbines did, resulting in better top-to-bottom turnover.

"Properly applied" means having the correct D/T ratio, which is generally 0.6–0.7 for the lower impeller, and slightly smaller for the upper impeller. The reasons for this will be explained more fully in Chapter 13. For now we will note that if the impellers are too small, they will "bore a hole" in the middle, resulting in no movement at the vessel wall, and if they are too large, they will cause mass rotation of the tank contents, similarly to a helix.

In this simulation, the hydrofoil saves 60% of the energy within the first 10 hours and 20% after that. It also uses much less torque, so the capital cost of the

**Table 8.1** Mixing progress, pitched blade turbine vs. hydrofoil.

| Time (h) | Viscosity at 1/s (cP) | Power law exponent | Hydrofoil shaft speed (rpm) | Pitched blade turbine (rpm) | Hydrofoil power draw (Hp) | Pitched blade turbine power draw (Hp) | Power ratio, pitched/hydrofoil | Torque ratio, pitched/hydrofoil |
|---|---|---|---|---|---|---|---|---|
| 0 | 1.21E+06 | 0.135 | 20 | 22 | 67.4 | 214 | 3.18 | 2.89 |
| 1 | 881 000 | 0.0917 | 13.8 | 14 | 27 | 65.2 | 2.41 | 2.38 |
| 2 | 160 000 | 0.159 | 11.9 | 11 | 21.2 | 38.4 | 1.81 | 1.96 |
| 3 | 120 000 | 0.17 | 11 | 7.98 | 5.58 | 8.97 | 1.61 | 2.22 |
| 4 | 90 107 | 0.21 | 10.1 | 7.3 | 4.25 | 6.78 | 1.60 | 2.21 |
| 5 | 45 000 | 0.3 | 9.2 | 6.54 | 2.79 | 4.39 | 1.57 | 2.21 |
| 6 | 20 000 | 0.5 | 9.2 | 6.38 | 2.42 | 3.61 | 1.49 | 2.15 |
| 7 | 10 000 | 0.7 | 9.2 | 6.25 | 2.26 | 3.17 | 1.40 | 2.06 |
| 8 | 5 000 | 0.8 | 9.2 | 6.05 | 2.04 | 2.74 | 1.34 | 2.04 |
| 9 | 2 000 | 0.9 | 9.2 | 5.82 | 1.85 | 2.37 | 1.28 | 2.03 |
| 10 | 500 | 1 | 9.2 | 5.66 | 1.72 | 2.16 | 1.26 | 2.04 |
| 10+ | 500 | 1 | 9.2 | 5.66 | 1.72 | 2.16 | 1.26 | 2.04 |

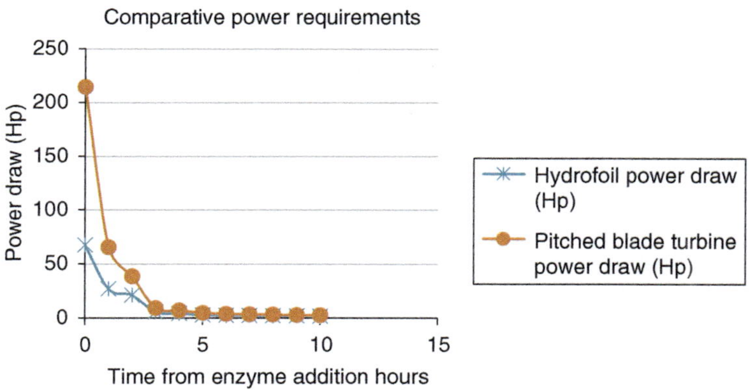


**Figure 8.14** Power comparison hydrofoil vs. pitched blade.

equipment will be less. So, a hydrofoil wins on both capital and operating costs. The earlier part of the simulation is more relevant for initial hydrolysis. The latter part of the simulation is more relevant for Hydrolysis II and subsequent operations. Though we used 500 cP for design in the latter part of the curve, the actual viscosity normally thins to less than 50 cP by the time full saccharification occurs. However, this has a negligible effect on agitator sizing in production equipment, as even at 500 cP, the flow will be turbulent.

## Hydrolysis II: Viscosity Reduction

Because of the drastic reductions in viscosity, some, though not all, producers opt to break up the hydrolysis reaction into another step, which we dub Hydrolysis II. Typically, this is done after anywhere from 2 to 20 hours contacting time with enzymes for pretreated stock, and maybe 15–20 hours for unpretreated stock such as cotton linters, paper mill waste and municipal solid waste, which all have relatively intact cellular structure. At this point, the viscosity is low enough so that baffled tanks should be used, a hydrofoil impeller with a D/T ratio of about 0.4–0.5 is recommended, and the design can be based on a scale of agitation of about 3. (The author has conducted several tests of two to three hour duration at a scale of 2 which showed no settling. However, a scale level of 3 is recommended due to possibly much longer retention time in the tanks and the possibility of nondigestible contaminants [aka "dirt"] in the slurry). When in doubt, depending on the initial condition in these tanks, testing and scale-up should be used.

## Simultaneous Fermentation and Saccharification (SSF)

To save time, energy, tanks, and agitator costs, most people are opting to start the fermentation process while saccharification is still occurring. This is also beneficial to enzyme usage, as the sugars produced actually inhibit enzyme activity (Ref. [6]), so it is good to let yeasts or other organisms consume the sugar as it is being produced. Agitator design depends on the degree of saccharification that has occurred in a given vessel. The presence of fermentation does not materially affect the agitation requirements. Therefore the sizing will be based on the previously described principles. If the fermentation begins after the viscosity is mostly broken down, a scale of agitation of 3 should suffice. However, it is possible that increased agitation may improve the rate of fermentation, the yield and the maximum titer as long as the fermentation is not limited by sugar production rate. See the discussion in Chapter 7.

## Fermentation

Though increasingly rare in new plants, some existing plants have a separate fermentation step after saccharification is complete. Others may have SSF for C6 sugars but have an additional fermentation stage for C5 sugars, which may be separated or may just use different organisms or conditions of temperature and pH. Though a scale of agitation as low as 2 may work when $CO_2$ is vigorously evolving, we recommend a minimum scale of 3 so as to keep solids in suspension even when gas is not evolving. The reader is also referred to the discussion in Chapter 7 regarding the effect of increased agitation on rate of ethanol production, yield, and titer. It may be good to buy the most agitation the budget will allow.

## Lignin Slurry/Stillage

Stillage is a combination of biomass and lignin. Some processes may separate the lignin from the biomass and recycle the living cells. In either case, the job of the agitator is to keep solids in suspension.

Although lignin slurry looks like mud, it does not settle as fast. Testing done by the author shows no visible settling in two hours at a scale of agitation of 2. However, it may simply take longer to settle than that, and additional dirt may accumulate over time. Therefore, we suggest a scale level of 3 as a basis for design.

If the lignin slurry is concentrated by filtration, centrifugation or other means, into a viscous liquid or paste, we recommend a lab test plus scale-up using a scaling exponent of 1, as described in Chapter 11. For such a material, viscosity may be measured by a conventional viscometer as long as the phases do not separate.

### Miscellaneous Applications

Since not all processes are the same, there can be other agitated tanks not described here, such as reagent make-up tanks, various hold tanks, etc. For the make-up tanks, normally a scale of agitation of 3 is sufficient. The procedures of Chapter 5 are applicable.

For tanks containing process materials such as cellulosic slurry, lignin, etc., the procedure should be based on the same principles described in this chapter and Chapters 12 and 13 for similar materials and applications.

## Summary of Chapter

We have discussed a generic cellulosic process for converting cellulose to sugars and then processing these sugars to other products, such as ethanol. We have gone over most of the tanks with mechanical agitation and provided sizing guidelines for them where possible to do so within a few paragraphs. The exceptions are those tanks containing high concentrations of fibrous material. The design procedures for those are so specialized that they require two dedicated chapters, namely Chapters 12 and 13.

## List of Symbols

A Tank cross-sectional area
D Impeller diameter
F Mass of solids
T Tank diameter
t Time
Z Liquid height

## References

**1** Benz, G. (2021). *Agitator Design for Gas-Liquid Fermenters and Bioreactors*. Wiley InterScience. ISBN: 978-1-119-65049-2.

**2** Demirbas, A. (2017). Higher heating values of lignin types from wood and non-wood lignocellulosic biomasses. *Energy Sources, Part A: Recovery, Utilization, and Environmental Effects* 39 (6): 592–598. https://doi.org/10.1080/15567036.2016.1248798.

3 Benz, G. (2017). Impeller selection for lignocellulosic hydrolysis reactors. *Biofuels Digest Magazine*, 16 March 2017, Special to the Digest.

4 Benz, G. (2017). Hydrofoil impellers vs. pitched blade turbines in lignocellulosic slurries. *Biofuels Digest*, 8 May 2017, Special to the Digest.

5 Benz, G. (2018). Consider hydrofoil impellers for laminar flow mixing. *Chemical Engineering Progress* 114: 30–35.

6 Xiao, Z., Zhang, X., Gregg, D.J., and Saddler, J.N. (2004, Humana Press). Effect of sugar inhibition on cellulases and β-glucosidase during enzymatic hydrolysis of softwood substrates. *Applied Biochemistry and Biotechnology* 113–116: 1115–1126.

# 9

# Biodiesel and Renewable Diesel Application Guide

## Introduction

In this chapter we review the main agitator applications found in the biodiesel and renewable diesel manufacturing processes. Though the products are different and the processes used to produce them are different as well, we chose to put them in the same chapter on the basis that most readers interested in diesel fuel would find both products and processes to be of interest.

### Biodiesel Definition

As currently defined in the industry, biodiesel is an ester. It is made by a transesterification of fats and oils route with an alcohol, usually methanol. As such, its properties are not an exact match to petroleum diesel, and most diesel engines are not rated for use of 100% biodiesel. The full technical definition, according to the National Biodiesel Board (www.biodiesel.org) is:

> Biodiesel, n - a fuel comprised of mono-alkyl esters of long chain fatty acids derived from vegetable oils or animal fats, designated B100, and meeting the requirements of ASTM D 6751.

### Renewable Diesel Definition

Renewable diesel is an exact or nearly exact match to petroleum diesel on a molecular level. It can be used directly as motor fuel in any proportion. It is made from renewable resources. It potentially can be made by a variety of processes, such as fermentation of sugars, pyrolysis, hydrogenation of fats and oils, and several other routes (Ref. [1]). However, processes being commercialized at present mostly use hydrogenation of fats, so that is the process we will describe in this chapter. Fermentation routes are covered in Ref. [2].

*Agitator Design Technology for Biofuels and Renewable Chemicals*, First Edition. Gregory T. Benz.
© 2022 John Wiley & Sons, Inc. Published 2022 by John Wiley & Sons, Inc.

## Feedstock Sources

Pretty much any renewable source of lipids (fats) can be used as feedstock for both biodiesel and renewable diesel. That covers a lot of possibilities. Most can be classified as either plant or animal fats, and within those groups, there are both edible and inedible oils, which probably impacts the economics of using them as fuel.

Most of the animal fats come from meat processing facilities as waste trimmings. Some come from restaurants or food processors that use them for frying until their flavor begins to deteriorate.

Most of the plant-based fats come from oilseeds, and most of these are edible, though some fractions (such as inedible fractions of corn oil) may be separated from the edible fractions. But other plant-based sources of oil include palm trees, flax, olive trees, Jatropha, Castor, Tung, Shea, cottonseed, some species of fungi, algae, and many others. Not all of these are edible.

Some have proposed extracting oil from sewage treatment plants. The author is not aware if this is being done commercially.

Though the author does not wish to get into the food vs. fuel debate, that issue is an important consideration influencing the availability and economics of making these kinds of diesel fuels. Food-grade oil goes for a much higher price than diesel fuel, so it makes no sense to use it as diesel fuel unless the supply is greater than the demand. Likewise, there are many other higher-value uses possible for fats and oils. Examples include chemicals, waxes, cosmetics, leather treatment, furniture finishes, perfumes, soaps, etc. All these uses compete in the market for fats and oils. Even "waste" oils such as used frying oil and meat trimmings have commercial value; no one is giving them away.

Regardless of the source of oil, after it is made relatively pure, that is where we begin our description of the biodiesel and renewable diesel processes. We will treat each process separately.

## Simplified Generic Biodiesel Block Diagram

There are a number of processes in use for making biodiesel. Some are acid catalyzed. Some are base catalyzed. Some may use no catalyst. What they all have in common is that the raw lipids must be converted to an ester to be considered an ASTM-approved biodiesel fuel, and they all produce glycerin (AKA glycerol) as a byproduct. They all involve various purification and drying steps to produce the final products.

We will be using a base-catalyzed process as a proxy for all processes, in the belief that all will have similar agitation requirements. We will briefly go over the main elements of this process, which is described on a simple block diagram in

Figure 9.1. We will roughly follow the flow of oil to diesel, though there are side streams and recycle streams.

Our chart begins at the upper right hand corner, with the oil storage tank. This is the crude fat or oil as bought from the suppliers. It will generally be relatively clean but too viscous for use directly as motor fuel. It may need to be heated in some cases to keep it liquid, depending on the fat or oil source.

From there it flows to the transesterification reactor, where it is blended and reacted with the catalyst, which is separately prepared as indicated in the upper left hand corner. The catalyst is generally a mixture of sodium hydroxide (NaOH) and methanol. It could be other short-chain alcohols as well. The base could be other alkali, such as potassium hydroxide (KOH). In an acid-catalyzed process, a mineral acid is used.

The transesterification reactor is a two-phase system. Upon completion of the reaction, the effluent from the reactor goes to a phase separation tank, though this could all be done in the reactor if a batch process is used.

The heavy phase is glycerin, which is treated with acid to facilitate separation of free fatty acids (FFAs). When the FFAs are separated, the glycerin goes to a still which recovers the excess methanol and returns it to the catalyst prep area. The crude glycerin is further refined to make saleable purified glycerin. We will not discuss the glycerin refining process here.

The oil phase from the transesterification reactor is crude biodiesel. It is neutralized with acid (alkali in an acid-catalyzed process) and then washed with

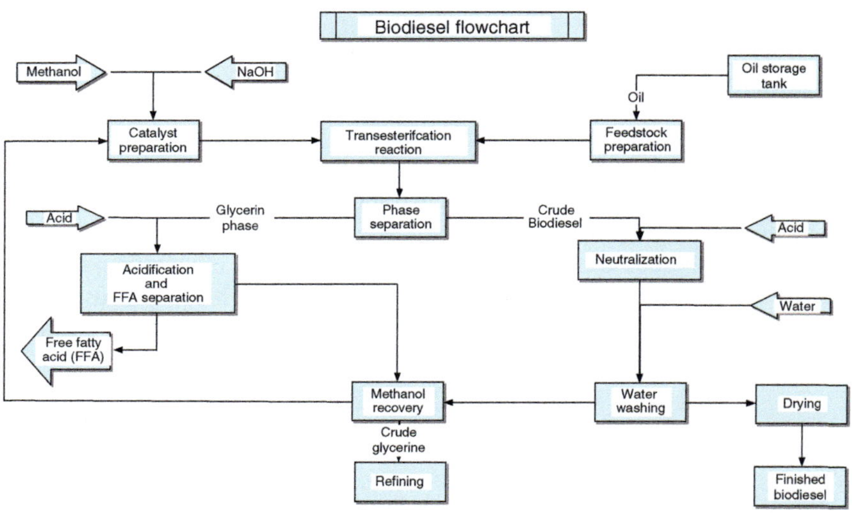


**Figure 9.1** Biodiesel flowchart.

water to remove salts and methanol. Some processes wash the crude biodiesel with absorbent solids to reduce the water that must be removed from both the methanol stream and the biodiesel stream. The washed biodiesel is dried, either by heating to evaporate water or by use of water absorbing material.

Finally, the finished biodiesel goes to storage and loadout of the product.

## Agitator Application Guide

### Crude Oil Storage

When crude oil is received from the suppliers, it may have trace solids (dirt) in it, and may have small amounts of water as well. In the case of some fats, the tank may need to be heated to keep them in a liquid state. Mild agitation may be required to promote temperature uniformity and avoid cold spots where the fats may solidify. It may also help avoid scorching if electric heating elements are used.

Generally, top entering agitators will suffice at a scale of agitation of 1 or possibly slightly less. Some facilities use no agitation in these tanks.

Though this author is generally not in favor of side entering agitators, the capital cost savings of using them over top entering agitators makes them attractive in some cases. Suggested sizing of side entering units may be calculated as follows:

$$\text{Motor size}(\text{kW}) = 7.5\text{E}-4 * \text{T}^3(\text{m}) \text{or} (\text{Hp}) = 3\text{E}-5 * \text{T}^3(\text{ft.}), \quad (9.1)$$

based on 350 rpm shaft speed

### Feedstock Preparation

This may involve blending of miscible liquids, such as other fats or oils from more than one storage tank, or other conditioning agents. A nominal scale of agitation of 3 should suffice if all additives are miscible and fall within the density difference ratio guidelines in Chapter 5.

If immiscible liquids are to be contacted, normally a scale of agitation of 5–7 is required. If it is a batch process and one of the impellers can be located just below the aqueous/oil interface, then a reduced scale of agitation of about 3–4 can be successful. This concept is illustrated in Figure 9.2.

### Catalyst Prep

This step involves blending alcohol (typically methanol) with an acid or a base. Depending on the concentration of the base, there may be a significant density difference to overcome. Scale of agitation should be selected based on the principles

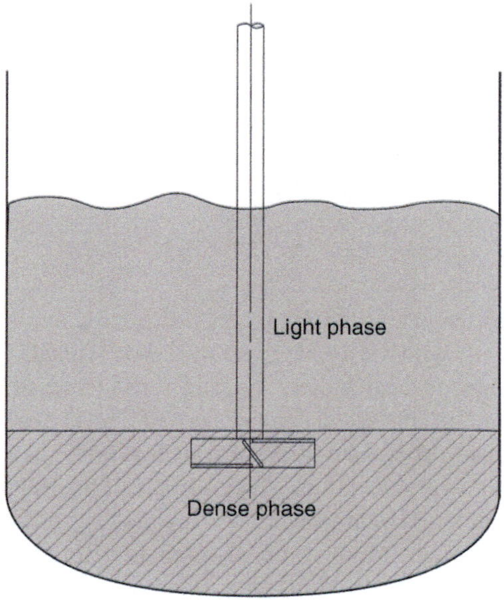


**Figure 9.2** Two-phase impeller placement.

of Chapter 5, but will generally be at least 3. If the process is continuous, a Static Mixer unit or similar product can be used. Only two elements should be needed, as the blending process will be in turbulent flow.

**Example 9.1** A 2 m diameter, 2 m SS tank will be used to prepare the catalyst. Final density is 1000 kg/m$^3$. Methanol, with a density of 786 kg/m$^3$ (per the Engineering Toolbox at 25 °C, www.engineeringtoolbox.com), will be blended with 50% sodium hydroxide, with a density of 1515 kg/m$^3$. Select an appropriate agitator to blend these two liquids together.

**Answer**

The density difference ratio, per Eq. (5.13), is $|1515-786|/1000 = 0.73$. Per Figure 5.10, we should use a scale of agitation of 7–8 for this application. We may be able to get by with less is we add the caustic in slowly or add it into the impeller rather than start from a stratified condition, but we will use a scale of 7 for this example first.

The working tank volume is about 6 m$^3$. We will repeat Eq. (5.12) here for clarity.

$$\text{Scale of agitation} = 600\left(P/\left(M*N\right)\right)^{0.5}, \text{where P is in kW, M is in kg and N is in rev/s} \quad (5.12)$$

If we choose a shaft speed of 125 rpm (2.08/second), we therefore get a power requirement of 1.7 kW. The nearest motor size is 2.2 kW. Based on a power number of 0.3 and a bulk density of 1000 kg/m$^3$, the impeller size would be about 0.91 m. This is about the maximum D/T ratio we would want to use, so the power cost is about minimum for this scale of agitation. In case of upset, such as inadvertently filling the tank with caustic, we may wish to upsize the motor to the next size, 3.7 kW.

## Transesterification Reactor

This step is really the heart of the process. The traditional way it is carried out is to use a simple turbine agitator, generally with a scale of agitation of 5–7. Typically, the reaction time is from two to six hours on a batch basis, and is speed dependent, going through an optimum. Therefore, variable speed is suggested (Ref. [3]). More recently, several facilities have begun using high shear dispersers, such as illustrated in Figure 9.3. These can cut the time down to less than 30 minutes, resulting in a cheaper tank and less in-process inventory. This indicates that the process is mass transfer limited, rather than reaction rate limited. However, it is expected that the power cost would be much higher, especially if the high shear unit is the only agitator in the tank, as it must still provide the equivalent to a scale of agitation of 5 minimum in order to contact the phases. If a turbine agitator is used to create overall tank motion and a high shear unit is placed between the turbine agitator shaft and the vessel wall, less overall power is needed. One caution: the particular chemistry should be tested with high shear units first in the lab, to make sure an emulsion will not form, which would make it difficult or impossible to separate the glycerin from the biodiesel.

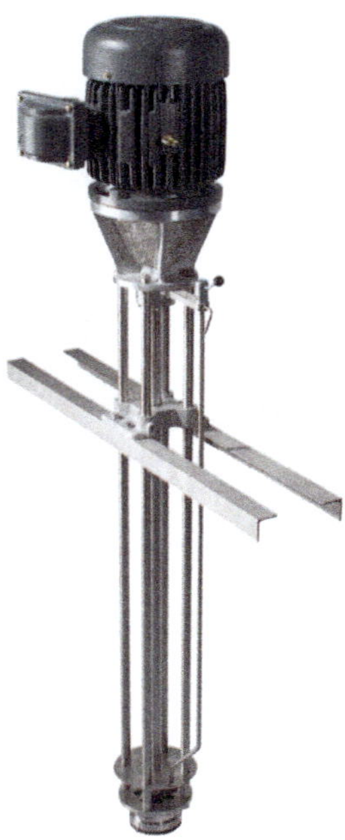

**Figure 9.3** In-tank style high speed disperser. *Source:* Greerco Homo mixer, permission of NOV.

This reaction can also be carried out on a continuous flow basis. If conventional turbine agitators are used, it can be carried out using tanks in series, an agitated column (Figure 9.4) or several other configurations described in Ref. [4]. Likewise, it would be possible to use a series of Static Mixer units at a higher than normal pressure drop to promote small droplets, or various high shear devices, such as the mechanical high-shear unit shown in Figure 9.5, ultrasonic dispersers, cavitating dispersers, etc. In Ref. [5], the author states that such high shear devices can produce

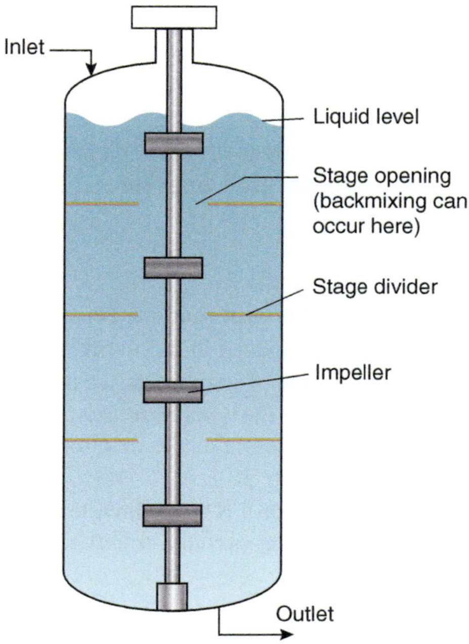


**Figure 9.4** Agitated column. *Source:* Courtesy Chemical Engineering Progress, blanket permission.

**Figure 9.5** Pipeline high shear mixer. *Source:* Greerco, permission of NOV.

biodiesel in two stages, with a typical power requirement of 60 Hp (45 kW) for a flow rate of 40 gpm (151.4 l/min). In our example problem, we will compare that power requirement to a batch reactor using a conventional turbine agitator.

**Example 9.2** As stated earlier, a total of 60 Hp applied to an inline high shear system can process 40 gpm of reactants into biodiesel. How does this compare with conventional agitation on a batch basis?

### Answer

Let us assume that we need to have two reaction tanks, such that one is reacting while the other is filling or draining. Thus, we need a capacity in each tank that is double the required batch time. Assuming a batch time of two hours, we need a capacity of 40 gpm * 120 minutes * 2, or 9600 gallons (36 336 l) per tank. Assuming a design density of 1000 kg/m$^3$ (actual density will be less, but the agitator may need to run in water occasionally for tank cleaning), our mass is 36 336 kg. If we apply Eq. (5.12), we get Scale of agitation = $600(P/(36\,336 * N))^{0.5}$, which leads to several combinations of power and shaft speed, as listed in Table 9.1. We show

**Table 9.1** Power/speed combinations for example problem.

| Shaft speed (rpm) | Shaft speed (1/s) | Scale of agitation | Shaft power required (kW) |
|---|---|---|---|
| 56 | 0.93 | 5 | 2.36 |
| 68 | 1.13 | 5 | 2.86 |
| 84 | 1.40 | 5 | 3.53 |
| 100 | 1.67 | 5 | 4.21 |
| 125 | 2.08 | 5 | 5.26 |
| 155 | 2.58 | 5 | 6.52 |
| 56 | 0.93 | 6 | 3.39 |
| 68 | 1.13 | 6 | 4.12 |
| 84 | 1.40 | 6 | 5.09 |
| 100 | 1.67 | 6 | 6.06 |
| 125 | 2.08 | 6 | 7.57 |
| 155 | 2.58 | 6 | 9.39 |
| 56 | 0.93 | 7 | 4.62 |
| 68 | 1.13 | 7 | 5.61 |
| 84 | 1.40 | 7 | 6.92 |
| 100 | 1.67 | 7 | 8.24 |
| 125 | 2.08 | 7 | 10.30 |
| 155 | 2.58 | 7 | 12.78 |

results for scales of agitation ranging from 5 to 7 and AGMA shaft speeds ranging from 56 to 155 rpm. Based on roughly square batch geometry, 56 rpm was chosen as the minimum to avoid needing too large of an impeller, and 155 rpm was chosen to avoid shaft design problems. The higher power/speed combinations will produce finer droplets, which may lead to a shorter batch time. However, it is clear that the total power consumed is much less than using a continuous high shear mixer, especially if the reaction tanks are drained into a settling tank or shut off while separating glycerin from biodiesel. So, it is a classic case of capex vs. opex.

The choice of batch vs. continuous flow requires knowledge of kinetics as well as economics. We will not be able to offer guidance on the kinetics. However, Ref. [4] describes how backmixing between stages of an agitated column may be taken into account for reactor design.

## Neutralization of Biodiesel

This is a two-phase process in which the biodiesel has excess alkali or acid neutralized. In the case of our model base-catalyzed process, dilute acid is added. Whether batch or continuous, an agitation scale of 5–6 is normally applied. Similarly to transesterification, high shear units and inline units have been applied, with reduced retention time. It is always advisable to test the system first to be sure emulsions will not form.

## Washing of Biodiesel

Mostly, this is done with water, to remove any water-soluble impurities from the biodiesel. The same agitation systems can be used as for neutralization of biodiesel, including the comments about high shear units. Some systems "wash" the biodiesel with absorbent solids. These are typically added as a powder. Suspension of these solids can be carried out with a scale of agitation of 3 unless they are granular, in which case solids suspension design procedures may be used. (This is not covered in this book). If the solids are hard to wet, lab testing and scale-up may be required. That would be on the basis of equal flux and a power/volume scale-up. We will cover scale-up in more detail in Chapter 11.

Some processes may use a fixed bed of solids. There is no agitation involved there, unless the solids are first made into a slurry and then pre-coated, as in a precoat filter. Though lab testing may be needed, most such slurries can be made up with a scale of agitation of about 6.

## Glycerol Neutralization

In an alkaline process, residual caustic from the catalyst will be present in the glycerin (aka glycerol) stream. Since this is mostly single phase, in a batch tank, a scale of agitation of 2–3 may be used. The resultant fatty acids may float to the top, where they can be decanted off. In a continuous flow system, a scale of 3–4 is suggested to keep the oil phase suspended until the stream goes to a settling tank or a centrifuge.

Similar agitation levels are suggested for an acid-catalyzed process.

### Crude Glycerol Storage

The purpose of an agitator in this tank is to keep trace solids in suspension as well as to keep stray fatty acids from accumulating at the surface. Normally a scale of 2 will suffice, and the agitator may be operated intermittently to save power costs.

### Finished Biodiesel Storage

This may not require any agitation unless there is a chance for some water to settle out. If a top entering agitator is used, the scale of agitation can be as low as about 0.5. Though the author is not fond of using side entering agitators, if used, they can be sized according to Eq. (9.2).

$$\text{Motor size}(\text{kW}) = 2.5\text{E} - 4\text{T}^3, \text{where T is in m, based on a shaft speed of 230 rpm.} \quad (9.2)$$

Though this book does not make specific recommendations regarding wetted part materials of construction, we comment here that most biodiesel applications will require stainless steel. Carbon steel, however, may be acceptable for finished biodiesel, as there should be no water present and the biodiesel itself is noncorrosive to steel. Care must be taken in the choice of any elastomers used in the shaft seals, as biodiesel is a strong solvent.

### Miscellaneous Applications

There may be several tanks used to make-up reagents used in the process, such as dilute acid or base, etc. These may be designed based on density difference ratio as described in Chapter 5. In general, most of these applications will work with a scale of agitation of 3. Small tanks using propeller agitators may get by with a scale of about 2, due to the relatively high speed and small D/T ratio, providing a jet-like stream which can often overcome density differences.

## Renewable Diesel Generic Process Description

Renewable diesel basically uses renewable feedstocks such as used fats and oils, new oils from plants, among others, to make a drop-in fuel that meets the same composition, pour point, and energy production as petroleum diesel. No modification of any kind is required to the diesel engine or its fuel handling system to use renewable diesel. That is because it is a true hydrocarbon, whereas biodiesel is an ester.

Though there are several possible ways to make renewable diesel, the principal way seems to be to hydrogenate the oil to decarboxylate the molecule, or remove oxygen (Ref. [6]). The principal byproducts are water and carbon monoxide. This is the process that will be described here.

## Flow Sheet

There are many possible ways to arrange a renewable diesel process. We will use a simplified description that will hopefully be sufficient as a guide for agitator applications in other processes. A simplified block diagram is provided in Figure 9.6.

We will follow the steps in order of material flow to the extent possible. Some steps, such as used oil storage and clean-up, may be performed at the supplier's facility prior to shipment to the diesel manufacturer.

The process starts out at suppliers' locations, and these may be collectors of restaurant oil and grease, rendering plants, oilseed processing plants, and other sources. These fats and oils will typically be held in a storage tank, which may need to be heated, depending on the melting point of the oils and the plant ambient conditions.

The used oil may not be very clean. It often has some water in it, as well as solids from animal parts, bits of vegetable matter such as particles of food from frying operations, etc. These must be removed before processing into renewable diesel fuel.

Removal is mostly accomplished by decanting, where the water and solids settle out. This can be done in a simple settling tank, or may be sped up by use of a centrifuge or centrifugal decanter. Filtration of the centrate will produce an even cleaner feed to the diesel manufacturing process.

Cleaned fats are stored upstream of the process in a storage tank. The fats in here should be free of water and particulate solids. Again, this tank may need to be heated.

The key to the process is the reactor. In this unit, the fats are reacted with hydrogen in the presence of a catalyst. Though process details vary, generally the temperature in the reactor is between 300 and 600 °C, and the pressure is rather high (50–100 bar-g), as per Ref. [6]. The reaction de-oxygenates or decarboxylates the feedstock and breaks the side chains off the triglyceride "skeleton" common to many lipids.

The discharge goes through a separator where hydrogen and other gasses may be removed. The gasses can go through a purifier to separate the hydrogen for re-use.

The remaining hydrocarbon mix goes into a fractionation column, where at least three main streams are produced: the diesel product, with average carbon chain lengths of around 13; naphtha or jet fuel, with chain lengths of 5–9 (or a little longer for some jet fuels), and a propane/light ends stream, with chain lengths from 1 to 5. The first two streams are liquid at room temperature. The last is a gas at room temperature and pressure, but is often stored under enough pressure to liquefy it. All of these fractions are saleable products.

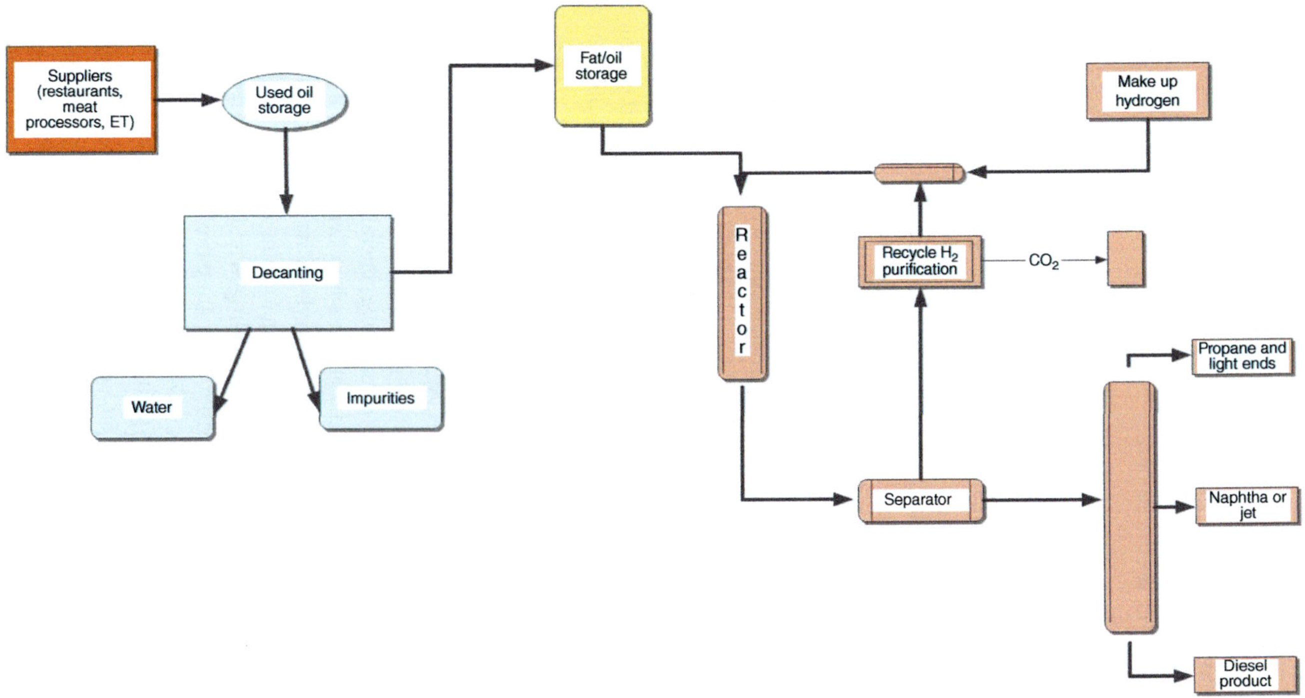


**Figure 9.6** Renewable diesel block flow diagram.

## Agitator Applications

### Used Oil Storage

Since the used oil may not be free of solids or water, in order to have a uniform feed to the decanter vessel or centrifuge, agitation is needed. It may also be needed to promote temperature uniformity if high-melting fats are used along with a heating jacket or submerged coil. Top-entering units should be designed with a scale of agitation of 2–3. Though not recommended, side entering agitators may be selected based on Eq. (9.3). Aspect ratio for side entering units should be limited to a maximum of 1.4 : 1.

$$\text{Motor size}(\text{kW}) = 1.5\text{E}-3 * \text{T}^3(\text{m})\text{or}(\text{Hp}) = 6\text{E}-5 * \text{T}^3(\text{ft.}), \tag{9.3}$$

based on 350 rpm shaft speed

### Decanted Water Tank

If this stream goes straight to a waste treatment facility, no tank may be needed. If a tank is used, a scale of agitation of 2 should be sufficient.

### Decanted Impurity or Sludge Tank

Sometimes this stream is combined with the water stream. In such a case, a scale of agitation of 3 should be used. If it goes to a separate tank, the agitation design depends on the % solids, which affects the rheology. If decanted by means of a centrifuge, for example, the % solids can be quite high, with fluid properties resembling a paste. In such a case, if a tank is used, lab testing plus scale-up is required, see Chapter 11. Such a paste may not go to a tank at all. It may go to a filter press and then to a dryer, to be sold as animal feed. It may also go to an extruder for similar purposes.

Generally, if the solids slurry is less than 30% solids, a scale of agitation of 3 will suffice.

### Clean Fat/Oil Storage

Presumably at this point the oil has no solids or water to speak of. Many will choose not to agitate this tank. If agitation is used, a scale of 1 should suffice. If the tank is heated to keep the fats liquid, agitation is definitely needed. Side entering units may be selected using Eq. (9.2).

### Reactor

As in biodiesel, the reactor for renewable diesel is the heart of the system. There are actually a number of ways to configure the reactor besides just batch versus continuous. Some systems are not mechanically agitated. For example, one could configure a packed column with catalyst on the surface of the packing, with a gas-continuous phase and liquid running down over the packing. This could be done in several stages to improve the residence time distribution. However, this is a book about agitator design, so we will focus on reactor designs involving mechanical agitation.

Mechanical agitation can be set up as batch or continuous. There are several design variations in each. Basically, the continuous flow mode will usually consist of tanks in series with the same range of agitation design as in a batch tank, or it can be an agitated column as illustrated previously in Figure 7.4 (not repeated in this chapter). So, we will focus on batch reactor designs. There are several variations possible which we will discuss here.

The most common batch reactor design is one which sparges hydrogen in at the bottom through a sparge ring, and re-incorporates any hydrogen which escapes at the surface by means of a common pitched blade turbine placed about one-half to one impeller diameter below the surface, so as to create a vortex which sucks the gas in from the head space. The tank baffles are generally cut off roughly at the level of the top of the pitched blade turbine to further enhance the vortexing action. This arrangement is illustrated in Figures 9.7 and 9.8. The difference

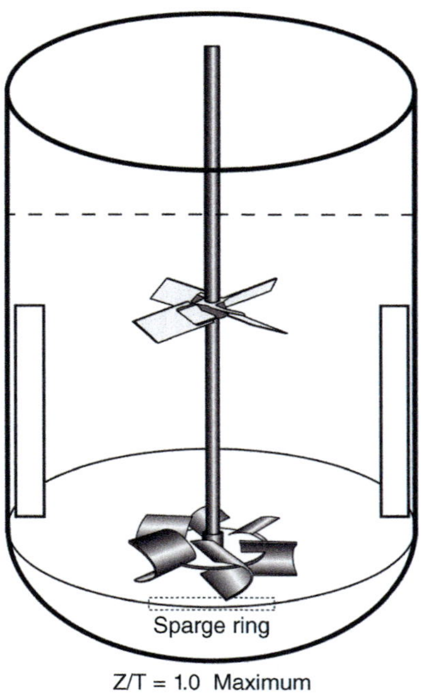


**Figure 9.7** Batch reactor Z/T = 1.

Figure 9.8 Batch reactor Z/T = 1.4.

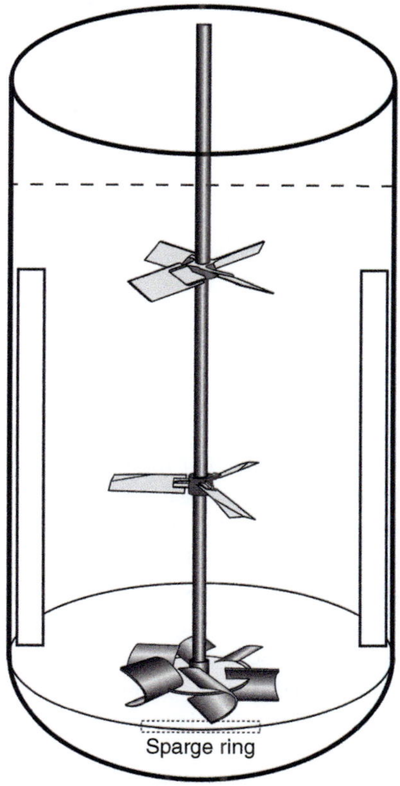


between the two is the aspect ratio, or Z/T value, which determines the number and types of impellers. Though four tank baffles are used, these figures only show two of them for clarity.

Note that this is not a flow-through system. The intent is to absorb all of the hydrogen without venting any. If the hydrogen is not 100% pure, a small amount of venting may be necessary if there are any gasses which will not react and will not condense. The reaction time can be several hours and may be carried out at 300–600 °C, depending on feedstock and catalyst (Ref. [6]). Pressure can be 50–100 bar-g.

The simple arrangement shown in Figures 9.7 and 9.8 is generally limited to a Z/T ratio of 1.4, as it is hard to get gas from the surface to travel back down to the bottom of the tank with a tall aspect ratio. Guidelines for impeller type and power split may be found in Table 9.2.

**Table 9.2** Impeller guidelines for batch reactor.

| | | |
|---|---|---|
| Z/T range | 0.6–1.0 | 0.8–1.4 |
| Number of impellers | 2 | 3 |
| Lower impeller type | Deep concave | Deep concave |
| Lower impeller power % | 75 | 60 |
| Middle impeller type | N/A | Hydrofoil |
| Middle impeller power % | N/A | 15 |
| Upper impeller type | Pitched blade | Pitched blade |
| Upper impeller power % | 25 | 25 |

The lower impeller should be a deeply concave disc turbine, as this will maximize power input when an initially high gas flow may be present. Typically, the gas flow starts out somewhat high and the pressure may be low. As the reaction goes to completion, the pressure rises and the net gas flow falls off. Eventually, the only gas being introduced to the liquid is what the upper impeller pulls in through its vortex. To the author's knowledge, the best vortexing impeller is the common pitched blade turbine. In vessels taller than a Z/T of 1.0, it is suggested to add a third impeller to transport some of the gas to the bottom and make sure that the liquid is circulated from top to bottom so it has a chance to react with the gas coming in through the vortex. That impeller should be a hydrofoil for best pumping action.

Normal practice is to choose a *motor* size based on a P/V of 1 W/l or 1 kW/m$^3$. The impellers are sized for fully baffled, ungassed power based on 85–90% of motor nameplate, even though under normal operating conditions they will draw less power than that. This is because if the tank is inadvertently filled higher than design level, and no gas is being fed in or drawn in by a vortex, the impellers could actually be operating in the ungassed condition. D/T ratio is not very critical, but should generally be kept in the range of 0.3–0.4. The upper pitched blade turbine should have a Froude number ($N^2D/g$) of at least 0.2.

**Example 9.3** Suppose we have a reactor with dimensions of 3 m diameter by 5 m SS, with a working volume of 25 m$^3$, and a specific gravity of 0.8. The tank has 2 : 1 elliptical heads. The depth of the bottom head = T/4, or 0.75 m, and the volume equals 0.131T$^3$, or 3.54 m$^3$. Therefore, the volume in the straight side is 25–3.54, or 21.46 m$^3$. This requires a straight side usage of 3.04 m, for a total Z value of 3.04 + 0.75, or 3.79 m. The Z/T value is thus 1.26. Per Table 9.2, we need 3 impellers.

Our motor size is nominally 25 kW, based on 1 kW/m$^3$. The nearest standard motor size that is not smaller than that is 30 kW, so that is what we will use. Based on Table 9.2 and 85% loading, we will design for about 15.3 kW in the lower impeller, which will be a deeply concave type, 4.5 kW in the middle impeller, which will be a hydrofoil, and 7.5 kW in the upper impeller, which will be a standard pitched blade turbine. Now we need to choose shaft speed and size the impellers to get an idea of what will give us a reasonable D/T ratio. We will use a power number of 1.8 for the lower radial turbine, 0.3 for the middle impeller, and 1.37 for the upper pitched blade turbine. Different shaft speeds will give different diameters. We will work through one AGMA speed and then provide a table for several more.

Let us choose 125 rpm, or 2.0833/second. We solve for D using Eq. (9.4).

$$P = N_P \rho N^3 D^5 \text{ or } D = \left(P / \left(N_P \rho N^3\right)\right)^{0.2} \tag{9.4}$$

For the lower radial impeller, we get D = (15 300 W/(1.8 * 800 kg/m$^3$ * (2.0833/second)$^3$))$^{0.2}$ = 1.03 m.

Similarly, the middle hydrofoil diameter would be 1.16 m and the upper pitched blade turbine diameter would be 0.95 m. Table 9.3 summarizes the results.

We would eliminate the 68 rpm option because the D/T ratio of the middle impeller is too large. The rest of the selections are OK, though shaft critical speed may dictate the design. With suspended metallic catalysts in use, a steady bearing is not an option. The final choice comes down to mechanical feasibility and cost. Note that the actual impeller sizes vendors may quote may be different, for two main reasons: (i) They may be using impellers with somewhat different power numbers, and (ii) Their gear drives may have actual shaft speeds that are different from nominal AGMA speeds. The important thing is that they input the desired power.

## Alternate Concept 1: Recirculating Gas Sparge

If the Z/T ratio is greater than about 1.4, or if the liquid level is variable such that an upper pitched blade turbine cannot be kept within 0.5–1.0 diameters below the

**Table 9.3** Hydrogenator sample problem summary.

| AGMA speed (rpm) | Lower impeller diameter (m) | Middle impeller diameter (m) | Upper impeller diameter (m) |
|---|---|---|---|
| 68 | 1.49 | 1.67 | 1.36 |
| 84 | 1.31 | 1.47 | 1.20 |
| 100 | 1.18 | 1.32 | 1.08 |
| 125 | 1.03 | 1.16 | 0.95 |
| 155 | 0.91 | 1.02 | 0.83 |
| 190 | 0.80 | 0.90 | 0.74 |

surface (we don't have a design for raising and lowering the upper impeller automatically or even manually!), a viable option is to use a compressor to recirculate the gas from the vapor space down to the sparger, as shown in Figure 9.9. Since the compressor is responsible for re-introducing gas that has escaped to the head space, an upper pitched blade turbine is not used. Fresh/make-up hydrogen can be introduced into the recirculation piping. Motor power is still generally set at 1 kW/$m^3$. When multiple impellers are used, the lower impeller should generally be a deep-concave turbine in a batch system, as the effective gas flow will reduce as hydrogen is consumed. The impeller system should be designed for the ungassed condition. In a continuous system, the gas flow can be fairly constant, so the impellers can be designed for the gassed condition. Power split is not very critical, as this is a chemical reaction, not a biological one, so a high rate of reaction in one area and a low rate in another causes no problems as long as the total reaction rate meets target. Putting more power in the lower impeller theoretically has a slight advantage, as the liquid head increases the absolute pressure at the bottom of the tank. However, with a back pressure of 50–100 bar-g, liquid head is insignificant. As a compromise, we suggest the impeller guidelines in Table 9.4. The lower radial impeller should be placed T/4 to T/6 off bottom. The upper ones

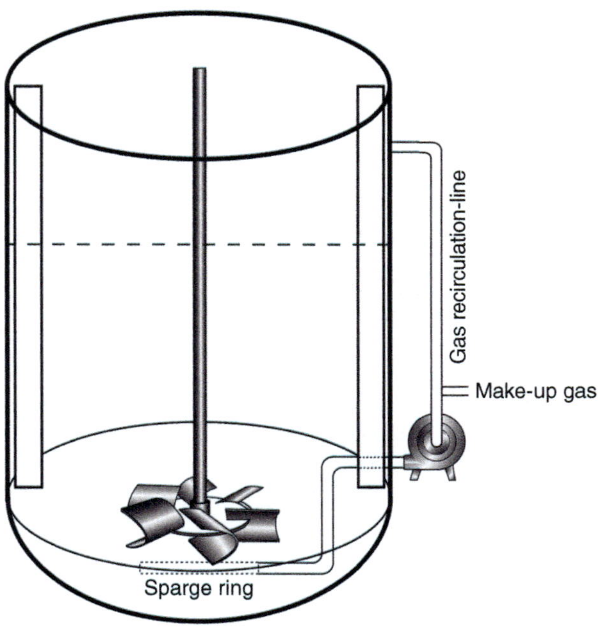


**Figure 9.9** Recirculating gas sparge.

**Table 9.4** Impeller guidelines for recirculating gas sparge.

| Z/T | Lower impeller type | Upper impeller type | Relative power split |
|---|---|---|---|
| 0.8 | Deep radial | None | 100 |
| 1.8 | Deep radial | Hydrofoil | 80/20 |
| Each 1.0 increment | Deep radial | Hydrofoil | 80/20/20. . . |

can be spread out so that the uppermost one is at 2/3Z to 0.75Z off bottom, with any intermediate impellers evenly spaced.

## Alternate Concept 2: Self-aspirating Stirrer

An alternate to re-incorporating gas from the vapor space by means of a vortexing impeller is to use an impeller deeper in the tank, specifically designed to create a suction effect, drawing the gas from the vapor space thought a hollow shaft which has openings above the liquid surface and behind the inner sections of the blades, so as to introduce the gas through the impeller at the point where its suction effect is the greatest. This concept, with an example impeller shown in the patent illustration Figure 9.10, does a very good job of sucking gas down in small vessels. However, in tall vessels, it must be placed fairly close to the surface, as there are limits to the absolute suction head it can produce. Theoretically, we can calculate the maximum velocity head using Eq. (9.5), using impeller tip speed as the velocity:

$$\text{Velocity head} = v^2 / 2g \tag{9.5}$$

Actual head will be less. Manufacturers do not generally publish detailed performance characteristics of proprietary impellers, but Ref. [8] studied one impeller system and found that the actual head for the system studied was about 15–20% of the theoretical head. If we generously assume that the impeller in Figure 9.5 is better that the impeller studied in Ref. [8], we can do a "what if" calculation based on, say, 25% of theoretical. If we, for example, look at a 1 m diameter impeller rotating at 125 rpm, our expected stalling head (where gas just goes down to impeller level but is not drawn through it) would be: $0.25 * (\pi * 1\,\text{m} * 125\,\text{rpm}/60)^2/2 * 9.8\,\text{m/s}^2 = 0.546\,\text{m}$. At this level, gas would probably not commence flowing. We would need to put it closer to the surface to get gas flow, or rotate it at a higher shaft speed. This depth is similar to what we would choose if using a 1 m diameter pitched blade turbine, so there is no obvious advantage. The author has been told in a private communication that in fact, some installations have placed the impeller deeper below the liquid by using a high shaft speed and small impeller, with a high probability that the unit was

US006394430B1

(12) **United States Patent**
**Forschner et al.**

(10) **Patent No.: US 6,394,430 B1**
(45) **Date of Patent: May 28, 2002**

(54) **AUTO-ASPIRATING ROTATIONAL DISPERSION DEVICE**

(75) Inventors: **Peter Forschner**, Hasel; **Rainer Krebs**, Schopfheim; **Hans-Jürgen Waiβ**, Kandern, all of (DE)

(73) Assignee: **Ekato Rühr-und Mischtechnik GmbH**, Schopfheim (DE)

( * ) Notice: Subject to any disclaimer, the term of this patent is extended or adjusted under 35 U.S.C. 154(b) by 0 days.

(21) Appl. No.: **09/418,000**

(22) Filed: **Oct. 13, 1999**

(30) **Foreign Application Priority Data**

Oct. 13, 1998 (DE) .................................. 298 18 255 U

(51) **Int. Cl.**7 .................................................. **B01F 3/04**
(52) **U.S. Cl.** .......................................... **261/87**; 261/93
(58) **Field of Search** ........................... 261/84; 87, 91, 261/93, 28, 29

(56) **References Cited**

U.S. PATENT DOCUMENTS

1,526,596 A * 2/1925 Greenwalt ................. 261/93
1,579,355 A * 4/1926 Greenwalt ................. 261/93
3,485,484 A * 12/1969 Quinchon ................... 261/87
3,776,531 A * 12/1973 Ebner et al. ............... 261/87
3,813,086 A * 5/1974 Ebner et al. .................. 261/93
4,283,357 A * 8/1981 Sidery ........................ 261/87
4,611,790 A * 9/1986 Otsuka et al. .............. 261/87
4,925,598 A * 5/1990 Kivisto et al. ............... 261/87
5,013,490 A * 5/1991 Tanimoto et al. ........... 261/87
5,389,310 A * 2/1995 Leiponen ................... 261/87
5,660,766 A * 8/1997 Van Dyk ..................... 261/87
5,795,504 A * 8/1998 Berchotteau ............... 261/87

FOREIGN PATENT DOCUMENTS

GB 745457 * 2/1956 ................ 261/87
GB 1028483 * 5/1966 ................ 261/93
JP 52-32748 * 8/1977 ................ 261/93
SU 621383 * 8/1978 ................ 261/87

* cited by examiner

*Primary Examiner*—C. Scott Bushey
(74) *Attorney, Agent, or Firm*—Kubovcik & Kubovcik

(57) **ABSTRACT**

An auto-aspirating, rotational dispersion device for gases and liquids with a rotating hollow shaft employed for gas suction and gas channels communicating with said hollow shaft to openings in the gas channels that are positioned at intervals over the circumference of the device, at which openings the gas and liquid are mixed. The device is distinguished by the fact that the gas being dispersed flows in a manner separate from the liquid, from the hollow shaft through the gas channels and is mixed with the liquid outside of the device.

**11 Claims, 2 Drawing Sheets**

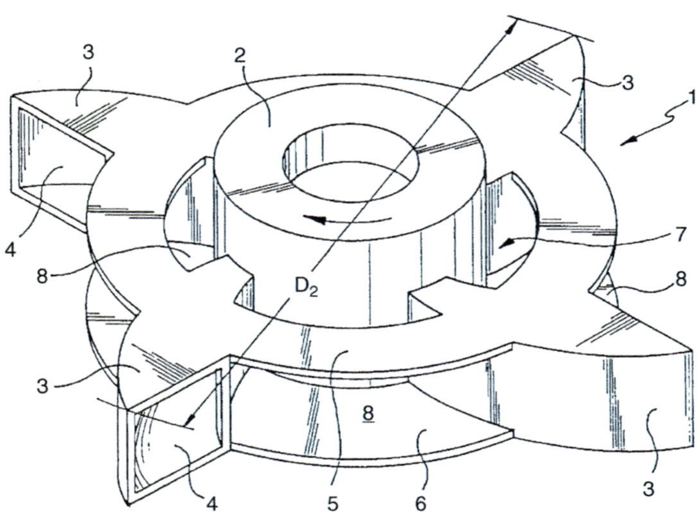


**Figure 9.10** Self-aspirating stirrer impeller. *Source:* Forschner et al. (2002)/U.S. Patent/Public domain.

operating above the first natural frequency. This is not a standard industrial practice on agitators above about 2.5 kW motor size, and is not recommended by this author. Steady bearings are probably not an option due to suspension of metal catalysts, which would be very abrasive to steady bearing materials.

The author has also been told by another consultant that his former employer did $k_l a$ comparisons of their own self-aspirating impeller to the upper pitched blade turbine option, and the $k_l a$ was lower with the self-aspirated system at equal power. It is the author's opinion that such would be the case with any such system, but actual tests would be needed to prove it.

When such impellers are applied in such a way as to pull in a lot of gas, they disperse it in the plane of the impeller, and do not drive it to the bottom of the tank, so the bottom of the tank may not have much reaction going on if no gas is being sparged into a lower impeller. This phenomenon is clearly shown in a video by a manufacturer of such equipment: https://www.ekato.com/products/impellers/gasjet/ (accessed 6 December 2021). There is no gas at the bottom of the tank unless it is being sparged in, which will not be the case in the later stage of reaction. It should be noted that the video was of a benchtop reactor, where the self-aspirating impeller was probably only about 150 mm below the liquid surface. The lower impeller in the video is a deep concave turbine, as would be recommended by this author. But the combination of two radial impellers creates a staged flow pattern, as illustrated in Figure 9.11. This limits fluid mixing from top to bottom. Likely, the results could be improved by adding an axial impeller, either up or down pumping, between the two radial impellers. This might help with another potential problem: suspending of metal catalysts. With limited axial flow, the catalyst might not get up into the gas-rich region of the upper impeller. Overall, this author does not see the advantage of a self-aspirating stirrer system over the first one mentioned.

## Additional Considerations

When hydrogen reacts with other species, it often does so by splitting off a hydrogen atom, leaving a single atom of hydrogen, also known as a hydrogen free radical, as a temporary byproduct. If the reaction takes place on the agitator metal surfaces or within some pores, there can be cases where the pore size is too small to pass a hydrogen molecule but may pass a hydrogen atom. If this happens, stray hydrogen radicals will readily recombine to form a hydrogen molecule, which cannot escape. Such molecules can build up very high pressure within the pore, causing cracking and hydrogen embrittlement. This can lead to equipment failure. The author is not a metallurgist, so the reader is urged to choose materials carefully for the reactor. Reference [9] provides some useful guidance. It also provides 169 references for further research.

As noted previously, these reactors often operate at considerable pressure. This creates a thrust force on the shaft and possibly a seal sleeve, and these forces must be supported by bearings somewhere. For example, suppose we had a 90 mm

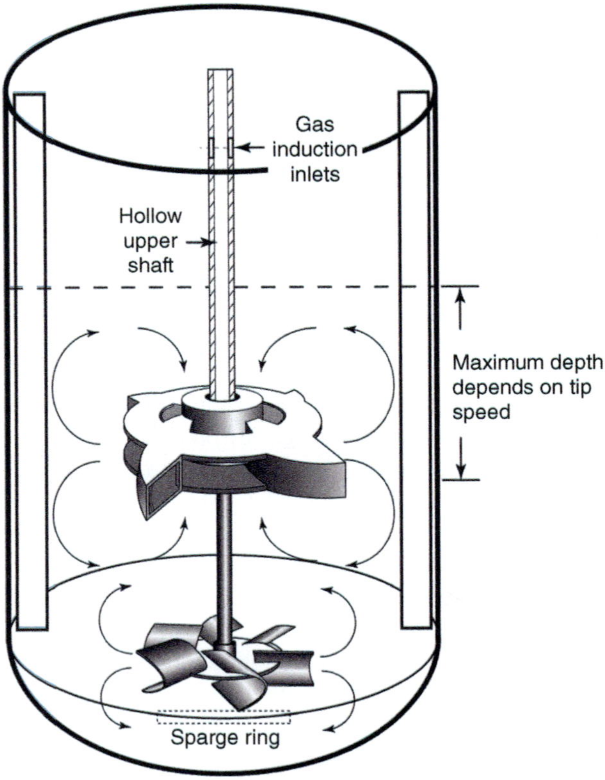


**Figure 9.11** Self-aspirating system.

diameter shaft and a tank pressure of 75 bar-g, or 7500 kPa. That would create a thrust force of 47 715 N, or the equivalent of a 4865 kg object being supported by the shaft if the machine were turned upside down. That is roughly the equivalent of two good-sized SUVs! That force may be taken by the gear drive bearings if they can handle it. If not, the load will have to be isolated from the gear drive. Chapter 16 discusses this in more detail.

## Diesel Product Tank

Inasmuch as the product is supposed to be pure diesel fuel with a low gel temperature and no water or solids, agitation may not be needed, except possibly to blend different batches to uniformity. This should require no more than a scale of agitation of 0.5–1. If a side entering agitator is used, Eq. (9.2) may be used to select the motor size.

Most renewable diesel applications will require stainless steel wetted parts. Carbon steel, however, may be acceptable for finished diesel storage, as there should be no water present and the diesel itself is noncorrosive to steel. Care must be taken in the choice of any elastomers used in the shaft seals, as they must be compatible with hydrocarbons.

### New Diesel Applications

As processes are constantly being developed or modified, the guidelines in this chapter may not cover all applications. Some new applications not covered may include fermentation of sugars to make renewable diesel. If the process is anaerobic, similar guidelines may be used as for ethanol fermentation. Aerobic fermentation is not in the scope of this book, but agitation procedures for aerobic fermentation are covered in Ref. [2]. Pyrolysis routes are covered in Chapter 10. Other processes may include catalytic upgrading of sugar, biomass to liquids, biogas to liquids and others. Although specific design recommendations are not available in this book, the reader is encouraged to seek similarities between applications in those processes to those in this book, and design accordingly. Use of scale of agitation will be an invaluable tool for any new process that is liquid motion controlled. Other, more complex applications may require lab testing and scale-up techniques, which are described in Chapter 11.

## Summary of Chapter

We have covered all of the major agitator applications found in generic biodiesel and renewable diesel processes. It is hoped that the breadth of applications covered is sufficient to allow the reader to apply these principles to new applications and processes, as the basic principles should be fully transferable.

## List of Symbols

| | |
|---|---|
| a | Interfacial area/volume |
| D | Impeller diameter |
| $k_l$ | Liquid film coefficient |
| $k_la$ | Overall mass transfer coefficient (1/time) |
| M | Mass of fluid |
| N | Shaft rotational speed |

$N_P$ Impeller power number
T Tank diameter
v velocity
V Volume of liquid
Z Liquid height

## Greek Symbols

ρ Liquid density

## References


1 McCormick, R. and Alleman, T. (2016). Renewable diesel fuel. *NREL Power Point Presentation* (18 July 2016). https://cleancities.energy.gov/files/u/news_events/document/document_url/182/McCormick___Alleman_RD_Overview_2016_07_18.pdf (accessed 1 January 2021).
2 Benz, G.T. *Agitator Design for Gas-Liquid Fermenters and Bioreactors*. Wiley InterScience. ISBN: 978-1-119-65049-2.
3 Hasan, A., Mashkour, M., and Mohammed, A. (2021). Impact of Mixing Speed & Reaction time on the biodiesel production from sunflower oil. *Journal of the Association of Arab Universities for Basic and Applied Sciences*. https://www.researchgate.net/publication/311843611 (20 January 2021).
4 Benz, G. (2014). Designing multistage agitated reactors. *Chemical Engineering Progress* 110: 30–36.
5 Janssen, M. (2010). Mechanical high-shear dispersion. *Biodiesel Magazine* (19 May 2010).
6 Sari, E. (2013). Green diesel production via catalytic hydrogenation/decarboxylation of triglycerides and fatty acids of vegetable oil and brown grease. PhD thesis in Chemical Engineering, Wayne State University.
7 The Engineering Toolbox. Velocity pressure head. engineeringtoolbox.com (accessed 20 February 2021).
8 Stelmach, J. and Musoski, R. (2017). Hydrodynamics in the blade region of a self-aspirating disk impeller. *Journal of Applied Fluid Mechanics* 10 (4): 1177–1188.
9 Bhadeshia, H. (2016). Prevention of hydrogen embrittlement in steels. *ISIJ International* 56 (2016): 24–36.

# 10

# Pyrolysis Application Guide

## Introduction

Pyrolysis is a process of heating a material in the absence of oxygen to break it down into simpler compounds. It is also known as destructive distillation. Products it can produce include alkane gasses, hydrogen, syngas (includes some carbon monoxide), liquid hydrocarbons (often called bio-oil these days), and charcoal (often called biochar when biomass is used as a feedstock). All of the products so produced have economic value.

Pyrolysis processes have few agitators. Some have none. In this chapter, we will give an overview of pyrolysis processes and then describe agitator applications where applicable.

## History

In its most basic form, pyrolysis is a very old process. It may even pre-date history. Likely, the earliest use was in manufacturing charcoal, which can burn evenly for a much longer time than wood. Later, people learned that it could be used for the gasification of coal, leaving coke as a residue and producing tar, oils, and industrial gasses such as methane and syngas if water vapor was added. One such product, coal oil, is essentially kerosene, though that name is usually applied to petroleum distillates. I can still remember my grandmother referring to store-bought kerosene or lamp fuel as coal oil; that was the generic term in her youth.

Pyrolysis was also used for destructive distillation of wood to make methanol, though that has fallen out of practice. However, pyrolysis of wood and other kinds of renewable biomass has undergone a bit of a renaissance recently as a potential source of engine fuel (bio-oil; similar to diesel fuel), gasses, and charcoal.

*Agitator Design Technology for Biofuels and Renewable Chemicals*, First Edition. Gregory T. Benz.
© 2022 John Wiley & Sons, Inc. Published 2022 by John Wiley & Sons, Inc.

## Feedstock Sources

As mentioned in the preceding section, coal and wood have been used as feedstock for pyrolysis processes for a long time. But almost any organic material can be used, liquid or solid. Petroleum, waste fats and oils, discarded plastics, municipal solid waste, sewage sludge, lawn clippings, leaves, crop residue, paper waste, forestry residue, discarded portions of food (animal or plant), etc. can all be used. Some even say that a big advantage of pyrolysis is that it is feedstock-agnostic. While this is partially true, different feedstocks require different time/temperature histories for optimal results, so some feedstock selectivity is usually required.

Not all of these feedstocks are from renewable sources. Renewable sources are generally some sort of biomass or waste product made from biomass, including paper waste, municipal solid waste, crop residue, and similar substances.

## Classification of Processes

Pyrolysis can be conducted over a range of temperatures and times. Higher temperatures require less time. But the product distribution also changes. The three main products are bio-oil, syngas, and biochar.

In general, pyrolysis processes can be characterized as slow, fast, or flash (which is kind of super-fast). These are described in more detail in Refs. [1] and [2]. Table 10.1 is an amalgamation of those references and other sources. Though the actual times, temperatures, and yields vary with process details and feedstock, this table illustrates that it is possible to manipulate product distribution to a considerable degree by choice of process.

We will briefly describe two processes in the next section.

**Table 10.1** Comparative pyrolysis regimes.

| Regime | Temperature range (°C) | Residence time (s) | Bio-oil yield | Syngas yield | Biochar yield |
|---|---|---|---|---|---|
| Slow | 300–850 | 300–6000 | 30 | 35 | 35 |
| Fast | 400–1000 | 0.5–10 | 75 | 13 | 12 |
| Flash | 700–1500 | Usually <0.5 | 5 | 85 | 10 |

Yields are in % of total output and are "typical."

## Two Typical Processes: Overview

Figure 10.1 shows a typical pyrolysis process for biomass, though it could also be used for other feedstocks, such as waste foods, paper, oils and fats, and plastics.

After unloading, the biomass may be stored to assure a steady supply to the process. Storage may be in some kind of vessel or simply spread out in a large pile, fed into the process with loading or conveying equipment. Liquid feedstocks will need some sort of vessel.

Since biomass may have some finite moisture content, which could interfere with pyrolysis and dilute the product stream, it is normally dried before being fed into the process. The drying step may come before or after the grinding step, depending on feedstock properties. The heat for drying normally comes from combustion of some of the reaction products (biogas or biochar).

Dried, ground feedstock is fed into the pyrolysis reactor. There it is exposed to heat in the absence of air for a certain time/temperature history, according to the desired product mix. The heat comes from burning some of the biochar or biogas, and some of the unburnt gas may be used to fluidize the bed of biomass, if a fluidized bed reactor is used. A variation of this process allows a limited amount of air into the reactor, letting partial combustion of the biomass supply the process heat. This method produces more carbon dioxide, which becomes part of the gas output as syngas.

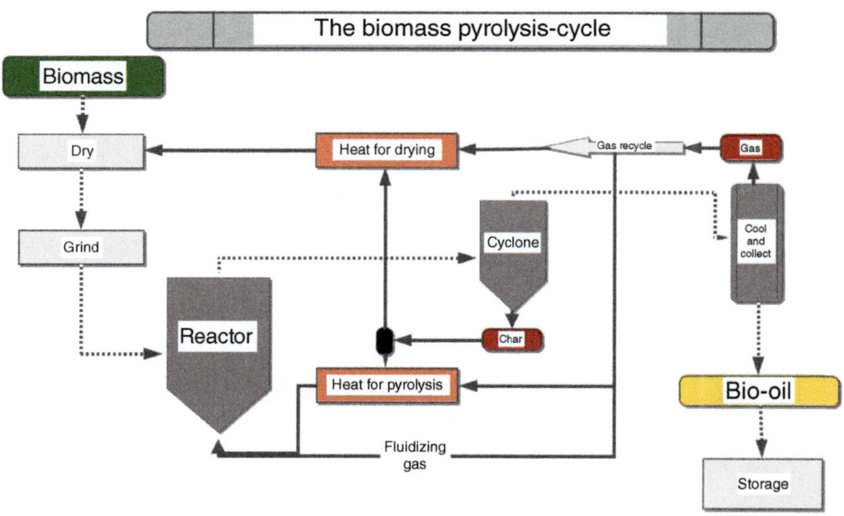


**Figure 10.1** Pyrolysis process 1.

A high-temperature cyclone separates the char from the vapors. The vapor stream is then cooled, condensing the bio-oil and separating it from the biogas. The biogas is partially used in the process; the rest can be used elsewhere in the plant or sold as a product. It can also be used as a feedstock for syngas-based fermentation.

The bio-oil is stored and may be further refined or fractionated for various fuel grades. The unfractionated oil is roughly equivalent to heating oil or diesel fuel in its flow and thermal properties.

A somewhat different process is illustrated in Figure 10.2. This schematic is based on the biomass to gas–biomass to liquid (BTG-BTL) process. In this process, the main products are bio-oil and steam. Very little char is produced, as most of it is consumed in the process. Pyrolysis is accomplished by contacting the biomass with very hot sand.

Biomass and hot sand are both introduced into the pyrolysis reactor concurrently. This results in charring the biomass and producing vapors, which are a mixture of bio-oil vapor and biogas. The sand and char mixture goes to a combustor that burns the char to heat the sand. In some processes, some of the char may be separated from the sand as an additional product. The hot sand goes back to the reactor. The combustion gasses are separated from ash in a cyclonic separator and the flue gas, which contains carbon monoxide among other things, is fed to an afterburner to provide heat for steam production. Water for cooling the char combustor is ultimately converted to steam.

The vapor stream from the reactor is cooled in the condenser, where it is separated into bio-oil and biogas. The gas is combined with the flue gas from the char

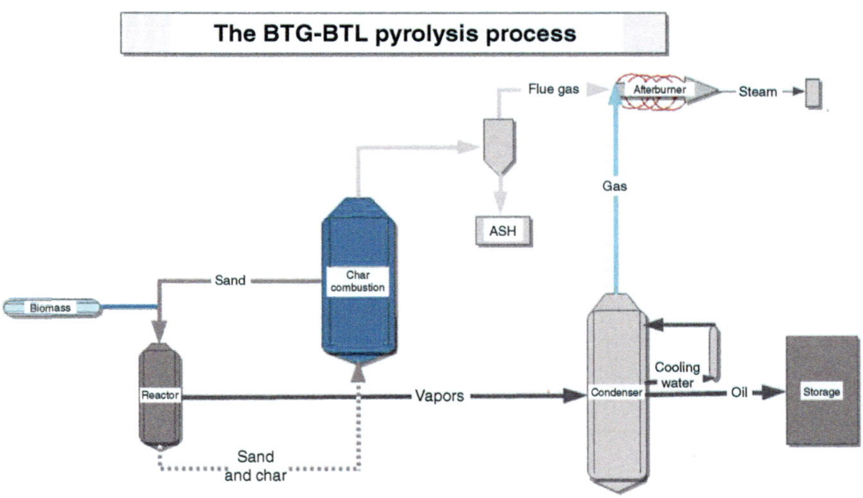


**Figure 10.2** Pyrolysis process 2.

combustor to act as fuel for the afterburner, which generates steam for heating or possibly power generation.

Bio-oil is sent to a storage tank, where it can be sold or further fractionated and processed into various molecular weight fractions.

## Agitator Application Guide

Unlike most processes described in this book, pyrolysis processes do not use agitators as a major part of the operation. Some may not use agitators at all. But some processes may use agitation or mixing at the beginning and/or the end of the process, so we will cover those here.

One of the advantages touted for pyrolysis as a means of production of renewable fuels and chemicals is that it is feedstock-agnostic. That is, one could potentially use any kind of organic matter as a feed to the process. However, this is only partially true. No process runs exactly the same on widely different feedstocks, and no process produces consistent results with inconsistent feed. So, for best results, a consistent feed must be supplied. Possibly, the source of biomass can be depended on to deliver a consistent feed. In the case of a single-feed plant, such as one fed chopped switchgrass, corn stover, or sawdust from a single species of tree, the feed may be consistent enough to need no up-front mixing. However, if the feed is inconsistent, for best results, it should be mixed to level out any such inconsistencies. How one goes about this depends on the general nature of the feed.

Broadly speaking, there are three cases to consider: 100% solids feed, 100% liquid feed, and processes adding liquids and solids together. We will cover these cases in turn.

### Solids-Only Feed

Solids mixing technology tends to be rather device-specific. Solids have been mixed in tumbling containers, horizontal ribbon blenders, horizontal paddle blenders, mixing conveyors, cone-shaped tanks with precessing augurs, and numerous other devices. The author's expertise is limited to liquid agitation, so we are unable to provide detailed guidance in this book for solids mixers. Mostly, sizing is done by a trusted vendor, though each vendor may come up with a very different approach. This is unlike the situation with fluid agitation equipment, where there is a considerable amount of similarity in the offerings of different vendors. Some guidance may be had by referring to chapter 15 of Ref. [3].

### Liquid-Only Feed

If various oils or fats are fed into the process, agitators can be designed using the principles of liquid motion described in Chapter 5, based on fluid property differences. In most cases, a scale of 3 will suffice, allowing for a small quantity of dirt in the mix.

### Solid Plus Liquid Feed

The strategy for this case depends on whether the mixture behaves as a damp solid or a liquid. In other words, can it be pumped or must it be conveyed? If it must be conveyed, the mixture needs to be considered a solids mixture, though it may be clumpy and sticky. Testing at a solids mixer manufacturer may be required.

If pumpable, the mixture can be considered a liquid. However, it may have a very high apparent viscosity and extremely non-Newtonian rheology if it is near the limit of what can be considered a liquid. It should be treated the same as a water-based biomass slurry in terms of characterizing rheology, lab testing, and scale-up. Use the procedures in Chapters 12 and 13.

### Bio-oil Storage Tanks

Essentially, this is the same application as finished biodiesel storage. Some may opt to use no agitation. If agitation is used, a scale of agitation of 1–2 should be sufficient. If a side entering agitator is used, sizing can be based on Eq. (9.2), which is repeated next for convenience. Tank geometry should not have a Z/T ratio of more than 1.4 if a side entering agitator is used.

$$\text{Motor size}(\text{kW}) = 2.5\text{E} - 4\text{T}^3, \text{where T is in m, based on a shaft speed of 230 rpm.} \tag{10.1}$$

## Agitation for Syngas or Biogas Fermenters

Biogas and syngas can be sold as products from pyrolysis processes. They can also be used as feed gasses for industrial fermentation. Microbes are available and in use for industrial processes that can consume methane, hydrogen, carbon monoxide, carbon dioxide, and other gasses that may be found in the off gasses produced by pyrolysis.

Agitation is needed to disperse these gasses into liquid (normally water) and enhance mass transfer. Liquid blending, heat transfer, and mechanical integrity are also big issues. The subject of gas–liquid agitation is far too complex to describe in a few paragraphs in this book. The author has written an entire book devoted to this topic, as listed in Ref. [4]. In general, agitators designed for gas–liquid contacting will require far more power than needed for the rest of the applications in this book, so power cost optimization can be very important.

## Summary of Chapter

We have described a couple of basic pyrolysis processes that can be applied to the manufacture of renewable fuels and chemicals. Though these processes do not always employ fluid agitation, we have endeavored to show where agitation may be applied and to provide sizing guidelines within the scope of topics in this book.

## List of Symbols

T Tank diameter
Z Liquid level

## References

1 Tomczyk, A., Sokolowska, Z., and Boguta, P. (2020). Biochar phicochemical properties: pyrolysis temperature and feedstock kind effects, published online February 2020. *Reviews in Environmental Science and Biotechnology* 19: 191–215. https://doi.org/10.1007/s11157-020-09523-3.
2 Goldfarb, J. (2018). Biochar and torrefied biomass short course Penn State bioenergy short course series. http://bioenergy.psu.edu/shortcourses/2018BiocharTorrefied/02_PrinciplesGoldfarb.pdf (accessed 29 October 2018).
3 Paul, E., Atiemo-Obeng, V., and Kresta, S. (2004). *Handbook of Industrial Mixing*. Wiley InterScience.
4 Benz, G. (2021). *Agitator Design for Gas-Liquid Fermenters and Bioreactors*. Wiley InterScience.

# 11

# Agitation Scale-up

## Introduction

Scale-up is an important topic in any industrial process. In this chapter, we will endeavor to explain how to scale-up agitation effects, especially in processes common in the biofuels/renewable chemicals industries. Because of the variety of tank geometries, flow regimes, and process results encountered, we will illustrate these concepts with a bevy of examples after the main body of the chapter, as well as some "demonstrator" examples within some sections.

## When Is Scale-up Necessary?

Basically, it is needed when we cannot calculate process results using normal procedures with enough certainty to assure successful design. Some examples of when we may wish to do this are listed as follows:

- Unusual tank geometry.
- Highly non-Newtonian rheology (yield stress scale-up will be covered in more detail in Chapter 13).
- Unknown relationship of process to agitation.
- Solids incorporation rate.
- Shear effects.
- Mass transfer.
- Droplet size.
- Particle attrition rate.

*Agitator Design Technology for Biofuels and Renewable Chemicals*, First Edition. Gregory T. Benz.
© 2022 John Wiley & Sons, Inc. Published 2022 by John Wiley & Sons, Inc.

## What Is Scale-up?

Putting it succinctly, scale-up is a procedure intended to produce equal results in one scale to those observed in another scale. Generally, the goal is to duplicate results in a larger scale, though sometimes it is desirable to model equipment in a smaller scale. In that case, the process is called scale-down. The same equations and principles work in either direction.

What makes scale-up quite challenging is that different kinds of results scale differently. As we will see, this makes it difficult or impossible to duplicate more than one result in another scale. One generally has to choose the result that is most important and compromise elsewhere.

## Kinds of Results

Generally we can divide results into process results and physical results. Although most of the time what we really want to scale-up are process results, agitation only produces physical results directly. We hope to produce equal process results by scaling certain physical results that seem to correlate with the process results. In fact, a great deal of work has been done to correlate physical results to process results, but the correlations are often process-specific. We will delve into some examples of each kind of result.

### Process Results

Though it is not possible to provide an exhaustive list of possible process results in this book (as there are simply too many processes), some of the more common ones are listed as follows:

- Rate of production.
- Yield.
- Product purity.
- Mass transfer rate.
- Product distribution.
- Cell growth rate (applies to fermentation).
- Cell viability (applies to fermentation).
- Crystal nucleation rate.
- Crystal growth rate.

### Physical Results

Some of these are direct measures of agitation. Others are possible to correlate with agitation parameters, an example being heat transfer coefficient. A list of some typical physical results follows. It, too, is not exhaustive, but the most common ones hopefully are covered.

- Characteristic fluid velocity.
- Maximum shear rate.
- Average shear rate.
- Mass transfer coefficient.
- Degree of solids suspension.
- Bubble size distribution.
- Droplet size distribution.
- Solids size distribution (applies when shear is enough to affect this).
- Particle attrition rate (in high shear applications).
- Eddy size distribution.
- Heat transfer coefficient.
- Heat transfer rate.
- Cooling/heating time.
- Vortex depth.
- Rate of solids incorporation (i.e. mass/area-time).

## Process Result Scale-up

Scaling-up of process results can be very complex. To do so successfully requires an intimate understanding of the relationships between all process variables that can affect those results. Therefore, scaling of process results resembles a design procedure more than a series of simple procedures or rules of thumb. Moreover, it may not be possible to actually duplicate all relevant process results in two different scales. We will provide some examples of what may or may not be possible in the next section.

### Process Result Scale-up Examples

1) Aerobic fermentation: We can scale-up some things but not others. For example, we can:
   - Scale-up on the basis of equal mass transfer rate/volume.
   - Maintain DO (dissolved oxygen) concentration at an acceptable minimum level.

   But we cannot:
   - Maintain equal DO distribution throughout the vessel, or even mimic what it was in a smaller scale.
   - Blend additive ingredients in as quickly in a large scale as can be done in a small scale.
   - Keep the shear rates or shear effects the same in two different scales.

2) Competitive-consecutive reactions: We basically cannot perfectly scale these up, as to do so would require equal blend time in both scales. This would require extremely high power inputs and aerate the batch. We will elaborate on this more under the section "Scaling a Single Physical Result," where such scale-up corresponds to a scaling exponent of zero. Various reactant feeding strategies can be used to aid in controlling product distribution, but it will never be exactly the same in two scales.
3) Anaerobic fermentation: Though more research needs to be done on the effects of agitation on this process, the work of Galindo referenced in Chapter 7 suggests we may be able to do the following:
   - Maintain equal rate of production
   - Maintain equal product yield
   - Maintain equal titer

   However, the way most agitation systems are designed, this will not be the case. So, in the real world, likely all three of the preceding process parameters will decrease upon scale-up. Buying larger, more expensive agitators than is the norm will probably help. Also, any additions will take longer to be blended in as vessels get larger.
4) Lignocellulosic hydrolysis reactors: Though there is a paucity of information in the literature about how rate and yield might scale-up, this author's experience has conclusively demonstrated that it is possible to at least scale-up in such a way that if there are no dead spots in the small scale, there will be none in the full scale. This probably means the rate of reaction will be equal. The author has done studies in an 18–20 l vessel and observed the same rate of viscosity reduction as in a 6000 l scale, using the scale-up procedures we describe in Chapter 13. So, we believe it is possible to achieve meaningfully the same process result. However, additive blending will take longer upon scale-up, so it is best to add all reagents to the feed stream. Since this is a relatively slow reaction, it is not usually blend time limited.

## Process Result Scale-up Variables

The variables that can be manipulated during a process result scale-up depend on the process, but there could be quite a few. Manipulating some to achieve a desired result may change other results, so priorities must be set.

The more common variables that can be manipulated can loosely be grouped into four categories:

1) Agitation parameters:
   - Impeller diameter, D
   - Number of impellers

- Type(s) of impellers
- Shaft speed, N
- Impeller placement in the tank

2) Tank geometry:
   - Tank shape (cylindrical, spherical, square, horizontal versus vertical, tapered, etc.). For example, an acute cone can be useful for feeding a reagent that must be kept dilute to control product distribution in a fed batch reactor. The resultant tank shape somewhat resembles an ice cream cone.
   - Tank diameter.
   - Tank height; aspect ratio (Z/T).
   - Number of tanks used to achieve capacity.
3) Process conditions:
   - Temperature (including that of the heating or cooling medium).
   - Pressure.
   - Steady state continuous versus batch or fed batch.
   - Planned temperature or pressure changes during the process.
4) Reagent control:
   - Feed rate of each stream
   - Composition of each stream

One can readily see that process scale-up is about more than agitation. There are no general rules for how to do it. Only full knowledge of the process relationships will make it possible. Therefore, we will focus on the physical results that can be achieved mainly by manipulating agitation variables.

## How Far Can We Scale a Process Result?

There is no simple answer to that question. It may be necessary to conduct studies in several scales to have confidence in the ability to scale-up a process result. In the absence of such studies, a good rule of thumb would be a 10 : 1 volumetric ratio for simple processes and 5 : 1 for complex processes. This is a more severe scale-up restriction than needed for purely physical results.

## Physical Result Scale-up

Scaling-up one physical result is generally possible. Scaling two is usually possible over a limited range of size. Scaling three or more results is almost impossible. One has to choose which results are most important. Otherwise, the way to scale from a 1 l reactor to 1000 l would be to install 1000 1-l reactors in parallel!

### Physical Result Scale-up Variables

The variables we can use to scale a physical result generally include the following:

- Shaft speed, N.
- Impeller diameter, D.
- Type(s) of impellers.
- Tank geometry (Z/T or other measure).
- Fluid viscosity (for simulations or hydrodynamic studies).

## Scaling a Single Physical Result

We start with the simplest form of scale-up: a single physical result. Though there are several ways to achieve that goal, we will start with the simplest and most reliable.

### Simplest Method: Geometric Similarity

Scale-up calculations are greatly simplified by using geometric similarity. By this we mean that the ratio of all corresponding dimensions is the same number, which we will call R, or the scale ratio. In other words:

$$R = T_2/T_1 = Z_2/Z_1 = D_2/D_1 = C_2/C_1, \text{etc.} \tag{11.1}$$

The concept of geometric similarity is illustrated in Figure 11.1.

This relationship fixes all dimensions, including the types of impeller, which must be the same in both scales. That leaves only one variable to manipulate: the shaft speed, unless viscosity can be manipulated. (This usually cannot be done with the actual process, but sometimes it is done to establish Reynolds number effects, mostly for academic study but sometimes for simulation of the real fluid.)

So, the scale-up procedure for a single physical result, using geometric similarity, consists of setting the diameter using $D_2 = RD_1$, and then calculating the shaft speed that will deliver that result. The shaft speed calculation method depends on the result being scaled.

## How Far Can We Scale a Physical Result?

As long as only one result is important, it should generally be possible to scale-up over at least a 10 : 1 scale ratio (R), which equates to a 1000 : 1 volumetric ratio. The author has in fact successfully scaled-up as far as a 20 : 1 geometric ratio, equating to an 8000 : 1 volume ratio. If more than one result must be scaled, it may

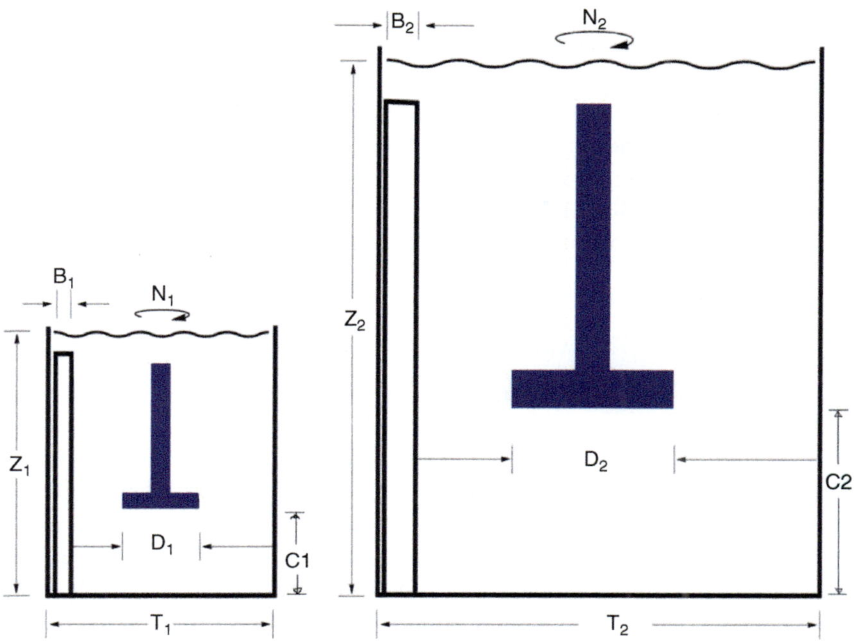


**Figure 11.1** Geometric similarity.

be impossible to go more than a 2 or 3 linear ratio, depending on the parameters. Since Reynolds number increases upon scale-up, most errors will be in the direction of a conservative design.

## What to Do When Geometric Similarity Is Approximate?

If we do not have perfect geometric similarity, but are close (say, Z/T ratio in both scales is within ±20%), we can define R as the cube root of the volume ratio:

$$R = \left(V_2 / V_1\right)^{1/3} \tag{11.2}$$

One can see that Eq. (11.2) gives the same results as Eq. (11.1) when geometric similarity is exact.

## What to Do When Geometry Is Not Close to Similar?

Sometimes all we have available in the lab or pilot plant is an aspect ratio (Z/T) close to 1, but the production vessel is tall and thin. For example, we might have a 1 : 1 lab or pilot scale vessel (also known as a "square batch") and a 3 : 1 production vessel. What we would do in this case is imagine dividing the tall vessel into zones and scaling-up each zone. For a simple process result, this would mean we basically calculate the shaft speed based on the geometry or volume of roughly a square batch (or 1/3 of the vessel for the example mentioned) using Eq. (11.2) for the *zone* volume (not the total volume), and then just add impellers for the rest of the zones. So, for our 3 : 1 example, we would scale for 1/3 of the volume but add two more impellers. This approach does not work fully for yield stress fluids; see Chapters 12 and 13 for detailed guidance on such fluids.

## Scaling Shaft Speed: The Scaling Exponent

As mentioned previously, geometric similarity leaves only one agitation variable to be manipulated upon scale-up. But how do we do that? If we have equal Reynolds numbers in both scales (which would normally require changing viscosity), we can use a simple exponential relationship, as expressed in Eq. (11.3). This equation will normally also be valid when the results are independent of Reynolds number, such as in fully developed turbulent flow. If there is a regime change upon scale-up (laminar to transition or turbulent; transition flow to turbulent), we would need to know how the desired process result may change as a function of Reynolds number and calculate the scaled-up speed iteratively.

$$N_2 / N_1 = (1/R)^n \quad \text{or} \quad N_2 = N_1 (1/R)^n, \quad \text{where n is the scaling exponent.} \tag{11.3}$$

The value of n depends on the result we wish to scale, as will be illustrated in the next section.

## Examples of Scaling Exponents, Their Derivation, and Their Consequences

We will go from the highest derived value of n to the lowest. High values of n mean lower shaft speeds as we scale-up.

1) n = 2: This corresponds to an equal Reynolds number scale-up if the viscosity does not change. This can be derived by noting that $N_{Re} = D^2N\rho/\mu$.
2) n = 1: This is likely the most common scaling rule. It results in equal characteristic velocity or scale of agitation, which can be derived by noting that $Q/A = N_QND^3/(\pi T^2/4) \propto ND$, since $T \propto D$ due to geometric similarity. It is also known as an equal tip speed scale-up, since tip speed = $\pi ND$. For turbulent flow, it is also called an equal torque/volume scale-up, since torque is power divided by $2\pi$ times shaft speed, or $N_P\rho N^3D^5/2\pi N \propto N^2D^5$, and volume is proportional to $D^3$, so we are

keeping $N^2D^2$ constant, leading to n = 1. This type of scale-up is used primarily for motion control applications. It can also be used for equal maximum shear rate.

3) n = (variable; 0.5 to 1.0). This is a special case for solids suspension, generally used for irregular tank shapes where the normal solids suspension equations may not be predictive. The exponent depends on settling velocity, with the exponent approaching 1 as the settling rate approaches zero, and the exponent approaching 0.5 as the settling rate approaches infinity. Inasmuch as the author is unaware of any applications in the biofuel industries where solids suspension is a serious problem (settling rates >0.001 m/s), and therefore can be scaled with an exponent of 1, we will not cover solids suspension in detail in this book. The fastest settling solids the author has seen in the biofuel industry are corn fiber, which settles approximately at the rate of 0.001 m/s. However, we have provided a graph in Figure 11.2 plotting the scaling exponent versus unhindered settling velocity, based on the information in Ref. [1]. Unhindered settling velocity can be measured by dropping particles down a long tube with the supernatant liquid and measuring the rate of fall. It should be noted that high solids slurries that have little free water are scaled as a motion control problem, i.e. using an exponent of 1. Extremely dilute slurries of corn fiber (<1%) may need an exponent of 0.95–0.98, but the author is unaware of any such application. Corn fiber in the corn wet milling industries is kept at a higher concentration and is designed with a scale of agitation of 3–4.

4) n = 2/3: Under turbulent conditions, this represents a power/volume scale-up. One can readily see this by noting that power $\propto N^3D^5$ and volume, for a given D/T ratio, is proportional to $D^3$. Therefore:

$$P/V \propto N^3D^2 \tag{11.4}$$

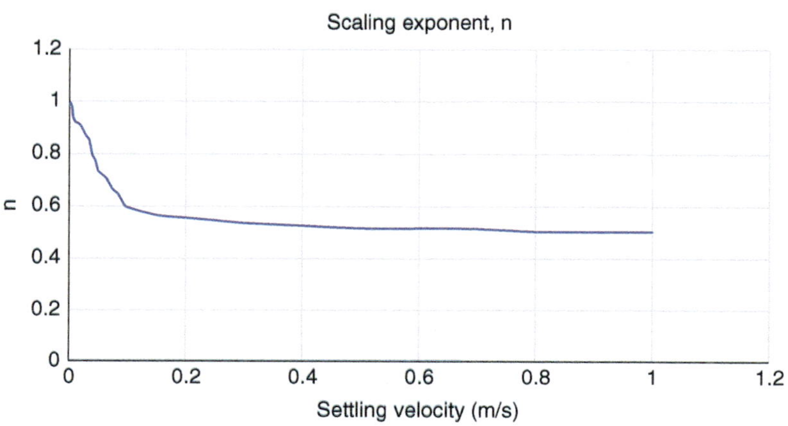


**Figure 11.2** Scaling exponents for solids suspension.

Generally speaking, a P/V scale-up is associated with an equal gas–liquid mass transfer coefficient, $k_l a$. It is also associated with an equal Sauter mean droplet size in liquid–liquid dispersions. There are theoretical explanations for why this should be so, mostly relating to turbulence theory and eddy size. We will not explore these derivations here.

Empirically, P/V has been found to be a useful scale-up method for achieving equal surface flux in dry solids make-down applications where the solids are somewhat to very difficult to draw in. Examples of some difficult solids include talcum powder, starches, kaolin clay, and sulfur. (Easy solids, such as granular sugar, salt, and sand, do not need this type of scale-up. They will simply fall in. They are treated as solids suspension problems.)

By equal flux, we mean equal mass of solids per tank cross-sectional area per unit time. Since the cross-sectional area per unit volume of the tank goes down by a factor of 1/R when scaling on the basis of geometric similarity, the time to incorporate the same amount of solids per unit of tank volume must be multiplied by R. Going the other way, if you have a production tank of 2 m diameter and you need to get the solids in over a period of 20 minutes, and you wish to find an agitation system that will do that based on a lab test in a 0.5 m vessel, you will need to get all the solids incorporated within 5 minutes in the lab scale.

5) n = ½: This exponent corresponds to equal Froude number ($N^2D/g$). It is proportional to the ratio of inertial to gravitational forces. At equal Reynolds number, or in a fully turbulent regime, it will provide a geometrically similar vortex depth. It will also provide an equal heat transfer coefficient to solid surfaces for most tank geometries. The author has not encountered any applications in the biofuels or renewable chemicals industries that would scale-up this way.
6) n = 0: An exponent of zero means the shaft speed does not change upon scale-up. This may be derived from the dimensionless blend time, Θ*N, which is a constant in either turbulent or laminar flow, but changes during transition. This is a rather extreme scale-up basis. It drastically increases power requirements compared to the other exponents. Yet it would be necessary in order to maintain the same product distribution with a competitive-consecutive reaction. However, this scale-up cannot be applied to a large volume range due to foaming and aeration problems. Incidentally, an exponent of zero also provides equal power/volume as long as both scales are in fully laminar flow. However, the author is unaware of any reason why it would be necessary to maintain equal P/V in laminar flow.

## Consequences of Using Different Scale Exponents

Choosing the right scaling exponent is of paramount importance. The choice of scaling exponent has a major impact on performance, energy consumption, and equipment cost. As the exponent chosen becomes smaller, the production scale speed increases. Power (in turbulent flow) is proportional to speed cubed. Torque is proportional to speed squared. As a rough rule of thumb, agitator cost is proportional to torque raised to the 0.6 power.

To illustrate the consequences, let us do a hypothetical scale-up from a 21.2 l volume in a 300 mm diameter tank to a 21 200 l volume in a 3 m tank diameter. Let us suppose that our smaller tank has a 100 mm diameter hydrofoil impeller, and that the desired process result was achieved at a shaft speed of 720 rpm. Our process liquid has a specific gravity of 1 and a viscosity of 1 cP. Let us assume the specific impeller used has a power number of 0.33 and a pumping number of 0.54. Since we have geometric similarity, we can use Eq. (11.1) to determine the scaling factor, R:

$$R = T_2/T_1 = 3\,\text{m}/0.3\,\text{m} = 10$$

Therefore using the same equation to determine full scale D, we get:

$$D_2 = RD_1 = 10 * 100\,\text{mm} = 1000\,\text{mm} = 1\,\text{m}$$

These details are summarized in Table 11.1.

Now we have only one variable left to determine: the shaft speed. We use Eq. (11.3) to do this:

$$N_2 = N_1(1/R)^n = 720(1/10)^n$$

Table 11.2 illustrates the consequences of different exponents. We show the scaled-up shaft speed as a function of n. We calculate the power based on a power number of 0.33 by using the following, where N is converted from rpm to 1/s:

$$P(w) = 0.33 * 1000\,\text{kg/m}^3 * 1\,\text{m}^5 * (N/60)^3.$$

For example, for the exponent of 1, we would have $N_2 = 72$ rpm, and therefore $P = 570\,\text{W} = 0.57\,\text{kW}$.

**Table 11.1** Scale-up exponent example problem.

| | Small scale | Full scale |
|---|---|---|
| Tank diameter | 300 mm | 3 m |
| Impeller diameter | 100 mm | 1 m |
| Volume | 21.2 l | 21 200 l |
| Liquid height | 300 mm | 3 m |
| Shaft speed | 720 rpm | To be determined |

**Table 11.2** Scale-up exponent example problem results.

| Exponent, n | N (rpm) | Power (kW) | Scale of agitation |
|---|---|---|---|
| 2 | 7.2 | 5.70E-04 | 0.3 |
| 1 | 72 | 0.57 | 3 |
| 2/3 | 155 | 5.7 | 6 |
| ½ | 228 | 18.1 | 10 |
| 0 | 720 | 571 | 30 |

For reference, we also calculate scale of agitation by noting that $Q = N_Q N D^3$ and scale of agitation = Q/(A * 6) when in English units of feet/minute, or Q/A * 32.8 if in m/s. Since $A = 3^2 * \pi/4 = 7.069\,m^2$, we get:

$$\text{Scale of agitation} = \left(N_Q N D^3 / 7.069\right) * 32.8 = \left(0.54\left(N/60\right) * 1\,m^3 / 7.069\right) * 32.8;$$ for our exponent of 1, at 72 rpm, we get scale of agitation = 3.

This, of course, is the same scale of agitation we had in the small vessel, because a scaling exponent of 1 delivers equal characteristic velocity.

One can readily see the major effect the exponent has on equipment size. Most biofuel/renewable chemical applications are liquid motion control problems and will successfully scale-up using an exponent of 1. Choosing an exponent of 2/3 to get equal power/volume will require 10 times the power input as an exponent of 1 for a 10 : 1 linear scale-up or 1000 : 1 volumetric scale-up. This would only be appropriate in two-phase mass transfer problems and certain solids incorporation applications. Exponents lower than 2/3 would be quite rare in general, and this author is unaware of any applications in the biofuel industry that require it. Can you imagine the frothing and foaming that would occur trying to turn a 1 m impeller at 720 rpm while trying to invest 571 kW in a 22.5 $m^3$ volume? This is not practical.

## Scaling of Two Physical Results

When scaling two physical results simultaneously, it is impossible to keep geometric similarity. We have to be able to manipulate two variables, whereas geometric similarity restricts us to only one, namely, the shaft speed. While shaft speed is still one variable we can manipulate, we need another. Possible things we can manipulate include impeller diameter, number of impellers, type(s) of impellers, and aspect ratio (Z/T). There are no general rules applicable. It depends on what things we wish to keep constant. In general, we can only do this over a small-scale ratio; perhaps a 3–5 volumetric ratio. This means several stages of pilot equipment may be needed to verify design as we scale-up, and it may ultimately be impossible to keep two variables fixed in a large vessel. Parallel trains may be needed.

## Example of Two Physical Results

One possible example may be to try to keep equal P/V and equal tip speed. This might be desired if we wish to keep a certain two-phase mass transfer coefficient while not increasing the maximum shear rate, which is proportional to impeller tip speed. An equal P/V scale-up combined with geometric similarity results in increasing tip speed upon scale-up. So how do we reduce tip speed while keeping power constant? There are several ways, and each may have some drawbacks, as shown next:

- Increase D/T and reduce shaft speed. This normally will work unless the D/T increase changes the flow pattern too much. For example, increasing D/T from 0.3 to 0.4 is probably okay, but from 0.3 to 0.6 will create a flow pattern change for a hydrofoil from axial flow to mixed flow. Also, the lower shaft speed means more torque and a more expensive agitator.
- Increase the number of impellers. This can be applied over a limited range. Too many impellers can mean the power draw is not additive; there is a spacing factor on power draw of less than 1 for multiple axial flow impellers. For low-viscosity liquids, the impeller spacing should normally be at least 1 impeller diameter.
- Increase aspect ratio. This lets us use more impellers while keeping spacing reasonable. However both the vessel and the agitator become more expensive.
- Change impeller style to one with a higher power number but a substantially similar flow pattern. For example, if we started with a low solidity three-bladed hydrofoil with a power number of 0.3, we could go to a four-bladed hydrofoil with a power number of about 0.36, or a high-solidity hydrofoil with a power number of 0.5–1.0, which would allow us to use a smaller impeller diameter. We may even consider using a pitched blade turbine with a power number of 1.37. These changes would allow us to use a smaller impeller diameter. For example, an impeller with a power number of 1 would have a relative diameter of $(0.3/1)^{0.2}$ or 0.786 times the diameter of a low-solidity hydrofoil.

## Three-Physical-Result Strategy

This is almost impossible. Fortunately, the author has yet to encounter a case where it was necessary.

## Several Example Problems

**Example 11.1** Motion Control, Turbulent, Geometric Similarity
Let us suppose we have a 1 m diameter tank, with (1) 400 mm pitched blade turbine located 300 mm off-bottom. It achieved the process objective successfully at a shaft speed of 200 rpm, with a liquid level of 1100 mm and a volume of 0.813 $m^3$ (the tank has an ASME dished bottom, so the volume is less than a cylindrical batch). The specific gravity is 1.18 and the viscosity is 50 cP. We wish to scale this up to a geometrically similar tank of 4 m diameter. We assume the process result is liquid velocity controlled. What will the impeller size, speed, and power draw be?

**Answer**
First, we must establish whether the flow is turbulent. Let us use SI units, so the viscosity becomes 50/1000 or 0.05 Pa-S. Therefore, we can calculate the Reynolds number as follows: $N_{Re} = D^2N\rho/\mu = (0.4\,m)^2 * (200/60) * 1180\,kg/m^3/0.05\,Pa\text{-}s = 12\,586$, so the flow is turbulent. No corrections are needed for Reynolds number. Next, we need to determine the impeller diameter and off-bottom clearance. Our scale ratio is 4 m/1 m = 4. Therefore, $D_2 = 4D_1 = 4 * 400\,mm = 1.6\,m$. The off-bottom clearance is 4 * 300 mm = 1200 mm = 1.2 m. The shaft speed, $N_2$, is $(1/R) * N_1 = 200/4 =$ 50 rpm = 0.8333/second. From Appendix 2, the turbulent power number of a pitched blade turbine is 1.39. Therefore, in SI units, $P = N_P\rho N^3D^5 = 1.39 * 1180\,kg/m^3 * (0.8333/second)^3 * (1.6\,m)^5 = 9952\,W = 9.95\,kW$. Since 50 rpm is not a standard AGMA speed, we may wish to adjust the shaft speed to 56 rpm and slightly reduce impeller diameter to keep within an 11 kW standard motor size.

**Example 11.2** "Almost" Geometrically Similar Motion Control
Let us take the same pilot data from Example 11.1, but scale it up to a tank of 5 m diameter, with a 2 : 1 elliptical bottom head and a liquid level of 4.5 m. Thus, this batch has a smaller aspect ratio than what we had in Example 11.1. How do we scale it up?

**Answer**
According to this website (Calculate the volume of Pressure Vessel Heads|LZR FIT Tools, accessed 8 April 2021), the volume of the elliptical head is 16.3 $m^3$. The depth of this style head will be one-fourth of its diameter, or 1.25 m. Therefore the liquid in the straight side occupies (4.5 − 1.25) = 3.25 m, for a straight side volume of 63.8 $m^3$. The total volume is thus (63.8 + 16.3) = 80.1 $m^3$. Since geometric similarity is not exact, we will define the value of R using Eq. (11.2): $R = (80.1/0.813)^{1/3} = 4.62$. Thus, our full scale impeller diameter equals 0.4 m * 4.62 = 1.85 m and our shaft speed is 200/4.62 or 43.29 rpm. The nearest AGMA speed is 45 rpm, so we will use this. The D/T has a minor effect on

power number. The power number at a D/T ratio of 0.35 is 1.37. It is 1.39 at a D/T of 0.4. Our D/T ratio is now 0.37. Thus, we will use a power number of 1.38. Therefore, Power = 1.38 * 1180 * $(45/60)^3$ * $1.85^5$ = 14 887 W = 14.9 kW. We would thus use a standard motor of 18.6 kW rating.

**Example 11.3** Very Different Geometry, Motion Control
Let us suppose that we achieved a successful result blending liquids of two different densities by using a 600 mm hydrofoil impeller operating at 190 rpm in a 1500 mm diameter flat-bottomed vessel with a liquid level of 1400 mm and a volume of 2.47 $m^3$. Although the final blended density is 1050 kg/$m^3$, we will design for the density of the densest liquid, which is 1200 kg/$m^3$. This is because it is possible and even likely that operators will someday fill the vessel with the denser liquid. The production vessel is 4 m diameter, 13 m straight side with a liquid level of 12.5 m. It has an ASME dished bottom head, with a volume capacity of 5.57 $m^3$ and a depth of 0.703 m. Therefore, the straight side volume is based on (12.5 − 0.703) = 11.8 m, which contains 148.3 $m^3$, bringing the total volume to 153.9 $m^3$. The Z/T ratio is 12.5/4 = 3.13. According to Table 5.1, we need three impellers. How do we scale this up?

**Answer**
Since the geometry requires a different number of impellers, we will scale-up by zones, that is, volume per impeller, and then just add the additional impellers. Based on three impellers, our volume per impeller is 153.9/3 = 51.3 $m^3$. Using Eq. (11.2), R = $(51.3/2.47)^{1/3}$ = 2.75. Therefore, our full scale impeller diameter = 600 mm * 2.75 = 1650 mm = 1.65 m, and we need three of them. The shaft speed required is 190 rpm/2.75 = 69.1. With a D/T of 0.41, according to Appendix 1, the power number is 0.27. Therefore, the power draw for three impellers is 3 * 0.27 * 1200 * $(69.1/60)^3$ * $1.65^5$ = 18 158 W = 18.2 kW. The nearest standard motor size with any safety factor is 22 kW. The nearest standard AGMA speed is 68 rpm, which is close enough to leave the impeller size unaltered. However, at equal thrust ($N^2D^4$), we could adjust the impeller size by 0.8% or to 1.66 m. But agitation is really not quite that precise!

**Example 11.4** Viscous Motion Control Scale-up Using Characteristic Fluid Velocity

The preceding examples were all in turbulent flow, so impeller performance characteristics did not change as a function of Reynolds number upon scale-up. This problem will be different, in that Reynolds number will have an impact on design and performance. As we shall see, the most rigorous design method requires an iterative approach.

Let us suppose we have a lab vessel in which we are adding water, with a viscosity of 1 and a density of 1000 kg/$m^3$, to a sugar solution with a density of

$1300\ kg/m^3$ and a viscosity of 10 000 cP. The vessel is a flat-bottomed tank with no baffles, and it has a diameter of 0.5 m with a liquid level of 0.6 m. It has two 0.2 m diameter hydrofoil impellers located 0.125 and 0.4 m off-bottom. It was found that a minimum shaft speed of 650 rpm was required to overcome the tendency of the water addition to form a stratified layer on top. We need to scale this up to a 3 m diameter tank with a liquid level of 3.6 m, so we have geometric similarity. What impeller size and shaft speed should the full scale system have, and how much power will it draw, bearing in mind we must design for the fluid properties of the sugar solution?

**Answer**

We believe this is a fluid velocity controlled problem so that is the basis on which we will scale-up. However, the lab vessel is not operating in turbulent flow, and the Reynolds number will change on scale-up, which will change the pumping number. Therefore, our approach will be iterative. But first, we need to calculate a characteristic velocity in the lab scale. Although the concept of scale of agitation requires us to resolve the problem to an equivalent square batch geometry with a single impeller, we do not need to do that for a geometrically similar problem. It is sufficient to define a characteristic velocity the same way in both scales. We will just use the Q/A value for each impeller based on the tank cross-sectional area. But first, we must calculate the impeller Reynolds number and pumping number, converting viscosity to Pa-s by multiplying by 0.001:

$$N_{Re} = D^2 N\rho / \mu = 0.2^2 * (650/60) * 1300/10 = 56.33. \tag{11.5}$$

This is just above laminar flow.

According to Appendix 1, the pumping number for a hydrofoil impeller with a D/T ratio of 0.4 and a Reynolds number of 56 is about 0.23. Therefore, the pumping capacity of each impeller evaluated individually is:

$$Q = N_Q N D^3 = 0.23 * (650/60) * 0.2^3 = 0.0199\,m^3/s \tag{11.6}$$

The cross-sectional area of a 0.5 m diameter tank is $\pi * (0.5\,m)^2/4 = 0.196\,m^2$. Therefore, our characteristic velocity, Q/A, equals $(0.0199\,m^3/s)/0.196\,m^2 = 0.102\,m/s$. We will try to match that in the larger tank.

As a first step, we will use geometric similarity: $R = 3/0.5 = 6$. Therefore, $D_2 = 0.2 * 6 = 1.2\,m$, and the off-bottom locations are 0.75 and 2.4 m, respectively. For our first iteration, we will use an exponent of 1, giving $N_2 = N_1/R = 650/6 = 108.33\,rpm = 1.81/second$. Now we calculate Reynolds number: $N_{Re} = 1.2^2 * 1.81 * 1300/10 = 339$. This is in transition flow. Per Appendix 1, the resultant pumping number is about 0.29. Therefore the pumping rate is $0.29 * 1.81/second * (1.2\,m)^3 = 0.907\,m^3/s$. Our tank cross-sectional area is $7.069\,m^2$. Therefore our characteristic velocity, Q/A, is 0.907/7.069 = 0.128 m/s. This is almost 26% higher than what we calculated in the lab, so we do not need to turn this fast. We will iterate until we match. By trial and error, we found that 90 rpm would give a Reynolds number of 282, a pumping number of about 0.28, and a resultant characteristic velocity of 0.103 m/s, which is in good agreement with our target value. The nearest standard AGMA speeds are 84 and 100 rpm. To save power, we will use 84 rpm and adjust impeller diameter keeping $N^2D^5$ constant, so our adjusted impeller diameter will be 1.25 m. This will result in a Reynolds number of 284, a pumping number of 0.26, and a characteristic velocity of 0.101 m/s, which is acceptable. The power number, per Appendix 1, is 0.4. Therefore, our power draw for two impellers may be calculated thusly:

$$P = 2 * 0.4 * 1300 * (84/60)^3 * (1.25\ \text{m})^5 = 8709\ \text{W} = 8.7\ \text{kW}. \tag{11.7}$$

An 11 kW motor will suffice. Had we adjusted to 100 rpm, a 15 kW unit would have been required.

Note that for this type of problem impeller placement and how the water is added, both are important. A dip tube may result in less agitation required.

This problem illustrates how Reynolds number changes upon scale-up can affect results. Our scaled-up speed was 17% lower than a simple scaling exponent of 1 would indicate, saving significant power, torque, and cost. However, some would opt to be more conservative and use the direct scaled speed.

**Example 11.5** Non-Newtonian Motion Control Scale-up

Now let us see what may happen if we scale-up a shear thinning (pseudoplastic) liquid. Let us suppose we have a pilot vessel of 0.3 m diameter, dished bottom vessel with a 0.4 m liquid level, which equates to a volume of 0.0261 m$^3$. In this vessel we have two 0.185 m diameter hydrofoil impellers, located 0.05 and 0.3 m off-bottom. In order to achieve full motion and avoid dead spots in the vessel, we had to rotate these impellers at 320 rpm. The fluid density measures 1080 kg/m$^3$. A curve fit to a plot from a rheometer indicated a power law model [reference Eq. (3.22)] with an M value (viscosity at a shear rate of 1/s) of 150 000 cP and a power law exponent, n′, of 0.20, which is quite shear thinning.

**In other words:**

$$\mu_a = 150000(dv/dx)^{(0.2-1)} = 150000(dv/dx)^{-0.8} \tag{11.8}$$

How do we scale-up to a 2.5 m diameter vessel with a liquid level of 3 m and a volume of 12.73 m$^3$ in such a way as to assure complete motion? Note that the geometric similarity is not exact.

**Table 11.3** Metzner-Otto constants for impellers.

| Impeller type | $K_M$ |
|---|---|
| Propeller | 10 |
| Low solidity hydrofoil | 10 |
| High solidity hydrofoil | 11 |
| Pitched blade turbine | 11 |
| Radial turbines | 11.5 |
| Chaos (reversing pitch) turbines | 40 |

### Answer

We can solve this problem in several different orders. But we will follow the order in Example 11.4. We will calculate a characteristic velocity in the small scale first. To do this we must calculate an apparent viscosity. We will use the relationship dv/dx = KN, with K from Table 3.1, repeated here as Table 11.3. From that table, we get a shear rate of 320 rpm * 10 = 3200/minute = 53.333/second. Plugging in to Eq. (11.8), we get $\mu_a = 150\,000(53.33)^{-0.8} = 6230\,\text{cP}$ = 6.23 Pa-s. Our Reynolds number is thus $0.185^2$ * (320/60) * 1080/6.23 = 31.6, which is transition flow but almost laminar. According to Appendix 1, our pumping number is about 0.19. Therefore, the single-impeller pumping rate is 0.19 * (320/60) * $0.185^3$ = 6.42E-3 $m^3/s$. The cross-sectional area of the tank is $\pi * 0.3^2/4 = 0.0707\,m^2$, therefore the characteristic velocity is 6.42E-3/0.0707 = 0.0908 m/s.

Now we can scale the geometry. Since the geometry is not exact, we use the cube root of the volume ratio to define R:

$$R = (12.73 / 0.0261)^{1/3} = 7.87$$

We will calculate as though we had geometric similarity with an equivalent vessel diameter. Therefore, we use $T_2 = 7.87 * 0.3 = 2.36$ m rather than the actual 2.5 m. This would have an equivalent cross-sectional area of 4.38 $m^2$. Our impellers will be 7.87 * 0.185 = 1.46 m diameter. Our first trial for a shaft speed will be based on a scaling exponent of 1, or 320/7.87 = 40.66 rpm. This corresponds to a shear rate of 6.78/second. That gives us an apparent viscosity of 32 454 cP or 32.45 Pa-s, giving us a Reynolds number of 44.87, which is transitional but almost laminar. According to Appendix 1, the pumping number based on equivalent D/T is still about 0.19. That means the individual impeller pumping is 0.4 m/s, and the characteristic velocity is 0.0915 m/s. That is basically the same as what we had in the pilot scale, so we need not change the speed to get equal motion control. We can either adjust to 45 rpm or 37 rpm; the scaled speed is midway between the two. If we use 37 rpm, let us change the impellers to 1.52 m diameter. The new shear rate is 6.17, the new viscosity is 34.98 Pa-s, the new Reynolds number is 41.1, and therefore the power number is 1.3 by interpolation. The power draw is thus: $P = 2 * 1.3 * 1008 * (37/60)^3 * 1.52^5 = 4987\,W = 4.98\,kW$. This would heavily load a 5.5 kW motor, so we would recommend a 7.5 kW motor.

Note that, for a Newtonian fluid, using a scaling exponent of 1, the Reynolds number is directly proportional to the scaling factor, R. Thus, for viscous Newtonian applications, a scaling exponent of 1 can be conservative, as the pumping number increases with scale and Reynolds number. But for shear thinning liquids, the viscosity increases as the speed decreases, which results in less change of Reynolds number, so a scale-up using an exponent of 1 often has little extra conservatism. In Chapter 13, we will see that fluids whose rheology is dominated by a high yield stress often have little or no change in Reynolds number upon scale-up.

## Summary of Chapter

We have explored some of the promises and limitations of scale-up. We have discussed the use of geometric similarity and the scaling exponent as a sound basis for scaling-up single physical results. We have shown that the scaling of two or more results is difficult and sometimes impossible. For complex processes with multiple important results, the possible scale ratio may be small, and production design may require multiple agitated tanks as opposed to just one large one. Scaling of viscous fluids is more complex and often requires an iterative approach.

## List of Symbols

| | |
|---|---|
| A | Tank cross-sectional area |
| a | Interfacial area/volume |
| C | Impeller off-bottom clearance |
| D | Impeller diameter |
| dv/dx | Shear rate |
| $k_l$ | Liquid film coefficient (gas–liquid) |
| N | Shaft speed |
| n | Scaling exponent for shaft speed |
| $n'$ | Power law exponent |
| $N_{Fr}$ | Froude number |
| $N_P$ | Power number |
| $N_Q$ | Pumping number |
| $N_{Re}$ | Reynolds number |

| | |
|---|---|
| $N_{Th}$ | Thrust number |
| P | Power draw |
| Q | Impeller pumping rate |
| R | Scale ratio |
| T | Tank diameter |
| Z | Liquid level |
| V | Liquid volume |

## Greek Letters

| | |
|---|---|
| $\rho$ | Density |
| $\mu$ | Viscosity |
| $\mu_a$ | Apparent viscosity |
| $\Theta$ | Blend time |

## Reference

1 Corpstein, R., Fasano, J., and Myers, K. (1994). The high efficiency road to liquid-solid agitation. *Chemical Engineering Magazine* (1 October). (Sourced by reprint. Original pagination unknown to author.)

# 12

# Basic Rheology and Viscometry

## Introduction

This chapter is about a few simple models of fluid behavior that are applicable, in the author's experience, to fluids commonly found in the biofuels industry, as well as how to measure characteristics of such fluids. We begin by reviewing the common models described in Chapter 3, but without the figures. (Refer to Chapter 3 if you want to visualize the models graphically.) We will then go into the various methods of measuring viscosity, especially for difficult fluids such as high-solids biomass slurries. Finally, we will show real-world data and treatment for some biomass slurries the author has tested.

## Review of Common Rheology Models

Though there are many models of fluid behavior, some quite complex, not all are needed for fluids encountered in the industries we are focused on in this book. For example, time-dependent fluids such as thixotropic (thins with time), rheopectic (thickens with time), and viscoelastic (has normal stresses as well as shear stresses) are not normally encountered. Therefore we will confine ourselves to the few simple models introduced in Chapter 3, which are reviewed once more here so you do not have to return to Chapter 3 to look them up.

### Newtonian

Many low-viscosity fluids are Newtonian, based on the conceptual model of fluid flow developed by Sir Isaac Newton. These fluids are those that have a shear stress that is directly proportional to shear rate (also called velocity gradient). Mathematically, this can be represented as:

*Agitator Design Technology for Biofuels and Renewable Chemicals*, First Edition. Gregory T. Benz.
© 2022 John Wiley & Sons, Inc. Published 2022 by John Wiley & Sons, Inc.

$$\sigma_s = \mu * \gamma \tag{12.1}$$

where μ, the coefficient on shear rate, is also known as viscosity. This can be arranged to:

$$\mu = \sigma_s / \gamma \tag{12.2}$$

Intermediate- and even some high-viscosity fluids can be Newtonian if they are true solutions, especially if they are solutions of small molecules. For example, honey and corn syrups are often Newtonian even at 30 000 cP. Some silicone solutions are Newtonian even at 20 000 000 cP, but this is quite rare. Most suspensions of more than 5–10% solids exhibit at least some non-Newtonian behavior. High-solids slurries can be very non-Newtonian. Usually, they are shear thinning, but some may be shear thickening, also known as dilatant. They often follow the power law model, described next.

Low-solids slurries operating in a water-like fluid are usually Newtonian. High-solids slurries may not be Newtonian, and those containing both an aqueous and an oil phase usually are not. But the overall viscosity for most biosolids slurries is less than 1000 cP. Many are less than 50 cP. The exception is fibrous materials, which can be very viscous, even semi-solid. Special techniques are often needed to measure and characterize their behavior, which will be described later in the chapter under the section "How to Measure Viscosity for Multiphase Macro Fluids Using an Agitator as a Viscometer."

## Pseudoplastic (Power Law)

These fluids are often called shear thinning because the slope of the shear stress versus shear rate curve decreases as the shear rate increases. Therefore, the apparent viscosity decreases as shear rate increases. However, some power law fluids are shear thickening. This is rare to nonexistent for fluids in the biofuels and renewable chemicals industries. The behavior of pseudoplastic fluids may be described by a power law model:

$$\sigma_s = M * \gamma^n \tag{12.3}$$

The exponent n is called the power law exponent. When it has a value of 1, the equation reverts to Newtonian. When it is less than 1, it is shear thinning. When it is greater than 1, the fluid is shear thickening. Starch slurry above 40% solids can exhibit shear thickening behavior, but this is not a normal situation in this industry.

The apparent viscosity can be expressed by dividing by the shear rate on both sides of the equation:

$$\mu_a = M * \gamma^{(n-1)} \tag{12.4}$$

where M is the viscosity coefficient, or the viscosity at a shear rate of 1 (usually in units of 1/s).

Many biosolids slurries exhibit this behavior at high-percent solids (typically above 20%). Fibrous materials may exhibit this behavior at solids concentrations as low as 2–3%, though higher concentrations are more common, which may require a more complex rheology model.

## Bingham Plastic

These kinds of fluids exhibit a yield stress. That is, until the fluid is exposed to a certain minimum shear stress, it will not move. Examples include some high-solids slurries, toothpaste, peanut butter, lignocellulosic slurries, and any fluid that can hold some of its shape after being disturbed. Above the yield stress, Bingham fluids exhibit a constant slope. Mathematically:

$$\sigma_s = \sigma_y + M * \gamma \tag{12.5}$$

and

$$\mu_a = \sigma_y / \gamma + M \tag{12.6}$$

At low shear rates, most high-solids lignocellulosic slurries seem to be a Bingham plastic, as the flow behavior is dominated by the yield stress. This manifests itself in an interesting way when measuring torque of an agitator impeller. In laminar flow, at low shaft speeds, the torque required to rotate the impeller is constant, i.e. independent of shaft speed. The same thing would happen with spindle viscometer readings if the slurry did not phase separate at the spindle surface, though it often does. At higher shear rates or shaft speeds, additional torque may be observed as the other viscous effects of the fluid come into play. The next model is a bit more comprehensive than the Bingham model at dealing with fluid behavior above the yield stress.

## Hershel–Bulkley

These fluids combine a yield stress with pseudoplastic behavior above the yield stress:

$$\sigma_s = \sigma_y + M * \gamma^n \tag{12.7}$$

and

$$\mu_a = \sigma_y / \gamma + M * \gamma^{(n-1)} \tag{12.8}$$

One may recognize that all three previous models are special cases of the Herschel–Bulkley model, making this the most versatile model commonly in use.

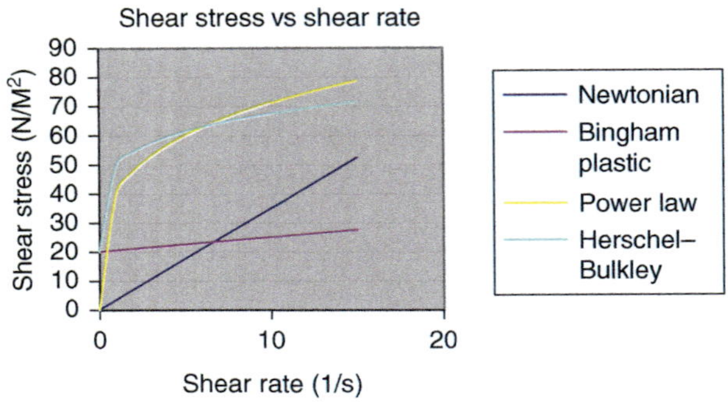


**Figure 12.1** Shear stress curve.

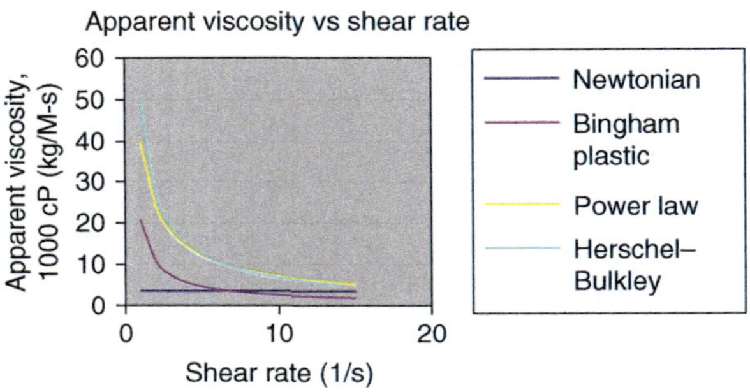


**Figure 12.2** Viscosity curve.

Most high-solids biomass slurries can be described well by the Herschel–Bulkley model. But measurements must be taken over a range of shear rates, including very low shear (typically less than 2/s) to pick up the yield stress term accurately, and must go to at least 15–20/s or even as high as 50/s to pick up good values of M and n.

These four models are described visually in Figures 12.1 and 12.2.

The effect of yield stress can be clearly seen in Figure 12.3, where some of the fluid is moving and some is not. Where the fluid is stationary, insufficient stress has been applied to overcome its yield stress. We will get into this in more detail in Chapter 13.

## Instrument Measurement Methods

As described in the earlier models, there is a relationship between the shear stress in a fluid and the velocity gradient, or shear rate, in that fluid. The purpose of a

**Figure 12.3** Yield stress.

viscometer is to enable us to quantify that relationship, although some viscometers only measure viscosity at one or two shear rates, which is basically useless for non-Newtonian fluids.

There are many kinds of viscometers on the market, using various principles to measure fluid characteristics. Examples include devices that measure pressure drop through an orifice or through a tube in laminar flow with known flow rate; the time it takes for a ball to fall through the liquid a certain distance (these viscometers actually measure kinematic viscosity, or $\mu/\rho$, rather than just viscosity); the time it takes for an aliquot of liquid to drain a certain distance through a funnel with a small orifice (also measures kinematic viscosity); and a bevy of rotational viscometers. Since rotational viscometers are the only ones that permit significant variation of shear rate, they are preferred for agitator design, and are the only types that will be discussed here.

At this point, we may need to define the difference between a viscometer and a rheometer. For the fluids of interest to us, there is not much practical difference. However, a viscometer, in principle, can only measure the shear stress versus shear rate relationship. A rheometer can measure other properties, such as time dependency (thixotropy or rheopexy), viscoelasticity, and maybe some others. Viscometers can be manually operated or computer controlled, with data acquisition. Rheometers are almost always computer controlled, with automatic ramping up and down in speeds, have a wide range of shear rates, and often have automatic temperature control as well. From a practical standpoint, simple rotational viscometers cost between $500 and 3000 in 2021 dollars. Rheometers can go for

$50 000 or more; some even cost more than $100 000. These costly rheometers are not necessary for the fluids we encounter in the target industries of this book. And, as we shall see, they do not even work for high-solids fibrous slurries or those with large particles. That renders them unusable for lignocellulosic slurries prior to significant hydrolysis.

## Spindle Viscometers

The most common rotary viscometers in use are similar to Figure 12.4 and are called spindle viscometers. This is because they employ spindles rotating in the fluid to create a reaction torque, which is used in the measurement process. Spindles can be disk-shaped or cylindrical, and come in different sizes, each of which has a different factor applied to the instrument readings to determine viscosity. A typical set of spindles is illustrated in Figure 12.5.

## Bob-and-Cup Rheometers

Spindle viscometers suffer from the limitation that the effective shear rate varies across the spindle, particularly with disk spindles. It also varies with container size and rheology, so data treatment methods are required to adjust for non-Newtonian

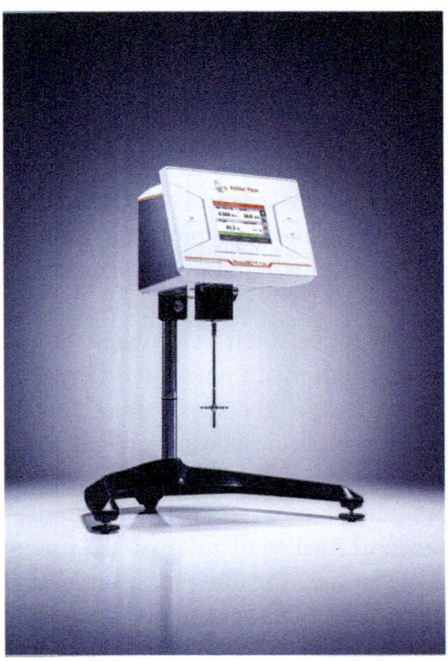

**Figure 12.4** Spindle viscometer.
*Source:* Courtesy, Anton-Paar.

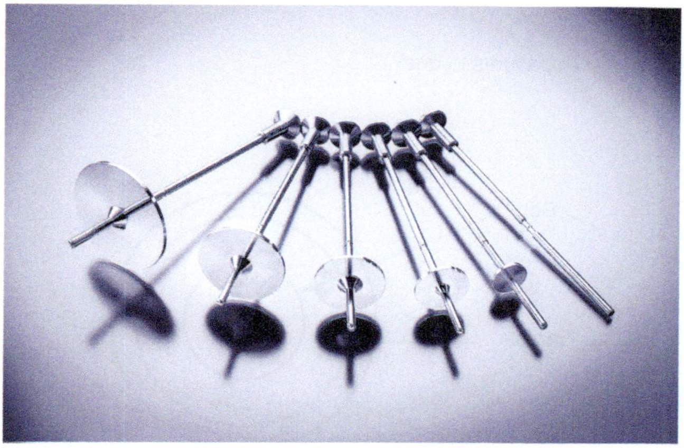

**Figure 12.5** Set of spindles. *Source:* Courtesy, Anton-Paar.

**Figure 12.6** Typical rheometer, with temperature control. *Source:* Courtesy, Anton-Paar Instruments.

fluids, as will be discussed under the section "Limitations of the Above Viscometers: Correction for Non-Newtonian Fluids." Most rheometers (such as in Figure 12.6) are equipped with a cup-and-bob arrangement (Figure 12.7), where the narrow gap size and cup geometry are designed to create a nearly constant shear rate that is independent of fluid rheology. This makes them more accurate at determining a fluid rheology model. Still, they have some limitations, too, which will be discussed.

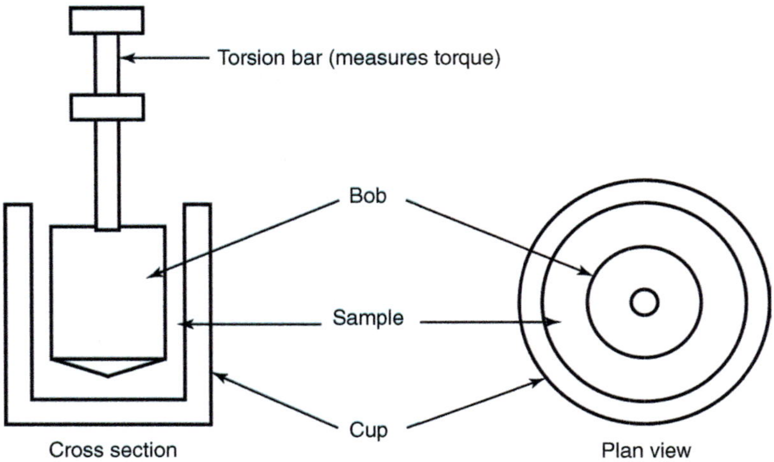


**Figure 12.7** Cup-and-bob arrangement.

## How Do Rotary Viscometers Work?

Brookfield and similar rotary viscometers are essentially torque meters. The raw instrument reading is proportional to the torque developed by the spindle operating in the fluid at a certain rpm. For a given spindle, the torque is proportional to the shear stress in the fluid, so the reading is also proportional to the shear stress. Viscosity is of course shear stress divided by shear rate, which is proportional to instrument speed or rpm. So, the spindle factor is based on the spindle number and rpm, and will give actual viscosity if the correct shear rate is used in calculations. The problem comes from the fact that the shear rate used to develop the spindle factor is based on Newtonian calculations. The actual shear rate for an open-spindle type of viscometer depends on the rheology of the fluid and the spindle geometry, which will be discussed in the next section.

While a cup-and-bob rheometer is still a torque meter, when distilled to its essence, the close clearance between the bob and the cup means the shear rate is nearly constant and independent of rheology, as stated earlier. Therefore, one can reliably use the direct instrument output to plot apparent viscosity versus shear rate.

## Limitations of the Above Viscometers: Correction for Non-Newtonian Fluids

Under this heading, we will discuss some of the problems endemic to rotational viscometers, and how to treat data from spindle-type viscometers when used in fluids for which they are suitable.

## Phase Separation

Many fluids of interest are slurries. This means they are actually composed of two phases: a solid phase and a liquid phase. These phases can separate, either by gravity or by the mild pumping or shearing action of the rotating spindle. Sometimes this problem can be ameliorated by rapidly testing the sample after agitating it to ensure uniformity and then getting readings quickly before any separation occurs. Some viscometers have an option to slowly lower the spindle as it rotates, though that effectively changes the spindle immersion and would have an effect on the reading accuracy, so the immersion should start above the scribed line indicating proper spindle immersion depth, and then the reading should be taken when the spindle is properly immersed.

## Particle Size Versus Gap

Although the close clearance between the bob and the cup in that type of rheometer confers the advantage of nearly constant and predictable shear rate, some slurries of interest in the biofuel industry have particles that are larger than the gap. In fact, most biomass slurries have such large particles prior to hydrolysis, as they are basically vegetation that has been put through a hammer mill, and it is not unusual to have length dimensions of 1–5 mm, which is greater than the gap. Obviously, a bob-and-cup viscometer will not work for such materials, though it may work on the hydrolyzed product, which is basically a lignin slurry in a sugar solution and has flow properties similar to a mud or clay slurry.

## Data Treatment

To properly evaluate viscosity data from a spindle-type viscometer, we have to go back to what it actually measures: torque. Then we must relate this to shear stress. Then we must calculate the shear rate so that we can divide shear stress by shear rate to get apparent viscosity. Unfortunately, the actual shear rate is not always the Newtonian shear rate the instrument viscosity factors are based on. The shear rate depends on the rheology. Fortunately, there is a relatively simple way to treat the data that is accurate enough for agitator design. It is generally described in Ref. [1] and the theoretical derivation is published in Czech in Ref. [2]. We repeat the fundamentals here, specifically for a Brookfield RVT viscometer. The same procedure can be applied to other viscometers as long as the spindle shear stress factors are known, and the spindle geometry is the same as the Brookfield spindles so that the shear rate correction can be made.

Referring to Table 12.1, we have identified the usual spindles associated with the Brookfield RVT, numbered 1–7. The symbol $K\tau$ is the multiplier that should be applied to the instrument reading, IR, to calculate the shear stress in Pa. In other words:

$$\sigma_s(\text{Pa}) = K_\tau * \text{IR} \tag{12.9}$$

**Table 12.1** Brookfield RVT factors for the Mitschka method.

| Spindle number | | 1 | 2 | 3 | 4 | 5 | 6 | 7 |
|---|---|---|---|---|---|---|---|---|
| $K_\tau$ | | 0.035 | 0.119 | 0.279 | 0.539 | 1.05 | 2.35 | 8.38 |
| $K\gamma$, at n = first column value | 0.1 | 1.728 | 1.431 | 1.457 | 1.492 | 1.544 | 1.366 | 1.936 |
| | 0.2 | 0.967 | 0.875 | 0.882 | 0.892 | 0.907 | 0.851 | 1.007 |
| | 0.3 | 0.705 | 0.656 | 0.656 | 0.658 | 0.663 | 0.629 | 0.681 |
| | 0.4 | 0.576 | 0.535 | 0.53 | 0.529 | 0.528 | 0.503 | 0.515 |
| | 0.5 | 0.499 | 0.458 | 0.449 | 0.445 | 0.442 | 0.421 | 0.413 |
| | 0.6 | 0.449 | 0.404 | 0.392 | 0.387 | 0.382 | 0.363 | 0.346 |
| | 0.7 | 0.414 | 0.365 | 0.35 | 0.343 | 0.338 | 0.32 | 0.297 |
| | 0.8 | 0.387 | 0.334 | 0.317 | 0.31 | 0.304 | 0.286 | 0.26 |
| | 0.9 | 0.367 | 0.31 | 0.291 | 0.283 | 0.276 | 0.26 | 0.232 |
| | 1 | 0.351 | 0.291 | 0.27 | 0.262 | 0.254 | 0.238 | 0.209 |

$K_\tau$ converts instrument reading to stress in units of Pa; $K\gamma$ converts rpm to shear rate in units of 1/s.

This assumes the instrument is of the kind that gives an instrument reading over a range of 0–100. On such instruments, a viscosity I refer to as "instrument viscosity," $\mu_i$ (because it may not be the actual viscosity), is obtained by multiplying the instrument reading by the instrument spindle factor, IF, which depends on spindle number and spindle speed. In other words:

$$\mu_i = IR * IF \tag{12.10}$$

The values of IF are supplied with the viscometer. We will use this information later for a more abbreviated procedure.

Some viscometer models read $\mu_i$ directly, rather than using a 0–100 scale. For such viscometers, it is necessary to get back to the actual instrument reading to use the procedure we are describing, so we would need to look up the factors and use a re-arranged version of Eq. (12.10) to calculate IR:

$$IR = \mu_i / IF \tag{12.11}$$

If we take a series of readings at different spindle speeds, hopefully covering most of the 0–100 rpm range, we can apply Eq. (12.9) to get the shear stress versus rpm. But to get viscosity, we must divide shear stress by shear rate, so we need to convert rpm to shear rate. That is where the rest of Table 12.1 comes into play. The actual shear rate of a given spindle equals the spindle rpm multiplied by the shear rate factor, $K\gamma$. In other words,

$$dv/dx = \gamma = K_\gamma * N(\text{spindle rpm}); \quad dv/dx \text{ in } 1/s \tag{12.12}$$

As noted in the table, Kγ is a function of spindle number and the power law exponent, n, for a power law fluid. For example, if we had a power law fluid with a power law exponent of 0.4, and used the number 4 spindle, Kγ would have a value of 0.529. So, if it were turning at 20 rpm, the shear rate would be 20 * 0.529 or 10.58/second. If our instrument reading were 75, the shear stress would equal the shear stress factor, 0.539, times 75, or 40.425 Pa. To get apparent viscosity, $\mu_a$, we would divide shear stress by shear rate, or 40.425 Pa/10.58/second giving an apparent viscosity of 3.82 Pa-s or 3820 mPa-s, which also happens to equal 3820 cP.

But where do we find the value of n? The answer comes about by curve fitting. If the raw readings are based on a traditional 0–100 scale, a log–log plot (or the curve fitting equivalent thereof) of the reading versus rpm will have a slope of n. If the line is not quite straight, the apparent slope at each point can be calculated and the data treated point by point. For instruments that display instrument viscosity directly, a similar plot can be made but the slope of the plot is n − 1. Either way, the value of n can be calculated for use in this procedure. Then either the detailed procedure can be followed or the simplified procedure after the next example section.

## Example

Suppose we have the following array of readings taken with a Brookfield No. 6 spindle on an RVT viscometer:

| RPM | Reading |
|---|---|
| 5 | 20 |
| 10 | 24.6 |
| 20 | 30.4 |
| 50 | 40 |
| 100 | 49.1 |

Let us determine apparent viscosity versus shear rate. We can begin by plotting the data on a log–log plot, as shown in Figure 12.8. Though we can measure the slope manually, regression analysis revealed that the value of n was 0.3 (yes, I designed it that way!). Therefore, we can treat the data using Eqs. (12.9) and (12.12) to create Table 12.2.

The test fluid is quite shear thinning. A little math reveals that the apparent viscosity versus shear rate can be described by Eq. (12.13):

$$\mu_a = 33\,290\left(dv/dx\right)^{-0.7} \tag{12.13}$$

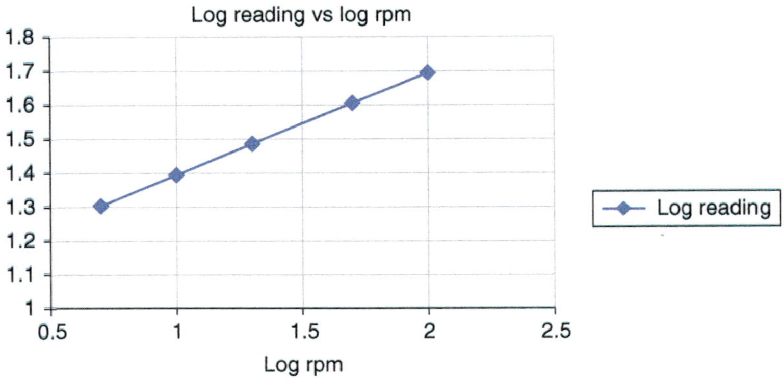


**Figure 12.8** Brookfield data plot.

## Simplified Method

Because the reason why the instrument viscosity is not the same as the apparent viscosity is due to the difference in actual shear rate versus the assumed Newtonian shear rate, we can directly correct the instrument readings and shear rates without calculating shear stresses, as long as we know the value of n. So, we can get the value of n either as the result of the raw readings or as n − 1 for units that read out instrument viscosity directly. Then we can simply multiply the instrument viscosity by the ratio of Newtonian shear rate (at n = 1), versus actual n based on factors in Table 12.1, and still apply Eq. (12.12) to get the actual shear rate. In other words:

$$\mu_a = \mu_i * K_\gamma(\text{at n} = 1)/K_\gamma(\text{at actual n}) \tag{12.14}$$

So, for our example problem, we would multiply the instrument viscosity by 0.238/0.629 = 0.378. When this method is applied to the same problem, we get Table 12.3, which has the values from above incorporated for comparison. There is less than a 2% deviation between the two methods. The instrument factor, IF, was found to be 10 000/N from the Brookfield unit manual.

The simplified method can be used for cylindrical spindles when the value of $K_\tau$ is not known, but the spindle factors or instrument viscosities are known, even for other viscometers besides the Brookfield RVT. One would just use the $K\gamma$ values for spindle 7, which is a cylindrical spindle.

If an error band of up to 10% maximum is acceptable, and in the conservative direction for shear thinning fluids, there is an even simpler method. The theoretical shear rate in a Newtonian fluid for an infinitely long cylindrical spindle operating in an infinite fluid may be calculated thusly:

$$dv/dx(1/s) = 4\pi N/60 \tag{12.15}$$

**Table 12.2** Example problem data treatment.

| RPM | Reading | Log rpm | Log reading | n | $K_\tau$ | Shear stress (Pa) | $K\gamma$ | Shear rate (1/s) | Apparent viscosity (Pa-s) | Apparent viscosity (cP) |
|---|---|---|---|---|---|---|---|---|---|---|
| 5 | 20 | 0.699 | 1.30103 | 0.3 | 2.35 | 47 | 0.629 | 3.145 | 14.94436 | 14944 |
| 10 | 24.6 | 1 | 1.39094 | 0.3 | 2.35 | 57.81 | 0.629 | 6.29 | 9.190779 | 9191 |
| 20 | 30.4 | 1.301 | 1.48287 | 0.3 | 2.35 | 71.44 | 0.629 | 12.58 | 5.678855 | 5679 |
| 50 | 40 | 1.699 | 1.60206 | 0.3 | 2.35 | 94 | 0.629 | 31.45 | 2.988871 | 2989 |
| 100 | 49.1 | 2 | 1.69108 | 0.3 | 2.35 | 115.385 | 0.629 | 62.9 | 1.83442 | 1834 |

**Table 12.3** Example problem alternative data treatment.

| RPM | Reading | Spindle factor * rpm | Instrument viscosity (cP) | n | Apparent viscosity (cP) | Kγ at n = 1 | Kγ | Shear rate (1/s) | Apparent viscosity cP (by long method) |
|---|---|---|---|---|---|---|---|---|---|
| 5 | 20 | 10 000 | 40 000 | 0.3 | 15 135 | 0.238 | 0.629 | 3.145 | 14 944 |
| 10 | 24.6 | 10 000 | 24 600 | 0.3 | 9 308 | 0.238 | 0.629 | 6.29 | 9 191 |
| 20 | 30.4 | 10 000 | 15 200 | 0.3 | 5 751 | 0.238 | 0.629 | 12.58 | 5 679 |
| 50 | 40 | 10 000 | 8 000 | 0.3 | 3 027 | 0.238 | 0.629 | 31.45 | 2 989 |
| 100 | 49.1 | 10 000 | 4 910 | 0.3 | 1 858 | 0.238 | 0.629 | 62.9 | 1 834 |

One may recognize that $4\pi/60$ is basically Kγ for $n = 1$, the Newtonian case, written as a decimal to three digits, $4\pi/60 = 0.209$. Note that this matches the value for spindle 7, which, although not an infinite cylinder, has an L/d of about 30 (I have not measured it) and is small compared to the usual 250–400 ml beakers used for testing, so it behaves a lot like an infinite cylinder in an infinite fluid. For such an infinite cylinder, the actual shear rate factor = Kγ (for $n = 1$)/n. Therefore, actual apparent viscosity is approximately instrument viscosity times n, at a shear rate of 0.209/n. This method was widely used before the Mitschka method was published. Comparing this method to the Kγ values for spindle 7 in Table 12.1, we see that the deviation is 0% for $n = 1$, 1.2% for $n = 0.5$, and 8% for $n = 0.1$. Though not as accurate as the Mitschka method, it errs slightly in the conservative direction. A lot of agitators have been designed using this method! Table 12.4 illustrates the method and compares shear rates to the longer method for $n = 0.3$ and using the data set from before but applying it to the spindle factors for spindle number 7.

## Calculating Apparent Viscosity for Agitator Impellers

Once we have a viscosity versus shear rate relationship for a fluid, we must apply that in agitator design. The most common way to do that is to use the Metzner–Otto relationship, which says that the average shear rate for an impeller is proportional to the shaft speed, with the proportionality constant, $K_M$, being called the Metzner–Otto constant:

$$dv/dx = K_M * N \tag{12.16}$$

This concept was also discussed in Chapter 3, and a table of constants was presented as Table 3.1 and is repeated as Table 12.5.

**Table 12.4** Example problem spindle 7 simplified data treatment.

| RPM | Reading | Spindle factor * rpm | Instrument viscosity (cP) | n | Apparent viscosity (cP) | $K\gamma$ at n = 1 | $K\gamma$ | Shear rate (1/s) | Shear rate by quick method (1/s) | % shear rate error |
|---|---|---|---|---|---|---|---|---|---|---|
| 5 | 20 | 40 000 | 160 000 | 0.3 | 48 960 | 0.209 | 0.683 | 3.415 | 3.48 | 2.0 |
| 10 | 24.6 | 40 000 | 98 400 | 0.3 | 30 111 | 0.209 | 0.683 | 6.83 | 6.97 | 2.0 |
| 20 | 30.4 | 40 000 | 60 800 | 0.3 | 18 605 | 0.209 | 0.683 | 13.66 | 13.93 | 2.0 |
| 50 | 40 | 40 000 | 32 000 | 0.3 | 9 792 | 0.209 | 0.683 | 34.15 | 34.83 | 2.0 |
| 100 | 49.1 | 40 000 | 19 640 | 0.3 | 6 010 | 0.209 | 0.683 | 68.3 | 69.67 | 2.0 |

**Table 12.5** Metzner–Otto constants.

| Impeller type | $K_M$ |
|---|---|
| Propeller | 10 |
| Low-solidity hydrofoil | 10 |
| High-solidity hydrofoil | 11 |
| Pitched blade turbine | 11 |
| Radial turbines | 11.5 |
| Chaos (reversing pitch) turbines | 40 |

The values of these constants came from several sources, including Ref. [3]. Reference [4] lists a value of $K_M$ of 3.4 for hydrofoil impellers. Several other references list it as around 10, so the value of 3.4 seems to be an outlier. It would predict a substantially higher apparent viscosity for a shear thinning fluid. The author on one occasion tested a high-solids lignin slurry using a bob-and-cup viscometer and then also tested the same slurry using an agitator as a viscometer, as described in the next section. The impeller used was a hydrofoil. The results matched up reasonably well using a $K_M$ value of 10, so this author recommends the values in Table 12.5. However, the uncertainly about what value of $K_M$ to use goes away when using an agitator as a viscometer, as we will see in the next section.

For now, let us illustrate how to use the Metzner–Otto concept with an example. Suppose we have a fluid that obeys the power law model, and the viscosity, in cP, can be expressed by $\mu_a = 50\,000(dv/dx)^{-0.55}$. Suppose we wish to operate a hydrofoil impeller at a shaft speed of 56 rpm in this fluid. What will be the apparent viscosity we must design for? Using a $K_M$ value of 10, we get $dv/dx = 10 * 56\,\text{rpm}/60 = 9.333/\text{second}$. Therefore, our apparent viscosity is $50\,000(9.333)^{-0.55} = 14\,637\,\text{cP}$.

## How to Measure Viscosity for Multiphase Macro Fluids Using an Agitator as a Viscometer

Many fluids of interest to the biofuel and renewable chemicals industries are complex macro fluids, with large particles (>1 mm; some >10 mm), a fibrous nature, particle-to-particle interactions, and, sometimes, a tendency to settle out during testing. For these and perhaps other reasons, conventional viscometers do not work well with them. But this author has successfully characterized rheology sufficiently for agitator design purposes on a wide variety of fibrous materials, both lignocellulosic and others, including such examples as wood pulp, cotton linters, beet pulp, switchgrass, various forms of cane, corn fiber, and other materials, with a variety of thermal, mechanical, and chemical pretreatments, with solids concentrations ranging from 4 to 30%. The technique is to use a suitably instrumented agitator as its own viscometer. We will go into considerable detail on how to do this.

As mentioned in Chapter 3, the power draw of any agitator impeller can be correlated as a function of Reynolds number, D/T, and other geometric factors. The power number becomes constant in turbulent flow. But in laminar flow, the power number is inversely proportional to Reynolds number, and geometric factors drop out. This is described by Eq. (12.17):

$$N_P = K_L / N_{Re} \text{ or } N_P * N_{Re} = K_L \tag{12.17}$$

In between laminar and turbulent flow, i.e. in the transition range, the power number changes in a nonlinear way that varies by impeller type, as illustrated in Figure 12.9. This figure shows that for some impellers, the power number can actually go through a minimum. Knowing the curve for a specific impeller gives us a way to use an agitator as a viscometer. This procedure was described generally in Ref. [5], and will be expanded upon here. We will go through a step-by-step procedure after listing and commenting upon some of the prerequisites for a successful viscosity study using this technique.

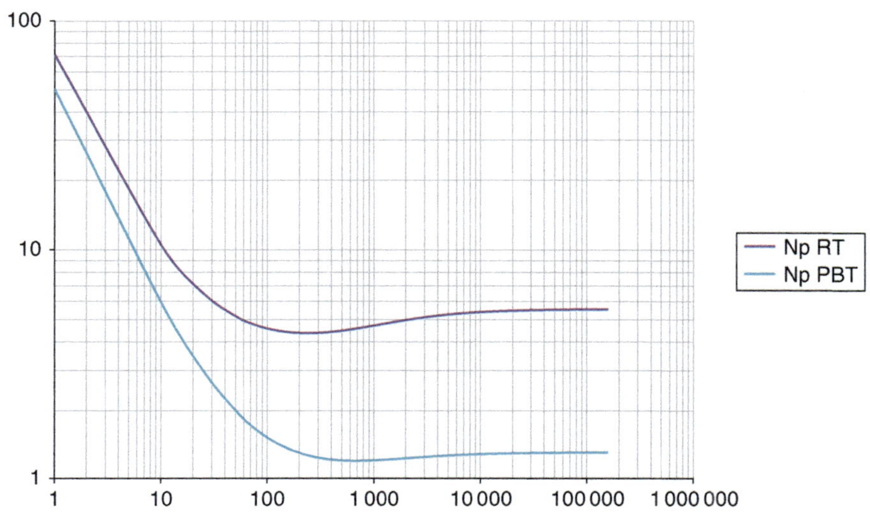


**Figure 12.9** Power number versus Reynolds number for Rushton turbine (RT) and pitched blade turbine (PBT).

### Based on Impellers with Known Characteristics

Though this might seem obvious, a little-known consideration is that the power draw characteristics of "similar" hydrofoil impellers may be different, and that lab-scale impellers may have different power draw characteristics from their full-scale counterparts. We will describe why after looking at some pictures.

Figure 12.10 is a rendition of a production-scale HE-3 hydrofoil impeller by Chemineer, a brand of NOV. It has a turbulent power number of 0.26–0.33 depending on D/T ratio. Figure 12.11 is a photograph of an 8″ Fusion Fluid hydrofoil taken by the author. Aside from the obvious difference of the fins, there are two other differences that could affect power draw. One is that the leading edge of the

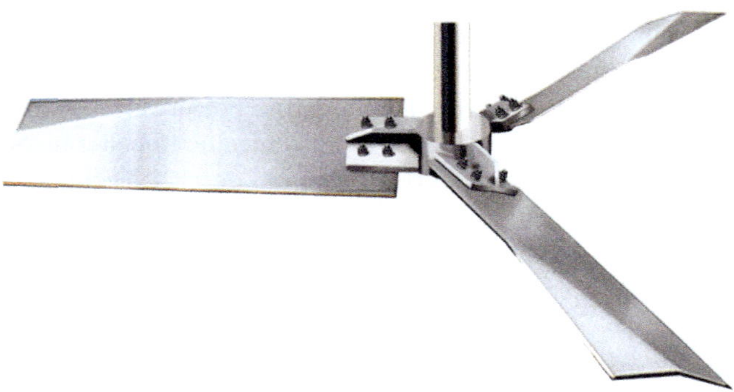

**Figure 12.10** HE-3 impeller. *Source:* Courtesy, NOV.

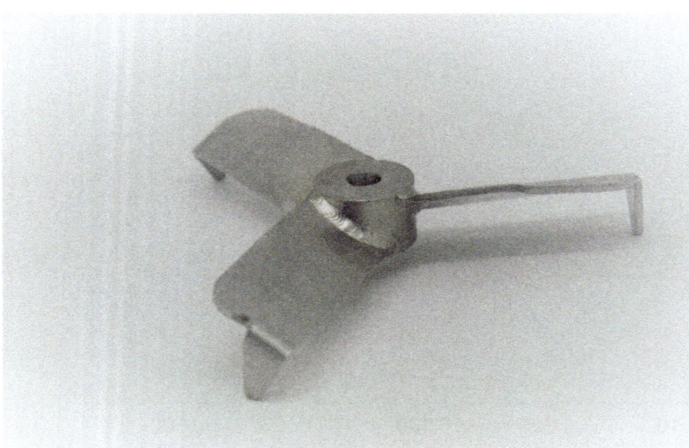

**Figure 12.11** Fusion Fluid 8″ lab hydrofoil.

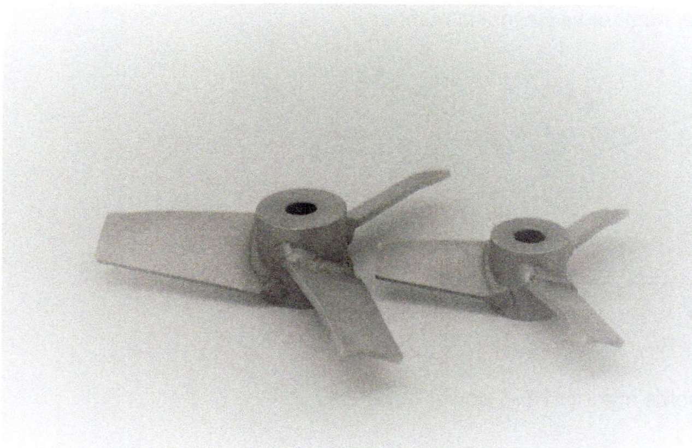

**Figure 12.12** Indco 7″ and 5″ lab hydrofoils.

blade is blunt, rather than profiled. This could affect the power draw, likely increasing the power number, though possibly less so in laminar flow than in turbulent flow. The other is that the thickness/diameter ratio is much higher in the lab impeller than in most production impellers, which will definitely increase the power number. That ratio is about 0.015 in the lab impeller, whereas it is usually less than 0.01 in production impellers and may often be much less than that. The author has not measured the power draw characteristics of this particular impeller.

Figure 12.12 shows two hydrofoil impellers manufactured by Indco, photographed by the author. They are 7″ and 5″ in diameter. Both have the same blade thickness, so the thickness/diameter ratio is higher on the smaller impeller. That thickness also seems to reduce the crispness of the bend line. The actual measured turbulent power numbers are 0.36–0.39 for the 5″ impeller and 0.27–0.32 for the 7″ impeller, depending on D/T. This is over a 20% difference in power number for the "same" kind of impeller. So, ideally, one would calibrate the impeller to be used in the tests before using it as a viscometer. A procedure for doing so will be included in an Appendix "Calibrating an Impeller."

## Simple If Reynolds Number <10

In Ref. [6], Mr. Zheng Ma studied the power number versus Reynolds number characteristics of several impellers, covering the full range of Reynolds numbers from creeping laminar ($N_{Re} < 1$) to fully turbulent ($N_{Re} > 100000$). These results were presented graphically at the Mixing conference in Ref. [7]. He found that hydrofoil impellers, such as the Chemineer HE-3, exhibited fully laminar characteristics, i.e.

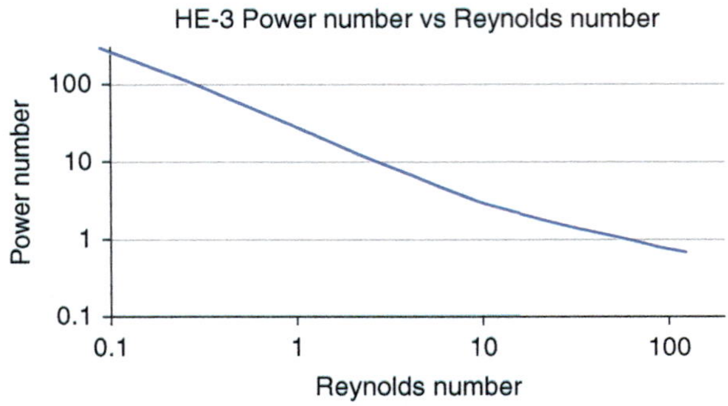


**Figure 12.13** Low Reynolds number HE-3 power numbers.

they obey Eq. (12.17), up to a Reynolds number of 10. Some other impellers started to deviate around a Reynolds number of 8. So, for hydrofoil impellers, the procedure is simple: measure the power with a given impeller and speed, calculate the power number, use Eq. (12.17) to calculate the Reynolds number, and, if it is 10 or less, use the definition of Reynolds number to calculate the apparent viscosity. Figure 12.13 plots the Ma data for a Chemineer HE-3 impeller from Ref. [6] at Reynolds numbers less than 100. For laminar conditions, $K_L$ was found to be about 27. Presumably, other three-blade hydrofoil impellers with a turbulent power number of about 0.3 will have a $K_L$ value that is quite similar.

## More Complicated and Less Accurate If Reynolds Number >10

Referring again to Figure 12.13, note that the curve deviates from linearity above Reynolds number of 10, though the error will not be huge up to a Reynolds number of 40. Between a Reynolds number of 10 and 100, a simple equation cannot be used to relate power number and Reynolds number. The user must manually relate the two using the empirically derived data set. Piecewise curves could possibly be fitted to computerize the calculations.

## Generally Not Usable If Reynolds Number >100

As we previously saw in Figure 12.9, above a Reynolds number of 100, the curve starts to flatten out too much to be usable. Moreover, for some impeller types, there is a dip in the curve, making the actual Reynolds number indeterminate for some power number values. For example, if a Rushton turbine were being studied and a power number of 5 were measured, we could not be sure if the Reynolds number were 50 or 1600. Likewise, if using a standard pitched blade turbine, a measured power number of 1.22 could mean a Reynolds number of 300 or 1800. Moreover, small errors in power measurement could lead to disproportionately large errors in Reynolds number calculations, and thus viscosity, for Reynolds numbers above 100. Thus, we recommend avoiding this technique for Reynolds numbers above 100, and strongly prefer to keep them below 40 for accuracy.

## Equipment Needed

To use an agitator as a viscometer, there are several prerequisites. Some have been described in Ref. [8]. We have a full list in the following text.

- A variable speed agitator with the capability to run at very low speeds so it can match the shaft speed of probable production equipment. This could mean speeds of 10 rpm or even lower in some cases. Most lab and small pilot agitators cannot run at such low speeds, as they are typically either DC-motor drives (Figure 12.14) or VFD-controlled AC drives. An outboard reducer (typically 4 : 1 ratio) can help, but the user must remember to log the recorded speed by dividing the speed on the instrument by the reduction ratio, and multiplying the measured torque by the reduction ratio. It has been the author's experience that lab units with electronic speed and torque measurement give reasonably accurate measurements down to about 10% of top speed (within a given speed range; some have dual speed ranges) and down to about 20% of maximum rated

**Figure 12.14** Electronic lab unit.

**Figure 12.15** Metallic traction lab unit.

torque. Below either of these conditions, errors rapidly increase, and can be 30% off in some cases. Metallic traction drives (Figure 12.15) are rare and expensive, and they do not generally have speed indication or torque indication, but they can operate all the way down to 0 rpm. That is why the author frequently uses them in his lab.

- The lab agitator should have a minimum torque capability of $0.12D^3$, where torque is in N-cm and D is in cm. For example, a 7″ impeller, or 17.78 cm diameter, would need a torque rating of at least 674 N-cm or about 60 in.-lb. This is based on the worst biomass slurries the author has seen. Some will not require as much torque.
- A means of measuring shaft speed. If a tachometer is not built in, a laser tachometer is recommended. In some cases, for low rpm, it may be necessary to count revolutions and time them with a stopwatch.
- A means of measuring torque. For low speed and loads, or for a metallic traction drive, a separate torque measuring device is needed. One way is to measure reaction torque on the vessel. This can be done with a low-friction turntable and a linear force measuring device, or by using a torque gage as shown in Figure 12.16.
- A vessel, which may be baffled or unbaffled as long as the Reynolds number is below 40, and preferably, with a D/T ratio of 0.5–0.75 so as to be similar to likely production equipment.

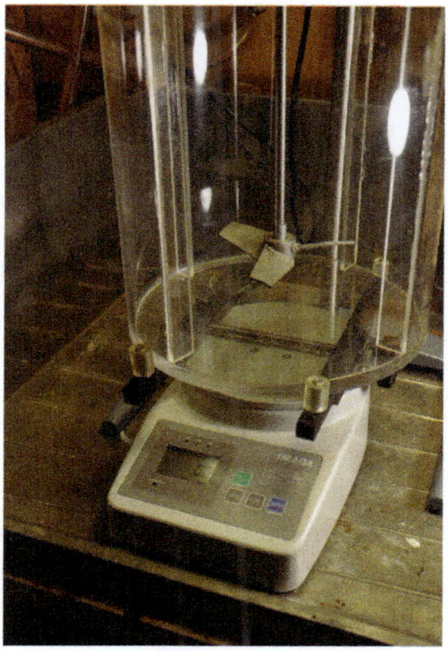

**Figure 12.16** Torque gage.

- An impeller, preferably hydrofoil, with known power number versus Reynolds number characteristics. The impeller chosen can be calibrated using the procedure in the Appendix "Calibrating an Impeller."

## Step-by-Step Procedure

1) Install the test impeller on the shaft of the lab agitator and position it so it will be fully covered with liquid.
2) If the torque sensor is external to the drive rather than a built-in electronic voltage/current algorithm, or if it is a strain-gage unit mounted on either the input shaft or the output shaft of the lab agitator, make sure it reads zero before any liquid is placed in the tank.
3) Pour the test liquid into the tank, covering the impeller by at least 1 impeller diameter. Bump the agitator a few times to dislodge any air pockets trapped under the impeller. If the slurry is made up by adding dry solids to water or other liquid, be sure it is well mixed and fully hydrated before running the viscosity test. Hydration may take 30–90 minutes. If in doubt, run the agitator while hydrating, measuring torque and waiting until the torque reading stabilizes.
4) If the slurry has any tendency to settle (such as corn fiber does), run the agitator at a sufficient speed to make the slurry uniform, then rapidly slow it down to the test speed for getting a power or torque reading. Increase the speed in

between readings each time to be sure it is well mixed before each viscosity data test.

5) Test at a series of speeds, going from at least as low as 10 rpm up to the maximum anticipated speed of full-scale equipment. I usually try to cover a range of about 5–100 rpm if possible. Sometimes I am able to go higher than 100 rpm.
6) For each test speed, record the speed and the torque. Some electronic meters read directly in power units.
7) For each test speed, calculate the power draw. This is most easily done in SI units: P (watts) = $\tau$ (N-m) * $2\pi N$ (rev/s). Most torque meters read in either N-cm or in.-lb. Convert to N-m by using 1 N-cm = 0.01 N-m and 1 in.-lb = 0.113 N-m.
8) Calculate the power number using the definition of power number: $N_P = P/\rho N^3 D^5$. In SI units, $\rho$ will be in kg/m$^3$, N will be in revolutions/second or 1/s, and D will be in m.
9) Do a first estimate of Reynolds number by dividing $K_L$ by $N_P$: $N_{Re} = K_L/N_P$. If this Reynolds number is in the laminar range (10 or less for hydrofoils; 8 or less for most other impellers), then this is accepted as the Reynolds number. If not, it will be necessary to go into a table or curve of power number versus Reynolds number to look up the actual Reynolds number. If the Reynolds number is greater than 100, the value cannot be trusted and the data point should be ignored.
10) With a known Reynolds number the apparent viscosity can be calculated by re-arranging the definition of Reynolds number: $\mu_a = D^2 N\rho/N_{Re}$. The viscosity in SI units will be in Pa-s. If desired, the user may convert to mPa-s or cP by multiplying by 1000.
11) Now for each valid data point ($N_{Re} \leq 100$) we have an apparent viscosity. We can plot this directly versus agitator speed or curve fit it to one of many possible viscosity models. It is recommended by this author to leave it as a function of agitator speed rather than shear rate, as that avoids the problem of uncertainty in the Metzner–Otto constant. However, some agitator design software requires the input to be a function of shear rate, so users of such software will need to convert any correlation into a function of shear rate. The shear rate may thus be estimated by using $dv/dx = K_M * N$.

## Dangerous to Apply to Other Kinds of Equipment

One might be tempted to use the resulting apparent viscosity relationship for other kinds of equipment, such as pipe flow, pump design, filtration, and other apllications. The author is aware of some who have done this successfully. However, it may be dangerous to do so for two main reasons:

1) Uncertainty in the Metzner–Otto constant.
2) Shear rates studied for agitator design may be much lower than for other kinds of equipment, resulting in considerable extrapolation.

The author recommends studying viscosity for other equipment by using such equipment as viscometers. For example, pipe flow and pressure drop could potentially be used as a way to measure viscosity for piping design.

## Examples of Data and Fit for Actual Biomass Slurries

In this section we will demonstrate data treatment and rheology model fit for three different actual biomass samples: pretreated sugarcane bagasse (abbreviated bagasse in the table), Municipal Solid Waste (abbreviated MSW), and root fiber pulp (abbreviated pulp). The percent solids and method of pretreatment will not be disclosed due to confidentiality. All runs used a 7″ (0.178 m) impeller with a $K_L$ value of 27. The raw data and the data treatment using the equations from this chapter are presented in Table 12.6.

The bagasse data were taken at an agitator manufacturer's mixing laboratory, using an agitator equipped with a strain gage on the agitator shaft. The MSW data were taken at a client job site, using a lab mixer with electronic speed and torque indication, which can become less accurate at low torque and rpm. The pulp data were taken in the Benz Technology International, Inc. facility, using a torque cell on the tank and a laser tachometer.

All samples visually had a yield stress. That is, they would retain their shape to some degree when disturbed, and would not settle down to a flat surface. They also had a tendency to form dead spots and "caverns," which will be discussed and visualized in Chapter 13. Therefore, we would expect to find the best data fits to be either Bingham Plastic or Herschel–Bulkley.

Examining the data from Table 12.6, we note that the bagasse data had almost constant torque, independent of shaft speed, other than measurement error that might be due to "clumpiness" of solids, resulting in some change in rheology as the clumps were redistributed by the rotating impeller. This behavior indicates a high yield stress and low viscosity above yield stress. In other words, the yield stress dominates the rheology at low shear rates.

The MSW had increasing torque with speed, but not proportional to speed, as would be the case with a Newtonian fluid.

The root pulp (Pulp) behaved similarly to the MSW, but was much thicker.

Data from Table 12.6 were fitted to power law, Bingham Plastic, and Herschel–Bulkley models using DataFit software by Oakdale Engineering, and the results are summarized in Table 12.7. As is often the case, we got pretty good fits from all three models. However, sometimes the model fits do not make sense from a mechanistic viewpoint.

**Table 12.6** Biomass sample rheology treatment.

| Sample ID | Density ($kg/m^3$) | Impeller diameter (m) | Shaft speed (rpm) | Torque (N-cm) | Power (W) | Power number | $K_L$ | Reynolds number | Apparent viscosity (cP) |
|---|---|---|---|---|---|---|---|---|---|
| Bagasse | 1010 | 0.178 | 9.3 | 235 | 2.2887 | 3405.4 | 27 | 0.00793 | 625 594 |
| Bagasse | 1010 | 0.178 | 19.3 | 222 | 4.4868 | 747.0 | 27 | 0.03615 | 284 776 |
| Bagasse | 1010 | 0.178 | 23 | 217 | 5.2266 | 514.1 | 27 | 0.05252 | 233 582 |
| Bagasse | 1010 | 0.178 | 31 | 223 | 7.2393 | 290.8 | 27 | 0.09284 | 178 095 |
| Bagasse | 1010 | 0.178 | 36 | 229 | 8.6331 | 221.5 | 27 | 0.12192 | 157 485 |
| Bagasse | 1010 | 0.178 | 42 | 231 | 10.16 | 164.1 | 27 | 0.16451 | 136 166 |
| Bagasse | 1010 | 0.178 | 49 | 236 | 12.11 | 123.2 | 27 | 0.21917 | 119 240 |
| Bagasse | 1010 | 0.178 | 61 | 225 | 14.373 | 75.8 | 27 | 0.35627 | 91 319 |
| Bagasse | 1010 | 0.178 | 73.5 | 230 | 17.703 | 53.4 | 27 | 0.506 | 77 473 |
| MSW | 1100 | 0.178 | 10 | 24.4 | 0.2555 | 280.8 | 27 | 0.09616 | 60 408 |
| MSW | 1100 | 0.178 | 20 | 26.2 | 0.5487 | 75.4 | 27 | 0.35821 | 32 432 |
| MSW | 1100 | 0.178 | 30 | 38.4 | 1.2064 | 49.1 | 27 | 0.5499 | 31 690 |
| MSW | 1100 | 0.178 | 40 | 42.7 | 1.7886 | 30.7 | 27 | 0.87916 | 26 429 |
| MSW | 1100 | 0.178 | 50 | 44 | 2.3038 | 20.3 | 27 | 1.3331 | 21 787 |
| Pulp | 1050 | 0.178 | 9 | 118 | 1.1121 | 1756.3 | 27 | 0.01537 | 324 599 |
| Pulp | 1050 | 0.178 | 12 | 124 | 1.5582 | 1038.1 | 27 | 0.02601 | 255 828 |
| Pulp | 1050 | 0.178 | 29.6 | 138 | 4.2776 | 189.9 | 27 | 0.14219 | 115 424 |
| Pulp | 1050 | 0.178 | 64 | 151 | 10.12 | 44.4 | 27 | 0.60751 | 58 412 |
| Pulp | 1050 | 0.178 | 132 | 172 | 23.776 | 11.9 | 27 | 2.26877 | 32 260 |

**Table 12.7** Biomass sample data fit results.

| | Power law constants | | | Bingham constants | | | Herschel–Bulkley constants | | | |
|---|---|---|---|---|---|---|---|---|---|---|
| **Sample ID** | **A** | **B** | **$R^2$** | **A** | **B** | **$R^2$** | **A** | **B** | **C** | **$R^2$** |
| Bagasse | 6.22E+06 | −1.033 | 0.999 | 5.80E+06 | −5723 | 0.998 | 5.68E+07 | −5.06E+07 | −0.996 | 0.998 |
| MSW | 2.50E+05 | −0.629 | 0.95 | 4.58E+05 | 13647 | 0.969 | 5.20E+05 | 3619 | 0.33 | 0.97 |
| Pulp | 2.19E+06 | −0.868 | 0.9999 | 2.83E+06 | 15000 | 0.999 | −1.00E+07 | 1.20E+07 | −0.967 | 0.9999 |

Power law model: $\mu_a = A * N^B$, where $B = n - 1$.
Bingham plastic model: $\mu_a = A/N + B$.
Herschel–Bulkley model: $\mu_a = A/N + B * N^C$, where $C = n - 1$.
$R^2$ = Correlation coefficient.

For bagasse, the power law fit was very good, but with a negative 1.03 exponent on shaft speed; it says viscosity is inversely proportional to shaft speed, which actually makes sense for a dominant yield stress characteristic. The Bingham model makes the most sense, though the negative viscosity component (B) is likely just a data error artifact. The Herschel–Bulkley fit involves subtracting two large numbers, and the value of B does not make mechanistic sense.

MSW results may be plagued a bit by instrument error, but are still usable. The power law fit is reasonable. The Bingham fit is reasonable and makes sense mechanistically, as there is a yield stress and a 13 647 cP viscosity slope above yield stress. The Herschel–Bulkley model has the best fit, but the "C" value of 0.33 indicates shear thickening behavior above the yield stress, which is quite improbable.

The pulp fits are very good for all three models. However, the Herschel–Bulkley "A" value implies a negative yield stress. That would imply that the fluid has stored energy that it can deliver to drive the impeller, rather than the impeller driving the fluid! So, though the data fit is good, the results are impossible.

So which model should one use if multiple models fit well? This author would use the one that gave the highest predicted apparent viscosity when designing production equipment, but would ignore models that are physically impossible. Though not recommended, these data fits can be converted to viscosity versus shear rate if $K_M$ is known for the test impeller, by substituting $(dv/dx)/K_M$ for N in the viscosity fit equations.

These data fit models are presented graphically in Figures 12.17–12.25

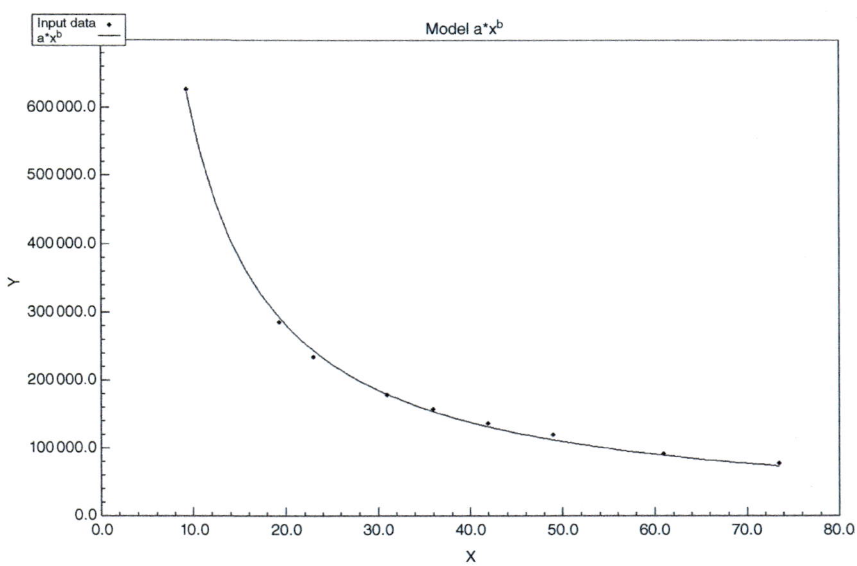


**Figure 12.17** Bagasse power law data fit.

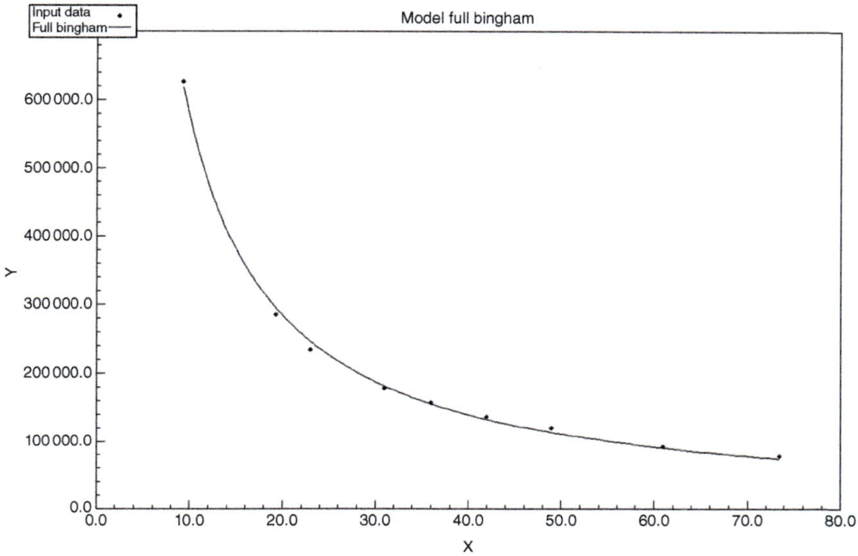


**Figure 12.18** Bagasse Bingham data fit.

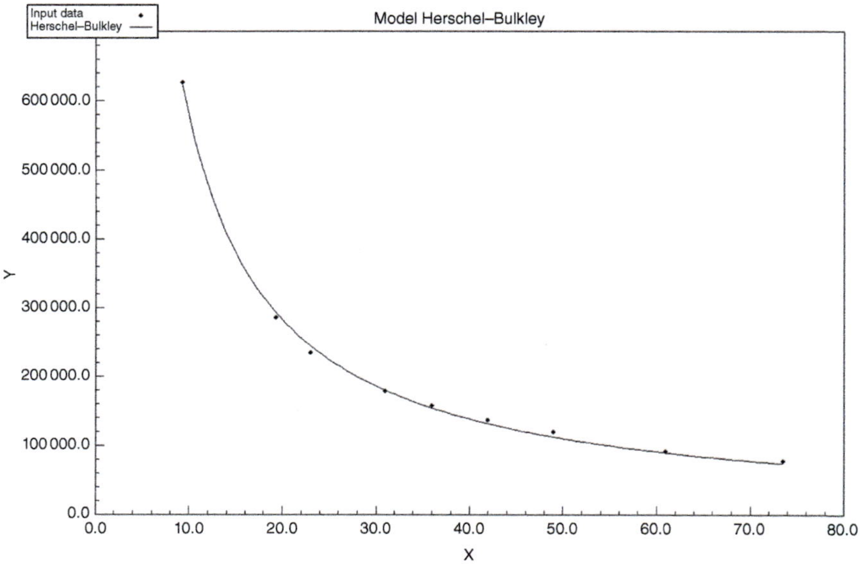


**Figure 12.19** Bagasse Herschel–Bulkley data fit.

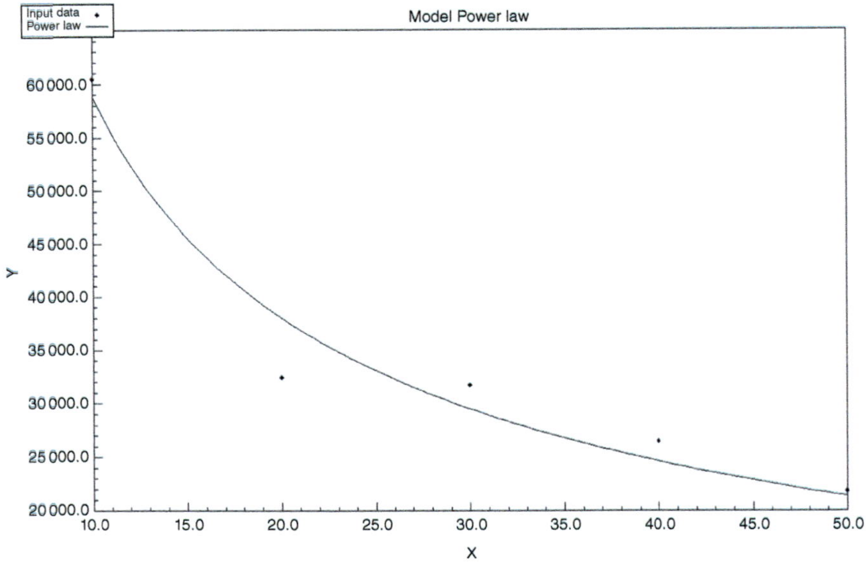


**Figure 12.20** MSW power law data fit.

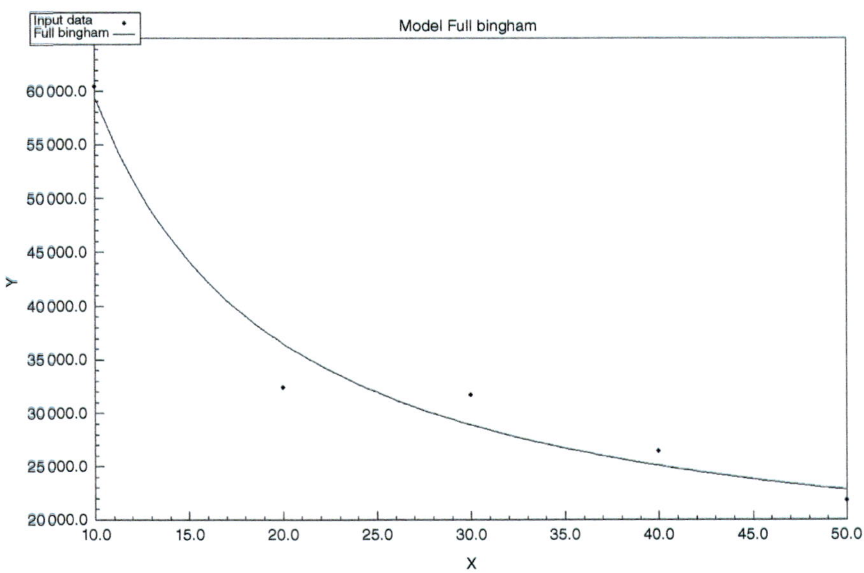


**Figure 12.21** MSW Bingham data fit.

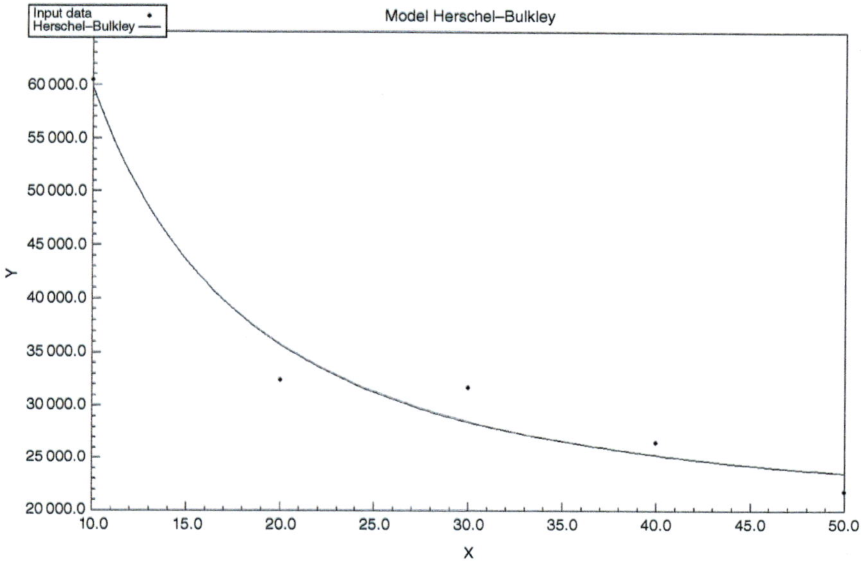


**Figure 12.22** MSW Herschel–Bulkley data fit.

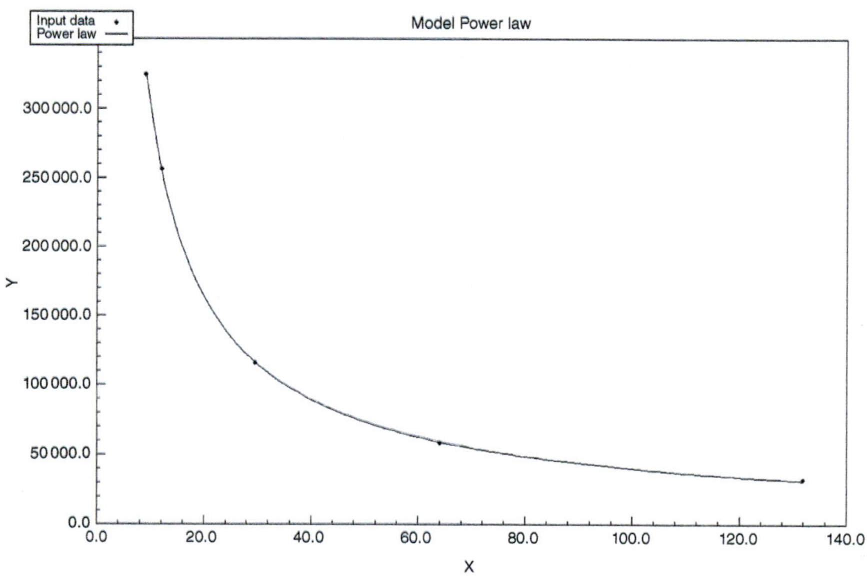


**Figure 12.23** Pulp power law data fit.

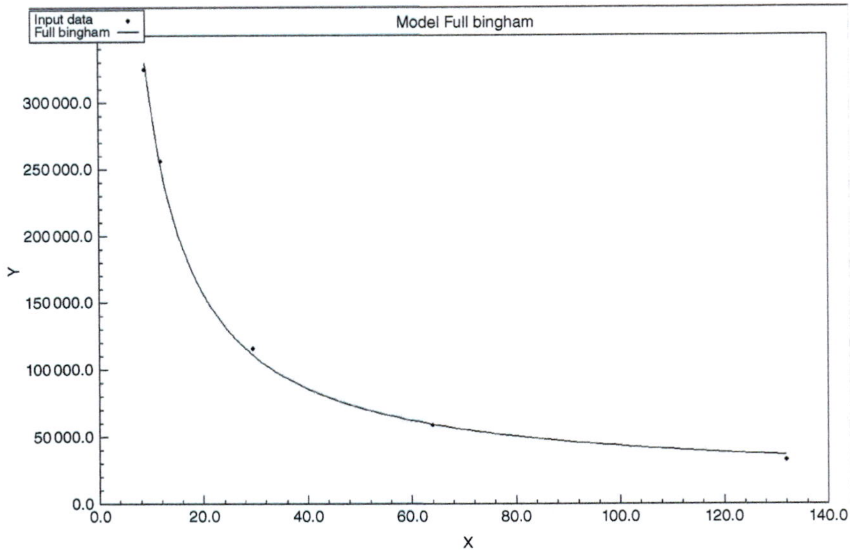


**Figure 12.24** Pulp Bingham data fit.

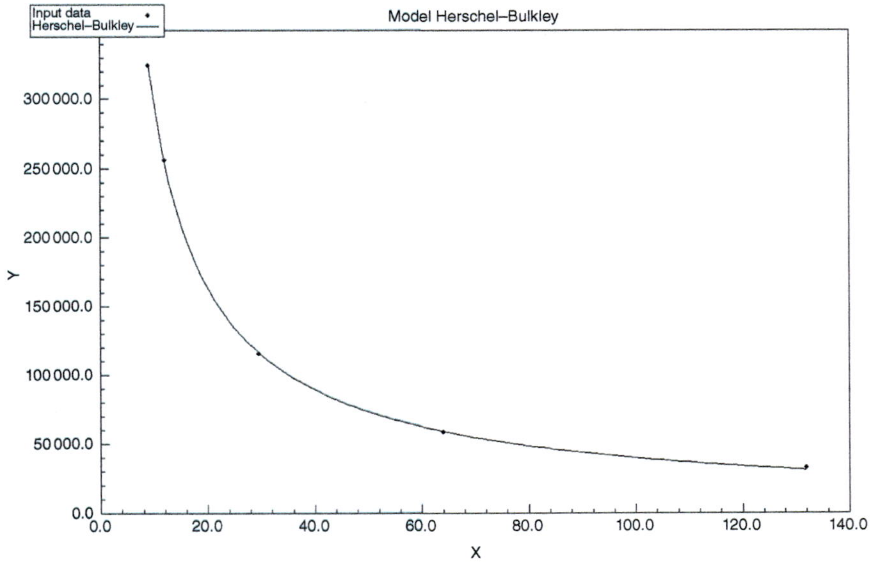


**Figure 12.25** Pulp Herschel–Bulkley data fit.

## Summary of Chapter

In this chapter we described some common rheology models used to characterize fluids commonly encountered in the biofuels/renewable chemicals industries. We described several kinds of viscometers and how to use them to characterize viscosity for non-Newtonian fluids. Of particular importance was the section on using an agitator as a viscometer, especially for fluids not amenable to testing with spindle-type viscometers. These fluids include fibrous materials, such as lignocellulosic feedstocks.

## List of Symbols

| | |
|---|---|
| D | Impeller diameter |
| dv/dx | Shear rate |
| IR | Brookfield instrument reading |
| IF | Instrument viscosity factor |
| $K_L$ | Laminar power number constant |
| $K_M$ | Metzner–Otto constant |
| $K_\tau$ | Brookfield RVT spindle torque/shear stress factor |
| $K\gamma$ | Brookfield RVT spindle shear rate factor |
| M | Power law model coefficient: viscosity at 1/s if seconds are used as the basis for the equation |
| N | Shaft speed |
| n | Power law exponent |
| $N_P$ | Power number |
| $N_{Re}$ | Reynolds number |
| P | Power draw |
| T | Tank diameter |

## Greek Letters

| | |
|---|---|
| $\rho$ | Density |
| $\sigma_S$ | Shear stress |
| $\mu$ | Viscosity |
| $\mu_a$ | Apparent viscosity |
| $\mu_i$ | Instrument viscosity |
| $\gamma$ | Velocity gradient or shear rate; same as dv/dx |
| $\tau$ | Torque |

## Calibrating an Impeller

Since laboratory impellers may not always be an exact match in all geometric details to their full-scale counterparts, it is best to calibrate the actual test impeller to be used when pressing a lab agitator into service as a viscometer. The following items are needed for the test.

- The test impeller: Normally this will be of 0.12–0.2 m diameter so that the geometry does not deviate severely from full-scale impellers.
- A test vessel: Benz Technology usually uses a vessel of about 0.3 m diameter. D/T should be 0.4–0.8. For laminar flow, baffles are optional. Baffles are recommended if any data are to be taken at Reynolds numbers of 50–100. Test results for Reynolds numbers over 100 are optional, but will not normally be used for viscosity testing.
- A Newtonian fluid or several Newtonian fluids of known viscosity: This could be a lab standard test fluid from Brookfield or others, but such fluids are expensive and the temperature must be tightly controlled to assure the standard will actually exhibit the reference viscosity. Other fluids that can be used include glycerin, silicone oils, and various grades of corn syrup. Be careful of the latter; it may form a skin on the liquid surface. Mix well before testing. These fluids should be calibrated on an accurate viscometer. A spindle type will be sufficient. The temperature in the viscometer must be the same as the temperature in the test vessel.
- The density of the test fluid must also be known or measured.
- A variable speed agitator drive: If it has a built-in tachometer, it may be useful to check it against a laser tachometer.
- For best speed accuracy, a laser tach should be used, at least for calibration of the agitator built-in tach.
- A means to measure reaction torque: Generally, built-in electronic algorithms are not accurate enough to calibrate the impeller. Shaft strain gages, vessel reaction torque gages, or a low-friction turntable with a force gage and a moment arm should be used for best accuracy.

The calibration procedure is as follows:

1) Place the test impeller in the vessel about 0.5–1.0 D off bottom. (Clearance has little effect on power in laminar or very low-transition flow.)
2) Pour the test fluid of known viscosity over the impeller until it is submerged by at least 0.5 impeller diameters, preferably 1 impeller diameter.
3) "Bump" the impeller by starting and stopping it several times to dislodge any air pockets that may have become trapped under it while filling. Typically this is done at a shaft speed of 50–100 rpm on lab-scale equipment.

4) Make sure the torque measuring device reads zero when the agitator is at rest.
5) Start the agitator at a low speed. Making sure the resultant torque is high enough to give good precision within the torque-meter's range. If it is not, increase the shaft speed until the torque is at least 10 times the precision unit of the measuring device. Record the speed and the torque.
6) If the previous steps result in a laminar Reynolds number, and if all product viscosity will be tested in the laminar range only, you are done. Just calculate $K_L$ for the impeller and record it, and do not extrapolate beyond laminar flow.
7) If the Reynolds number at the lowest acceptable torque is not laminar, but is below 100, take a few more data points to get to a Reynolds number of 100, or more if you desire a more complete curve. Then switch to a higher-viscosity test fluid. It is necessary to get some data in the laminar range.
8) With the higher-viscosity test fluid, repeat steps 5 and 6. Repeat until you get some useful data in the laminar range.
9) For each data point, calculate Reynolds number as follows: $N_{Re} = D^2N\rho/\mu$. This calculation works best in SI units: D is in m, N is in rev/s, $\rho$ is in $kg/m^3$, and $\mu$ is in Pa-s (AKA kg/m-s).
10) For each data point, calculate power draw from torque and speed: $P = 2\pi\tau N$. This, too, works best in SI units: P is in W, $\tau$ is in N-m (Newton-meters), and N is in 1/s. Convert the torque-meter output to N-m.
11) For each data point, calculate power number: $N_P = P/\rho N^3D^5$. Obviously this also works best in SI units.
12) Now you can plot $N_P$ versus $N_{Re}$. In the laminar range, $N_P * N_{Re} = K_L$.
13) Now the impeller is calibrated for use as a viscometer.

## References


**1** Mitschka, P. (1982). Simple conversion of Brookfield R.V.T. readings into viscosity functions. *Rheologica Acta* 21 (2): 207–209.

**2** Wein, O., Mitschka, P., and Wichterle, K. (1981). *Rotational Flows of Non-Newtonian Liquids*, Published (in Czech only) as Studie No. 5 by Academia. Prague: Publishing House of the Czechoslovak Academy of Sciences.

**3** Bakker, A. and Gates, L. (1995). Properly choose mechanical agitators for viscous liquids. *Chemical Engineering Progress* 91: 25–34.

**4** Post, T. (2005). Overcome the fear of shear. *Pharmaceutical Manufacturing* 4 (8): 26–37.

**5** Benz, G. (2010). Agitating fibrous materials. *Chemical Engineering Progress* 106: 28–32.

6 Ma, Z. (2014). Impeller power draw across the full Reynolds number spectrum. Master's thesis, University of Dayton.

7 Ma, Z., Myers, K., and Janz, E. (2014). Impeller power draw across the full Reynolds number spectrum. Poster Presentation, Mixing XXIV, Lake George, New York (22–27 June 2014).

8 Benz, G. (2018). Equipment needs to study mixing of fibrous materials. *Biofuels Digest* (16 April).

# 13

# Cellulosic Materials and Hydrolysis Reactor Agitator Design

## Introduction

This chapter will delve into the details of successful agitator design for cellulosic and other fibrous materials, especially for hydrolysis reactors. The reaction schemes have been broadly described in Chapter 8. Here we examine how rheology, tank geometry, and scale-up techniques are used for successful design. We emphasize discovery of the minimum threshold of agitation that will provide complete motion, as dead spots are apt to have a rather slow reaction rate; reagents penetrate by diffusion rather than by convection in those dead spots. For batch reactors, minimal but complete motion will suffice, as the biomass slurry will rapidly thin out as enzymes break down the cellular structure. For fed batch or continuous flow schemes, it is helpful to have a blend time less than or equal to the retention time. However, there are no well-established methods for calculating blend time in a multi-impeller system in a rheologically complex fluid. In the author's experience, a shaft speed 20–50% higher than that required to create full motion seems to suffice.

## Background: Yield Stress Fluids

In general, one of the defining characteristics of fibrous slurries, including lignocellulosic slurries, is that they have a yield stress. This means they will not flow until a certain threshold of stress is applied. In Chapter 12, we described some common rheological models that include a yield stress, such as the Bingham plastic model and the Herschel–Bulkley model. Other models are possible. Sometimes, a simple power law model seems to fit the data as well as models containing a yield stress. However, this does not necessarily denote the absence of a yield stress. Given time, the surface of a disturbed Newtonian or power law fluid will flatten out in a container except for forming a meniscus at the edges due to

*Agitator Design Technology for Biofuels and Renewable Chemicals*, First Edition. Gregory T. Benz.
© 2022 John Wiley & Sons, Inc. Published 2022 by John Wiley & Sons, Inc.

**Figure 13.1** An 18% solids angle of repose.

surface tension. However, fluids with a yield stress will not fully flatten out. When gravitational stresses fall below the yield stress, the material will flow no more and will retain its shape indefinitely. This leads to the fact that a nominally liquid slurry can actually have an angle of repose, which is a phenomenon normally associated with piles of solids. Figure 13.1 shows this phenomenon with a biomass slurry of about 18% solids. The pile is about 10 cm across and about 4 cm high, giving it an angle of repose of about 35–40°. The individual particle size is about 0.5–2 mm. Though this looks like just a damp pile of solids, it can actually be agitated with hydrofoil turbine impellers.

In essence, for complete agitation with no dead spots, we need to be able to overcome the yield stress throughout the vessel. Though several authors have published papers on how to predict this (Refs. [1–4]), it is difficult to measure the yield stress directly, and correlations that try to predict the size and shape of dead zones based on yield stress alone do not seem to work for varying impeller types. Therefore, we favor a "wet" lab test and scale-up approach to design.

## What % Solids Can Be Handled?

This is a question often asked, and the author has discussed it briefly in Ref. [5]. Here we revisit the topic.

In the early days of the development of biomass hydrolysis reactors, it was common to use rotating barrels (similar to continuous solids dryers), horizontal paddle mixers, and other mixing apparatus intended principally to mix solids. This is because high solids biomass resembles damp peat moss more than it does a liquid. However, after even a modest degree of hydrolysis, the biomass slurry liquefies. So, if used, the solids mixers would be needed only for the earliest stages of hydrolysis. The fact that such mixers are expensive and limited in volume to about 40 000 l or so has led the industry to look into cheaper, scalable alternatives. Today, simple turbine agitators are used for almost all hydrolysis reactors. They cannot handle materials that are basically damp solids, but can handle very viscous biomass slurries. The key is to configure the process so that there is at least a couple of percent free water. By that I mean water that is not bound up within biomass cells, but is outside the cells and available to fluidize the mass. It is necessary for the free water to fill up all of the interstitial spaces between solid particles as a minimum. In practice, it takes about 2–3% more free water than needed to fill in the spaces between particles before the slurry behaves as a liquid and can be agitated by a turbine impeller.

But what % solids does that correspond to? It depends on the kind of biomass and its history of pretreatment and how hydrolyzed it is. Untreated materials such as switchgrass and corn stover are limited to approximately 12–14% solids. The same materials, after pretreatment, may be able to be agitated at concentrations in the 18–22% range. When fully hydrolyzed, they can probably go above 40% solids.

But how does one maximize the % solids in the actual process? A common way is to backmix a portion of already hydrolyzed material with the inlet pretreated material. This can allow the pretreated material to be more than 30% solids, as the turbine agitator does not see the pretreated material alone. In practice, this can work with either a fed batch reactor or a continuous flow reactor with a backmixed first stage. Lab/pilot testing is always required to ascertain quantitatively what is possible for a given process. The author has successfully agitated some unhydrolyzed biomass at up to 40% solids. Small particle size and various kinds of chemical or steam explosion pretreatment allow for higher % solids than merely adding a coarse grind to the tank.

## Caverns

When a fluid has a yield stress, if enough torque is not applied, the moving zone, or "cavern," will not extend to the walls of the tank, resulting in dead zones that contribute nothing to productivity. P. Arratia and others used visualization techniques with dyes and photography (Ref. [1]) to show this phenomenon. Others have used different techniques, such as conductivity probes and salt solutions.

Figure 13.2 comes directly from Ref. [1]. The left image, "a," shows a cavern that has a diameter less than the tank diameter. The "b" image shows a cavern that touches the wall. We would definitely prefer the latter condition! The height of the cavern at the vessel wall is about 40–50% of the tank diameter, using a Rushton turbine, applied at a D/T ratio of about 1/3. Increasing agitation once the cavern reaches the vessel wall results in a cavern height proportional to the shaft speed, up to an aspect ratio of about 1 for the cavern. For axial impellers, the cavern

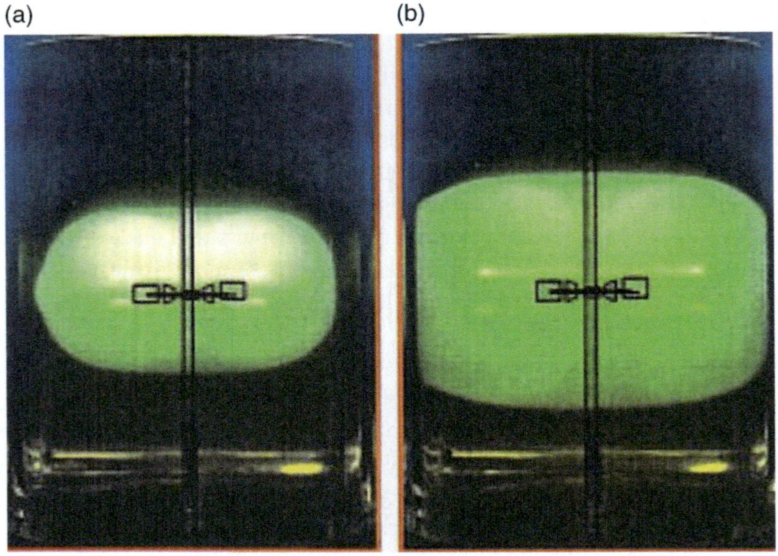


**Figure 13.2** Cavern visualization of two different cavern sizes (a) and (b). *Source:* With permission from Wiley InterScience.

**Figure 13.3** Biomass caverns.

height is closer to 75% of the tank diameter when the cavern just touches the tank wall.

Figure 13.3 is a photo of cavern formation in a biomass slurry in a transparent laboratory vessel. The motion was visualized by using a time exposure. Streaks indicate moving fluid, and we can see that it is largely in axial flow. There are stagnant areas along the bottom wall and very slow areas at the surface, which we will discuss later in this chapter.

Obviously, for a typical tall vessel, multiple impellers are needed. But multiple radial impellers are not the best choice. Axial flow impellers are normally used in all biomass applications. The relative size and spacing must be determined by lab testing.

### Estimating Cavern Size

In Ref. [2], the authors derived an equation for the cavern diameter, based on a cavern shape of a cylinder with an aspect ratio of 0.4. Their equation is as follows:

$$\left(D_C/D\right)^3 = \left(1.36/\pi^2\right) * \left(N_P \rho N^2 D^2/\sigma_y\right) \tag{13.1}$$

If we want to have the cavern reach the wall, this means that $D_C = T$.

With a few derivations, the minimum torque per impeller for full motion with the number of impellers set to (Z/0.4T) becomes:

$$\text{Minimum torque/impeller}\left(N-m\right) = 0.1156T\left(m\right)^3 * \sigma_y\left(\text{dynes/cm}^2\right) \tag{13.2}$$

Since zones do not fully intermesh, the actual minimum torque is probably 20–30% higher than this for multiple impellers. Also, per Ref. [6], getting full movement at the bottom requires more torque, as the tank bottom is another shear surface that must be overcome. It is also the author's experience that different impellers may require different torque inputs for the same process result. We discuss this later in this chapter.

The authors of Refs. [1] and [2] did not note any difference in torque requirement as a function of D/T for minimum movement for pure Bingham plastic fluids. Of course, less power is required at a larger D/T due to the lower shaft speed required to invest the same torque. (Power ∝ torque times shaft speed.) However, many yield stress fluids exhibit shear-thinning behavior above the yield stress, and benefit from larger D/T ratios to avoid "boring a hole" into the fluid. In this author's experience, high-solids biomass slurries require a D/T ratio of 0.5–0.65; a few extreme cases require as much as 0.8. Too small a D/T ratio will result in "boring a hole" and having no wall movement. Too large will result in mass rotation of the batch with no mixing.

The authors also did not examine different torque requirements for different kinds of impellers, as theory has it that at equal torque, equal cavern size will result. But, as we will see in the "Impeller Choice" section, this is not the case.

## Surface Phenomena

The phenomenon of yield stress and its angle of repose creates interesting phenomena at the liquid surface. The fact that such materials are self-supporting to some degree means that often the upward flow along the vessel wall created by the impellers will result in material climbing the walls, sometimes, at least in the lab scale, climbing 100–150 mm and exposing the upper impeller. A side view of this phenomenon is shown in Figure 13.4a and b. (The "a" and "b" pictures are from two

**Figure 13.4** (a, b) Rising surface.

different kinds of biomass slurries.) As the solids pile higher, eventually they will try to collapse inward due to their own weight. In so doing, they generally form what looks like a truncated cone or dome roof effect with a hole in the middle, as shown in Figure 13.5a and b. Though such a surface structure is often almost motionless, just creeping upward and inward until a section falls off and drops into the eye of the upper impeller, just below this structure there is often fairly strong agitation. If one manually collapses the structure in the lab scale with a spatula, a new one will quite rapidly rise up to replace it until a new equilibrium is established.

Because the yield stress remains the same upon scale-up, the height of such structures does not scale-up proportionately. These battlement-like structures will tend to be of the same absolute height or slightly smaller as the tank diameter gets larger. This is actually due to two causes: with the same yield stress, the maximum height the structure can reach before its own weight will cause it to flow is independent of scale. And the shape of the structure is important, too. With a small radius, a truncated cone or dome-like structure has a higher strength to weight ratio than one with a large radius. Conceptually, these points are illustrated in Figures 13.6 and 13.7, which show approximate surface shapes in a 0.3 m diameter lab vessel and a 3 m diameter production vessel, respectively. The end result is that formation of such structures at the surface is less of a problem as the tank diameter becomes larger. This means we should do our testing in larger lab vessels when possible. Although the author has successfully scaled from lab scale to large pilot/demo scale from vessels as small as 0.15 m diameter, he prefers to use 0.3 m or larger. Any scaling errors from small vessels will be in the conservative direction, but may result in oversized equipment. There are techniques to be used in the pilot protocol to compensate for this, which will be described in the "Pilot Protocol: Lab Scale and Large Pilot Scale" section.

**Figure 13.5** (a, b) Top view, truncated dome surface.

## Concept: Lab Test Plus Scale-up

Given the variety of biomass slurries, and the rheological complexity of them, this author believes that the best approach to design is a lab test plus scale-up. Though there may be some value in computational fluid dynamics (CFD) simulations, it is necessary to input the rheological properties to the CFD model, which means that at least a partial "wet" test is required anyway. Overall, lab testing plus scale-up takes much less time and is much less expensive than CFD simulations. We will therefore go into considerable detail on how to do the test, what equipment is needed, and how to scale-up the results. That is the principal concept of this chapter.

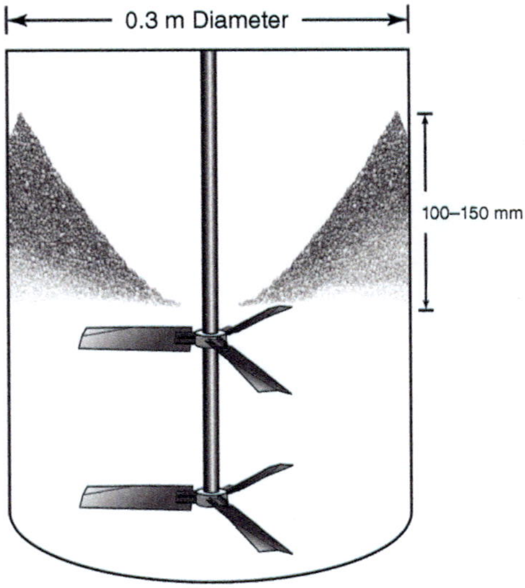


Figure 13.6 Surface shape, 0.3 m diameter vessel.

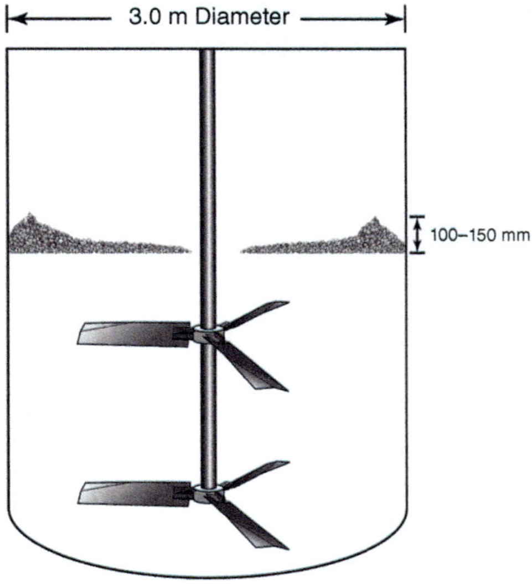


Figure 13.7 Surface shape, 3.0 m diameter vessel.

## Pilot Equipment Requirements

As an aid to those who wish to pilot agitation of lignocellulosic materials, we are providing guidance for the equipment necessary to conduct such studies. This section was inspired by Ref. [7].

There are two main purposes for the pilot agitation equipment: determining the minimum shaft speed required for creating full tank motion and using the agitator as a viscometer. The latter purpose was covered in detail in Chapter 12. Techniques for both of these purposes are also described in Ref. [8]. A complicating factor is the wide speed range required to achieve both of these purposes, as the viscosity must be measured at the same shaft speed range as the production-scale equipment. Production equipment might have speeds in the 10–45 rpm range, whereas a 5–10 gallon (20–40 l) pilot unit may need speeds of hundreds of rpm to achieve minimal wall movement. Table 13.1 illustrates suggested agitator power, torque, and speed range as a function of vessel diameter, for very difficult slurries at the edge of being solids rather than liquids. This table is *not* to be used to size production equipment. It is intended to give guidance for research/piloting purposes only, to handle the most difficult slurries that can be agitated with a turbine agitator, as opposed to a solids mixer.

Vessels should be transparent whenever possible as that is the best way to ascertain that the entire contents are being agitated, with no dead spots. A dished bottom is preferred. Flat bottoms will form fillets, which are best to ignore when taking data. This will be discussed in the next section “Pilot Protocol: Lab Scale and Large Pilot Scale.” The vessel should at least have an aspect ratio (Z/T) of close to 1, so that the effects of dual impeller torque split can be studied. It is not necessary to model a tall aspect ratio in the lab. For highly viscous slurries, there should be no tank baffles.

**Table 13.1** Suggested pilot agitator drive parameters.

| Vessel diameter (in./m) | Speed range (rpm) | Torque (in.-lb/N-cm) | Power (Hp/kW) |
|---|---|---|---|
| 12/0.3 | 10–800 | 50/565 | 0.75/0.55 |
| 18/0.45 | 10–600 | 150/1 700 | 1.5/1.1 |
| 24/0.6 | 10–420 | 400/4 500 | 3/2.2 |
| 30/0.75 | 10–350 | 800/9 000 | 5/3.7 |
| 36/0.9 | 10–280 | 1 400/16 000 | 7.5/5.5 |
| 42/1.07 | 10–230 | 2 200/25 000 | 10/7.5 |
| 48/1.2 | 10–190 | 3 200/36 000 | 15/11.1 |
| 54/1.4 | 10–190 | 4 600/52 000 | 20/15 |
| 60/1.5 | 10–155 | 6 300/71 000 | 20/15 |
| 72/1.8 | 10–155 | 10 000/113 000 | 30/22 |
| 84/2.1 | 10–125 | 17 000/194 000 | 40/30 |
| 96/2.4 | 10–100 | 25 000/290 000 | 40/30 |
| 108/2.75 | 10–84 | 36 500/412 000 | 50/37 |
| 120/3.0 | 10–84 | 50 000/565 000 | 75/55 |

If mixing of highly hydrolyzed slurries is anticipated, baffled vessels should be used for those studies. Use four baffles, T/12 wide, T/72 off wall.

Larger pilot vessels are often not available in transparent materials. In such a case, observations must be made through sight glasses and by looking down on the liquid surface. As a minimum, sight glasses should be about 150 mm diameter viewing area, and there should be at least one located T/4 off bottom and at 0.75Z.

Though we will discuss impellers later in this chapter, the author recommends hydrofoils exclusively for such agitation, though they could be studied at different solidity ratios and different base power numbers. The author has primarily used hydrofoil impellers with a turbulent power number of 0.3. For high solids fibrous slurries, the lower impeller should normally be 0.6–0.65T in diameter, and the upper impeller should be about 10–15% smaller in diameter. There should be two impellers for a batch Z/T ratio between 0.75 and 1.0. For use as a viscometer, impellers should have known power draw characteristics in laminar and transition flow. For lignin slurries, which are much less viscous, a D/T of 0.3–0.5 will suffice, and only 1 impeller per square batch is needed.

There needs to be a means of measuring shaft speed. Many lab agitators have a built-in tachometer, which is usually accurate enough. For some units, and for low speeds, a laser tachometer is recommended. For extremely low speeds, revolutions may be counted by hand and timed with a stopwatch.

For use as a viscometer, there must be a means of measuring torque. This was covered in detail in Chapter 12. For mixing studies, measuring and recording torque is optional.

## Pilot Protocol: Lab Scale and Large Pilot Scale

Although the goal and kinds of data to be taken are independent of scale, the actual construction of pilot equipment affects our ability to observe mixing, so the protocol is somewhat different depending on size and construction. The major difference is whether the tank is transparent. In most cases, transparent vessels are only available up to about 0.5 m diameter, though this author has seen one as large as 1.5 m diameter. For best scalability, test work should be done in as large a scale as possible, but with a strong preference given to transparent vessels. Most work done by the author in transparent vessels has been in 0.15, 0.3, and 0.45 m diameters. The author has done work in larger pilot vessels up to 6 m diameter, but these were stainless steel vessels equipped with sight glasses. How to deal with using sight glasses instead of a fully transparent vessel will be described next in step 3.

The step-by-step protocol follows:

1) Evaluate viscosity vs shaft speed using the procedures in Chapter 12. Transparent vessels are not needed for this, but are needed if possible for mixing tests.

2) Install impellers for fibrous slurries according to the size and position guidelines given under section "Pilot Equipment Requirements." If the test is for material that has already been hydrolyzed, just use a single impeller per square batch geometry, with a D/T ratio between 0.25 and 0.5.
3) Operate the agitator at a shaft speed that creates movement in the bottom of the tank and record the shaft speed. It is crucial to not overdo the shaft speed, especially for flat bottom tanks. It is very hard to get motion near the transition from the bottom to the sidewall in a flat bottom tank. Fillets will tend to form unless a great deal of torque is applied (see Figure 13.8). Assuming we will be using a dished bottom tank in the production equipment, which will have much less of a problem with solids fillet formation, our complete motion speed in a flat bottom tank can mimic what is required by ignoring the fillet. See Figure 13.9 for a dished bottom tank, which would have no bottom fillets at a shaft speed that would produce fillets about T/6 to T/8 high if operated in a flat bottom tank. A reasonable guideline for observation if one must use a flat bottom tank is to consider full bottom motion to be obtained if the stationary fillet height equals about 0.15T or less. Only if the production tank will be flat or sloped bottom should the pilot scale attempt to get complete motion down to the bottom corner, but this is done at considerable cost. See later comments under section "Tank Shape."

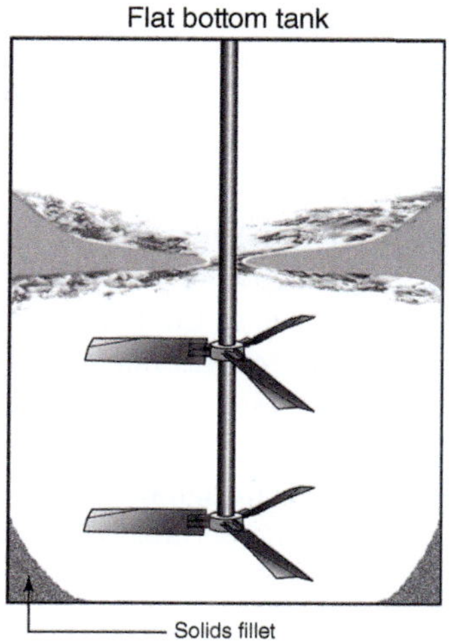


**Figure 13.8** Flat bottom tank.

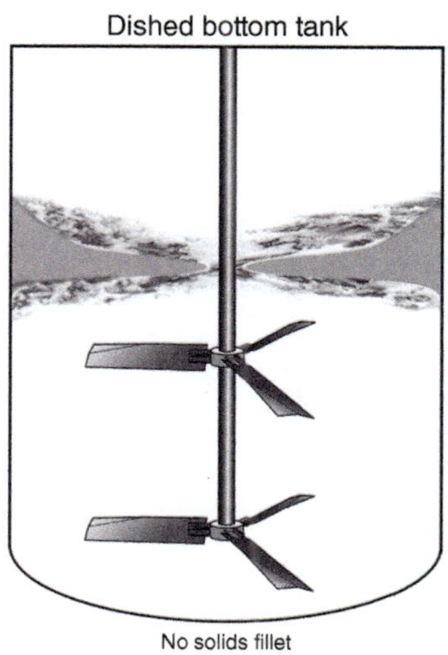


**Figure 13.9** Dished bottom tank.

In the case of an opaque tank, one must resort to the use of sight glasses. These should be at least 150 mm in diameter. It would be ideal if they were made of curved glass that matched the radius of the tank wall and were mounted flush with the wall. However, that is not how they are actually built. Sight glasses must be of the "flush mounted" type, rather than mounted on a nozzle, or they will be useless for monitoring slurry movement. However, even flush mounted sight glasses have lips and recesses that tend to create dead spots.

Referring to Figure 13.10, one can see that such sight glasses have a recess where they are mounted to the metal supports. (The tank contents are on the right of the sight glass in this illustration.) These recesses create dead spots (stagnant solids), which may not be moving even when there is good bulk movement throughout the tank. Even the center of the glass tends to have much less velocity than the bulk tank movement. This author has observed that when a small amount of pulsing or creeping, intermittent movement is observable at the middle of the upper sight glass, the tank contents, when viewed from the liquid surface, are moving well out to the vessel wall. Therefore, it is suggested that if any movement at all is observed in the middle of the sight glass, the tank contents are moving fully in that region of the tank.

4) Operate the agitator at a shaft speed that creates visible motion in the upper region of the tank. This is not necessary for lignin slurries with only one

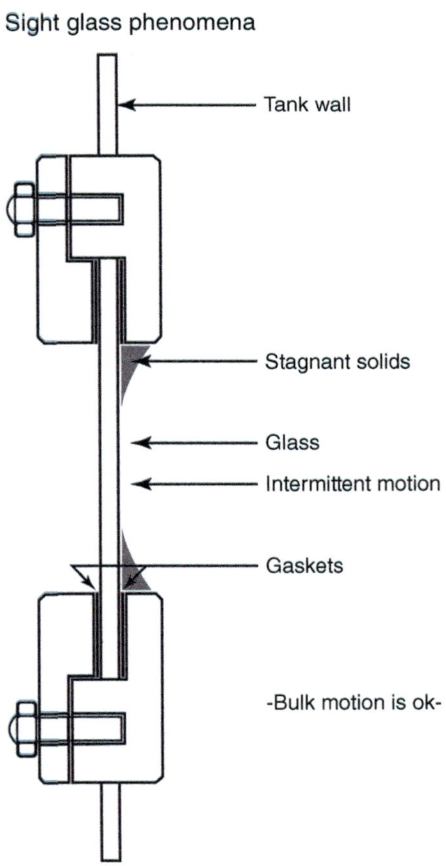


**Figure 13.10** Sight glass phenomena.

impeller. Record this speed. Ideally, this would be the same speed as required to create full motion in the bottom of the tank, but that is not always the case. It could be faster or slower, depending on the slurry. If the two speeds are within 30% or so of each other, we will adjust impeller relative sizes upon scale-up, as we cannot operate each impeller at a different speed in production equipment. If the speeds are too far apart, it may be necessary to change the upper impeller size and recheck the required speed.

In smaller tanks with high % fibrous solids, the surface may tend to form structures as shown in Figures 13.4 and 13.5. It is not necessary for the agitator to rapidly fold these in when testing in small, transparent tanks of less than about 0.5 m diameter. If there is visible movement below those structures, and if the structures themselves slowly fold in, it is enough. A further check can be made by pushing the structure over with a spatula. If slurry rapidly rises to replace that which was manually toppled over, it is sufficient.

For larger alloy tanks equipped with sight glasses, ideally there would be several along the straight side of the vessel. There should be at least one about 90% of the liquid height. One can observe both the sight glasses and the liquid surface in most cases if a tank top light-and-sight glass is installed. As long as the liquid surface is moving and/or there is at least intermittent motion at the middle of any sight glasses in the upper part of the tank, the speed is sufficient to create full motion.

5) If using a batch or fed batch process to emulate a continuous flow process, the above data may have to be collected several times as a reaction progresses, so that agitation requirements can be estimated in each part of a tanks-in-series design. For a fed batch reactor, the most difficult condition is generally when all solids have been added, unless they are added too quickly at first. After the solids have been added, the enzymatic hydrolysis reaction will rapidly thin the slurry.

## Tank Shape

As mentioned earlier, bottom contour has a big effect on movement. If a large tank is being contemplated, it may be difficult to get it with a dished bottom. An acceptable alternative would be to fabricate a modified flat or sloped bottom tank with a metal fillet that mimics the shape of fillet that the solids will tend to form. This is illustrated in Figure 13.11.

Non-circular tank cross sections and eccentric agitator mounts are to be avoided. If for some reason they cannot be, a lab model of the actual geometry should be built and the agitation studied in the step-by-step protocol.

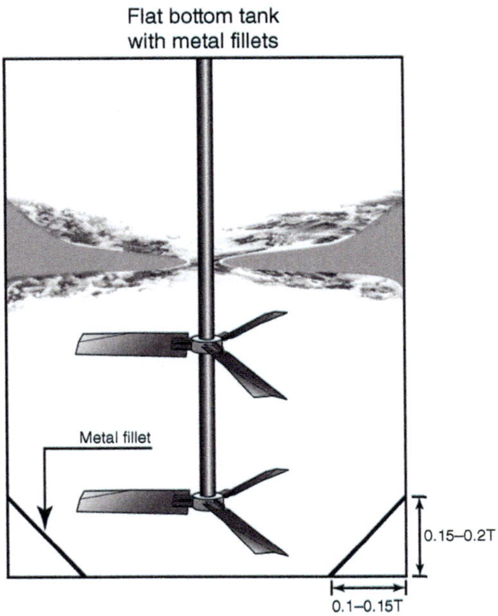


**Figure 13.11** Filleted tank construction.

## Internals

By now, with all this discussion of yield stress and dead spots, it should be obvious that any tank internals will be prone to create dead spots in yield-stress fluids. Therefore, all tank internals, even baffles, should be avoided. If the process is batch or semi-batch, hydrolysis will thin the fluid enough that baffles would be beneficial later in the batch. But the swirl typical of an unbaffled tank in a low-viscosity liquid is not as detrimental as stagnant areas during the initial stage of the process. For more efficient operation, even with batch processes, some will transfer the reactor contents after some time has elapsed to another vessel that is baffled. For example, after the initial unbaffled vessel is filled, they may allow it to react for 30–90 minutes, and then transfer it to one or more larger baffled vessels where it may still continue to react for several days.

## Torque Split

As mentioned in the pilot protocol, the shaft speed needed to create full motion in each impeller zone must be measured separately. This is because the torque required to create movement is greater at the bottom of the tank than along the sidewall. The logical explanation for this is that the bottom impeller must overcome yield stress along the bottom of the tank as well as its portion of the sidewall, whereas the upper impellers need only overcome yield stress along their portion of the sidewall. In fact, one can do a simple derivation of the theoretical torque required based on this concept. If we assume a flat-bottomed tank with two impellers per square batch (Z/T per impeller = 0.5), we can derive the following for the torque required to overcome yield stress on the tank bottom:

$$\int(0 \rightarrow r)\sigma_y * 2\pi r^2 dr = \sigma_y * 2\pi r^3/3 \tag{13.3}$$

For the sidewall, we can likewise derive the following:

$$\sigma_y * r * 2\pi r * r = \sigma_y * 2\pi r^3 \tag{13.4}$$

Thus, the total torque required of the bottom impeller is Eq. (13.3) plus (13.4), which equals $\sigma_y * 8/3\pi r^3$.

The upper impeller need only overcome the yield stress at the sidewall, same as Eq. (13.4). The total torque is thus $\sigma_y * 14/3\pi r^3$.

In other words, for a two-impeller system with a Z/T of 1.0, theoretically 8/14 or 4/7 of the torque should be invested in the lower impeller. In a taller geometry, the 4/7 ratio would extend to additional impellers, which will be equal in size and torque to the second impeller from the bottom. This would mean the upper impellers would be about 94% of the diameter of the lower impeller.

However, actual results determined experimentally differ from the theoretical derivation. Table 13.2 summarizes actual data from a number of tests conducted

**Table 13.2** Experimental torque split.

| Material | % solids | Enzyme treat time (h) | Torque split, bottom/top |
|---|---|---|---|
| A | 12 | 0 | 69/31 |
| A | 13 | 0 | 69/31 |
| B | 12.4 | 0 | 88/12 |
| B | 13.1 | 0 | 81/19 |
| B | 15.3 | 0 | 90/10 |
| B | 16.6 | 0 | 89/11 |
| B | 18.6 | 0 | 90/10 |
| B | 18.6 | 1.5 | 88/12 |
| B | 18.6 | 2 | 90/10 |
| C | 18 | 0 | 68/32 |
| C | 16 | 0 | 68/32 |
| D | 17 | 10 | 90/10 |

**Table 13.3** Torque split consequences.

| Aspect ratio (Z/T) | Relative torque without proper split | Relative torque with proper split, 70/30 basis | Relative torque with proper split, 90/10 basis |
|---|---|---|---|
| 0.5 | 1 | 1 | 1 |
| 1 | 1 | 0.71 | 0.56 |
| 1.5 | 1 | 0.62 | 0.41 |
| 2 | 1 | 0.57 | 0.33 |
| 2.5 | 1 | 0.54 | 0.29 |

by the author. He has conducted many more, but these should suffice to give an idea how much variation there is in the actual torque split requirements. In all cases, the actual portion of torque required by the lower impeller exceeded the theoretically derived 4/7 ratio. Some samples required as much as 90% of the total torque invested in the lower impeller. The author does not have an explanation for these results, but has seen such results over dozens of tests with multiple feedstocks and processes.

There are consequences to these results in full-scale design. It is vital that all pilot tests use at least two impellers in a square batch geometry so that the relative torque split can be determined. Scaling-up from a single impeller by just adding impellers for a greater Z/T ratio can result in seriously oversized equipment. Table 13.3 explores these consequences for several aspect ratios and a couple of typical torque split ratios. One can readily see that an assumption of an equal torque split will result in oversized, more expensive equipment, and the difference gets greater as the aspect ratio increases. We will discuss aspect ratio in more detail in the next section.

## Aspect Ratio

Because of the asymmetric torque split between upper and lower impellers, aspect ratio affects capital and energy costs of the agitator. The degree to which it affects them depends on the torque split. Table 13.4 gives examples of relative torque and power as a function of aspect ratio for 50/50, 70/30, and 90/10 torque splits. For an equal split, the total torque will be independent of aspect ratio, but the power will increase. For a 70/30 split, once we get past one impeller, the power is roughly constant, but the torque decreases as the aspect ratio increases. For the 90/10 split, both power and torque decrease with aspect ratio. Capex is closely related to torque, though a long shaft and more impellers may negate the savings in gear

**Table 13.4** Relative power and torque as a function of aspect ratio.

| | | Square batch power split basis | | | | | |
|---|---|---|---|---|---|---|---|
| | | 50/50 | | 70/30 | | 90/10 | |
| Z/T | Number of impellers | Relative torque | Relative power | Relative torque | Relative power | Relative torque | Relative power |
| 0.5 | 1 | 1.00 | 0.79 | 1.41 | 1.12 | 1.80 | 1.43 |
| 1 | 2 | 1.00 | 1.00 | 1.00 | 1.00 | 1.00 | 1.00 |
| 1.5 | 3 | 1.00 | 1.15 | 0.86 | 0.99 | 0.73 | 0.84 |
| 2 | 4 | 1.00 | 1.26 | 0.80 | 1.00 | 0.60 | 0.75 |
| 2.5 | 5 | 1.00 | 1.36 | 0.75 | 1.02 | 0.52 | 0.70 |

drive selection afforded by the lower torque level. Opex is closely related to power, so a taller vessel may be favored when the pilot data show an extreme torque split. That is why it is important to study this.

## Impeller Choice

Though lignocellulosic slurries often have very high viscosities (e.g. some have viscosities at a shear rate of 1/s over 1 000 000 cP), they are quite non-Newtonian, and this often dictates the choice of impeller. Conventional agitation design philosophy would say that a helical ribbon or helix impeller (no illustration shown here) should be used for such high viscosities. However, the author's experience dictates otherwise. Aside from the fact that helix impellers are much more expensive than turbines, cannot be scaled-up to large sizes, and require close tolerances on the vessel, they have another problem: they do not work in high yield stress materials. What happens with a helix impeller in a biomass slurry is that it shears at the wall and the whole batch rotates with the impeller, with very little actual mixing. However, it turns out that ordinary turbines applied at the correct D/T ratio can work very well. Then the question becomes, what kind of turbines? In Refs. [9] and [10], the author studied pitched blade turbines vs hydrofoil impellers in laminar flow. In all cases, the hydrofoils performed better than the pitched blade turbines.

This author has carried out a study based on lab data that used a mixed lignocellulosic slurry intended to be an average of several kinds of feedstock. It was found that even at the highest viscosity, the hydrofoil had an advantage. In fact, the hydrofoil impeller only required about half of the torque of a pitched blade turbine. As the batch thinned with time, the pitched blade turbine continued to require about twice the torque of a hydrofoil. These results are shown in a simulated 50000 gallon (189000 l) batch reactor profile, depicted in tabular form in Table 13.5 and graphically in Figure 13.12.

Remarkably, in the most viscous conditions, the pitched blade turbine actually had to rotate at a higher shaft speed than the hydrofoil to create minimal motion! Though cavern theory would predict that all impellers would need to input the same torque to overcome yield stress, empirically we have shown that is not the case. So, for now, the author strongly recommends hydrofoil impellers. In Table 13.5, the hydrofoil saves 36% of the energy of a pitched blade turbine design in the first 10 hours and 20% after that.

The author has not had the opportunity to test reversing pitch, "chaos" impellers such as shown in Figures 4.25 and 4.26. It is possible they may be satisfactory as well. The buyer or designer is encouraged to do testing with other impellers to find out.

## Use of Variable Speed to Save Energy

Table 13.5 and Figure 13.12 illustrate another point. In batch or fed batch reactors, the slurry starts off very thick but thins out rapidly. If one were to maintain constant shaft speed, the agitation would be excessive as time went on and much energy would be wasted. Table 13.6 presents the results of what would happen if we kept the shaft speed at 20 rpm throughout the batch, as opposed to using a variable frequency drive (VFD) to slow it down as needed. Using the VFD saves almost 90% of the total energy for a several day-long batch cycle.

## Scale-up Procedure

In most cases, lignocellulosic hydrolysis reactors are liquid motion controlled. This means they can be scaled-up using the principles of Chapter 11. The reader will recall that, for turbulent flow and geometric similarity, a scaling exponent of 1 can be used, as per Eq. (13.5):

$$N_2 / N_1 = 1 / R, \text{where } R \text{ is the geometric scale ratio.} \tag{13.5}$$

This equation, though generally thought to be for turbulent flow only, is really based on the assumption that the impeller pumping number, $N_Q$, is constant. And, in fact, it is constant for turbulent flow conditions ($N_{Re} > 10000$). But it is also constant for laminar flow ($N_{Re} < 10$) and nearly constant up to a Reynolds number of 40. It changes in the transition range ($40 \leq N_{Re} < 10\ 000$). So, if the flow is either laminar or turbulent in both scales, an exponent of 1 can be used. If the flow

**Table 13.5** Hydrofoil vs pitched blade comparison, mixed biomass.

| Time (h) | Viscosity at 1/s (cP) | Power law exponent | Hydrofoil shaft speed (rpm) | Pitched blade turbine shaft speed (rpm) | Hydrofoil power draw (Hp) | Pitched blade turbine power draw (Hp) | Power ratio, pitched/hydrofoil | Torque ratio, pitched/hydrofoil |
|---|---|---|---|---|---|---|---|---|
| 0 | 1.21E+06 | 0.135 | 20 | 22 | 254 | 446 | 1.76 | 1.60 |
| 1 | 881 000 | 0.0917 | 13.8 | 14 | 115 | 157 | 1.37 | 1.35 |
| 2 | 160 000 | 0.159 | 11.9 | 11 | 21.2 | 38.4 | 1.81 | 1.96 |
| 3 | 120 000 | 0.17 | 11 | 7.98 | 15 | 17 | 1.13 | 1.56 |
| 4 | 90 107 | 0.21 | 10.1 | 7.3 | 10.6 | 12.3 | 1.16 | 1.61 |
| 5 | 45 000 | 0.3 | 9.2 | 6.54 | 5.6 | 7 | 1.25 | 1.76 |
| 6 | 20 000 | 0.5 | 9.2 | 6.38 | 3.56 | 4.8 | 1.35 | 1.94 |
| 7 | 10 000 | 0.7 | 9.2 | 6.25 | 2.69 | 3.65 | 1.36 | 2.00 |
| 8 | 5 000 | 0.8 | 9.2 | 6.05 | 2.2 | 2.81 | 1.28 | 1.94 |
| 9 | 2 000 | 0.9 | 9.2 | 5.82 | 1.89 | 2.38 | 1.26 | 1.99 |
| 10 | 500 | 1 | 9.2 | 5.66 | 1.72 | 2.16 | 1.26 | 2.04 |
| 10+ | 500 | 1 | 9.2 | 5.66 | 1.72 | 2.16 | 1.26 | 2.04 |

Basis: 50 000 gallons, 248″ tank diameter, batch reactor.
Pseudoplastic rheology.

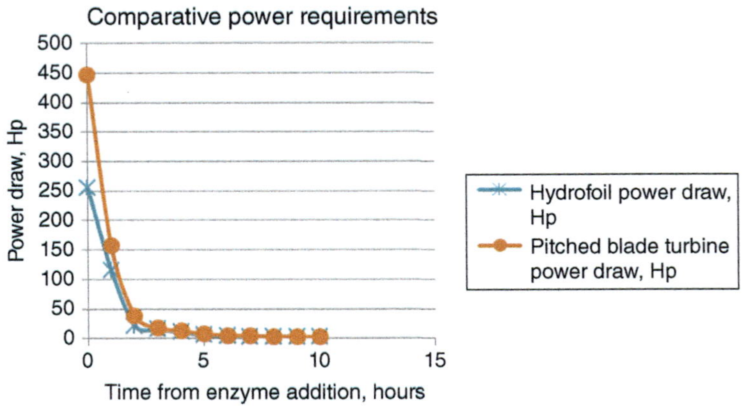


**Figure 13.12** Hydrofoil vs pitched blade turbine.

**Table 13.6** Energy savings with VFD.

| Time (h) | Scaled shaft speed (rpm) | Power draw at scaled shaft speed (Hp) | Power draw at constant 20 rpm (Hp) | Power ratio |
|---|---|---|---|---|
| 0 | 20 | 254 | 254 | 1.00 |
| 1 | 13.8 | 115 | 181 | 1.57 |
| 2 | 11.9 | 21.2 | 49 | 2.31 |
| 3 | 11 | 15 | 42 | 2.80 |
| 4 | 10.1 | 10.6 | 36.4 | 3.43 |
| 5 | 9.2 | 5.6 | 27.7 | 4.95 |
| 6 | 9.2 | 3.56 | 22.2 | 6.24 |
| 7 | 9.2 | 2.69 | 21 | 7.81 |
| 8 | 9.2 | 2.2 | 19.3 | 8.77 |
| 9 | 9.2 | 1.89 | 18.1 | 9.58 |
| 10 | 9.2 | 1.72 | 17.4 | 10.12 |
| 10+ | 9.2 | 1.72 | 17.4 | 10.12 |

regime changes, the scale-up process will need to be iterative to be rigorously correct. However, since $N_Q$ increases with Reynolds number, failure to account for the change is an error in the conservative direction. Some may opt to use this fact as a safety factor, and not make a correction. We view the correction as optional, especially since, with yield stress fluids and strongly shear-thinning fluids, the Reynolds number does not change as much upon scale-up as it does with Newtonian liquids. We will explore the consequences in our example problem.

Since the bottom impeller invests the most torque and has the most critical job, we will scale it up first, and then adjust for the speed ratio for top and bottom movement in the tank by factoring the upper impeller diameter. The step-by-step procedure follows.

1) Define the scaling ratio, R, as $T_2/T_1$. Do not use the cube root of volume ratio, as we need to keep two impellers per square batch, and we will adjust for aspect ratio by adding impellers if needed, not by making the main impeller larger or smaller.
2) Scale-up the lower impeller diameter by multiplying the small-scale impeller diameter by R; i.e. $D_2$ (bottom) = $D_1$ * R.
3) Scale-up the shaft speed per Eq. (13.5), applied to the shaft speed needed by the lower impeller to create full motion in the bottom of the tank.
4) If the upper impeller needed to rotate at a different speed from the lower one to create motion in the upper part of the tank, we will adjust its diameter so that, at the same speed as we scaled-up the bottom impeller, it will input the same relative torque. Since the lab scale will almost certainly be in laminar flow, we note that torque $\propto ND^3$. Therefore we scale the upper impeller by a factor of $R * (N_{upper}/N_{lower})^{1/3}$.
5) Adjust to the nearest standard shaft speed if needed.
6) Calculate power draw using standard procedures from Chapter 3, accounting for the rheology determined using the methods of Chapter 11.
7) Optional: adjust speed for Reynolds number effects on pumping number. This is an iterative process. We first calculate the Reynolds number in the lab scale, and then look up the lower impeller pumping number from Appendix 1 or 2 from this book, if appropriate, or from other specific sources for the impellers used, if available. Next, we calculate the Reynolds number in the production scale, and look up the corresponding pumping number. If it is the same as lab scale, no adjustment is needed. If it is greater, we can iteratively adjust the shaft speed so that $N_QND^3$ is the same as we would calculate using the lab value of $N_Q$. The calculation is iterative because the Reynolds number changes each time we change the shaft speed, which also changes the pumping number.

## Example

For this example problem, we will do both viscosity treatment and scale-up, as we will need the viscosity model to calculate power draw and to do the optional Reynolds number correction, if applicable. We will treat the viscosity first, and then do the scale-up.

By way of background, the example used here is based on actual lab testing done by Benz Technology International. The material tested was a biomass sample, pretreated by the customer's proprietary pretreatment method. It was desirable to push the % solids to the maximum possible and yet remain nominally liquid, so as to be possible to agitate with turbine agitators rather than solids mixers. The maximum % we tested was 21.2%, which was very close to the maximum possible; it may be that 21.5 or even 22% was feasible, but we had difficulty keeping the sample from climbing out of the tank at the walls due to yield stress. We had to use a spatula to break the surface formations, which let new material rapidly rise to the surface to replace it. A photo of the surface is shown in Figure 13.13.

Testing was done in a nominal 0.3 m diameter vessel with a 7″ (178 mm) hydrofoil impeller, which had a laminar Reynolds number times power number value of 27. Torque and speed were measured and recorded. The resulting data set and conversion to apparent viscosity are tabulated in Table 13.7.

Rheology models were fitted to these data using DataFit software by Oakdale Engineering. We looked at three models: Bingham plastic, Power law, and Herschel–Bulkley, as previously defined. The results are tabulated in Table 13.8 and presented graphically in Figures 13.14–13.16.

Interestingly, all three models showed an excellent data fit, as judged by the correlation coefficient, $R^2$, which exceeded 0.99 for all three. However, though the Herschel–Bulkley model makes the most sense from a mechanistic point of view, this particular fit results in a negative yield stress term (A/N), which is not physically possible. But the correlation essentially involves subtracting one very large number from another, which is problematic when the original torque data were only accurate to two significant digits. So, I would suggest using the power law fit for full-scale shaft speeds between 20 and 80 rpm, and the Bingham model below that, as the yield stress will dominate the viscosity at very low shaft speeds/shear rates.

**Figure 13.13** High solids biomass test.

**Table 13.7** Example problem viscosity data.

| Shaft speed (rpm) | Torque (in.-lb) | Torque (N-cm) | $N_P$ | $N_{re}$ | Apparent viscosity (cP) |
|---|---|---|---|---|---|
| 9 | 6.7 | 76 | 1137 | 0.024 | 198 783 |
| 13 | 7.1 | 80 | 578 | 0.047 | 145 835 |
| 22 | 7.4 | 84 | 210 | 0.128 | 89 816 |
| 28 | 7.7 | 87 | 135 | 0.200 | 73 431 |
| 46 | 8.2 | 93 | 53 | 0.507 | 47 600 |
| 62 | 8.4 | 95 | 30 | 0.899 | 36 177 |
| 80 | 8.5 | 96 | 18 | 1.478 | 28 371 |

Basis: 7″ (178 mm) impeller diameter, hydrofoil, $N_P * N_{Re} = 27$ in laminar flow.
Vessel ID = 0.29 m.

**Table 13.8** Example problem viscosity models.

| Model | A | B | C | M | $R^2$ |
|---|---|---|---|---|---|
| Bingham plastic | 1.73E+06 | 9600 | N/A | N/A | 0.9980 |
| Power law | N/A | N/A | −0.8826 | 1 390 000 | 0.9998 |
| Herschel–Bulkley | −1.28E+07 | N/A | −0.9840 | 14 100 000 | 0.9998 |

Bases:
Bingham plastic model: $\mu_a = A/N + B$.
Power law model: $\mu_a = M * N^C$.
Herschel–Bulkley model: $\mu_a = A/N + M * N^C$.
Viscosity stated in cP or mPa-s.
Comment: the Herschel–Bulkley "A" value implies a negative yield stress.
This is not physically possible.

Now we get into the scale-up. With the referenced material, we used a lower 7″ (178 mm) impeller located 38 mm off bottom, and an upper 6″ (152 mm) impeller located 180 mm off bottom, in a flat-bottomed tank with a nominal diameter of 0.29 m, with a liquid level of 0.27 m. Essentially, this was a square batch geometry. Complete motion in the bottom of the tank (except for a 38 mm tall solids fillet) was obtained at a shaft speed of 588 rpm. Complete motion in the upper part of the tank was obtained at a shaft speed of 513 rpm. We wish to scale this up to a production-scale tank of 5 m diameter, with a nominal aspect ratio of 1.5, or a 7.5 m liquid level. We will use a dished bottom tank to avoid the problem of a solids fillet. What shaft speed and impeller system will we need, and how much power will it require?

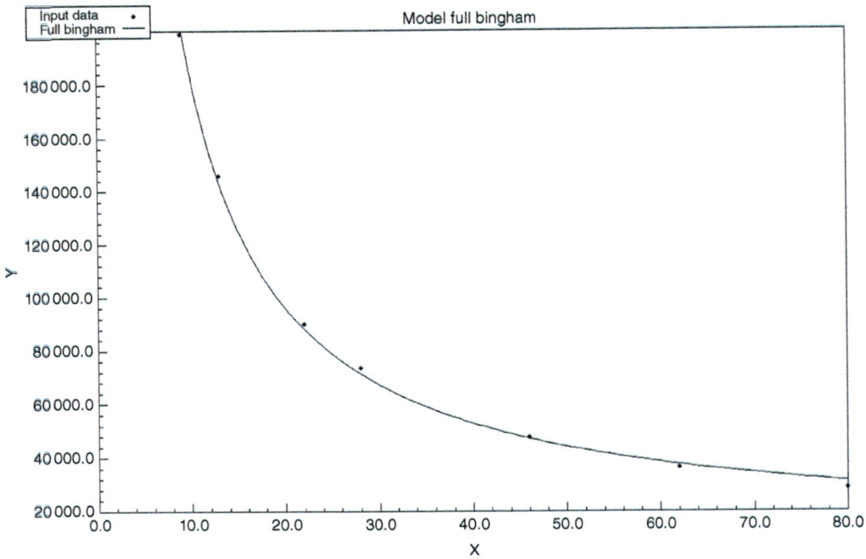


**Figure 13.14** Example problem, Bingham data fit.

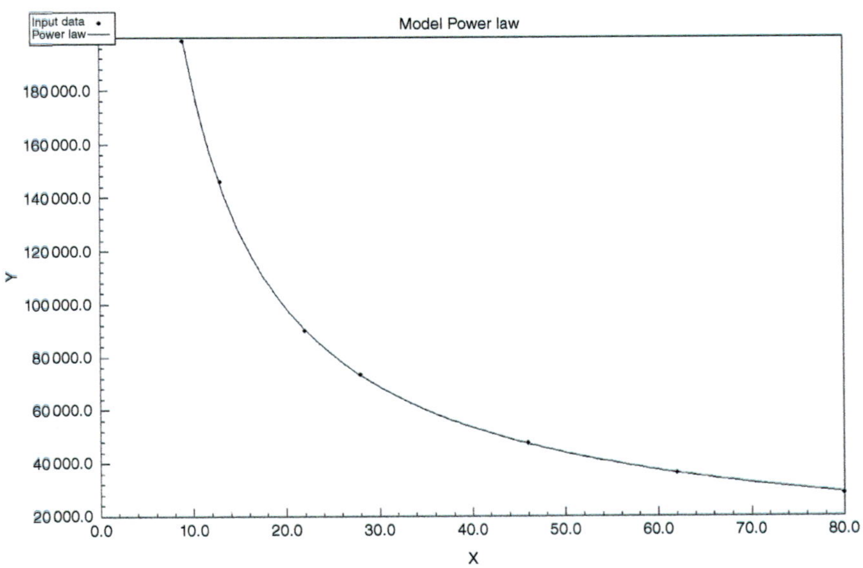


**Figure 13.15** Example problem, power law data fit.

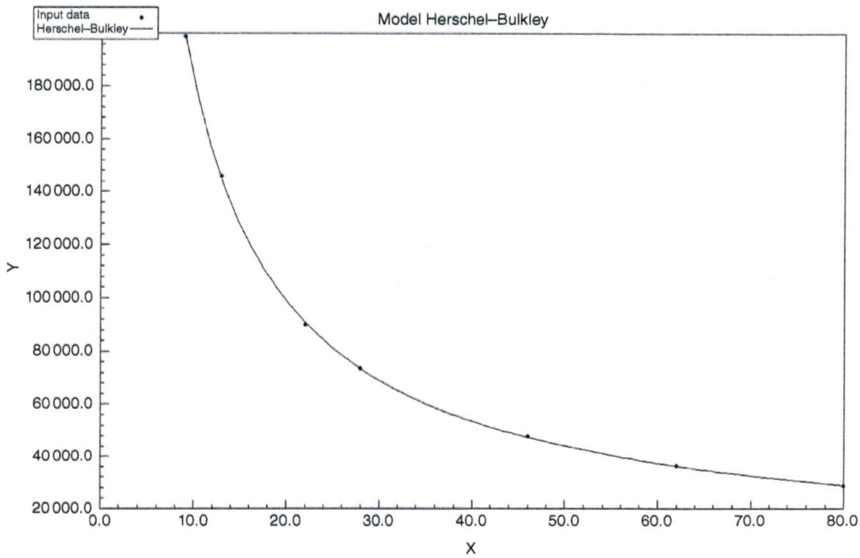


**Figure 13.16** Example problem, Herschel–Bulkley data fit.

We will follow the steps given earlier.

1) Scale ratio = R = $T_2/T_1$ = 5/0.29 = 17.24. Due to the aspect ratio change, we will be using three impellers.
2) Scale the bottom impeller by multiplying by R: $D_2$ (bottom) = $D_1$(bottom) * R = 178 mm * 17.24 = 3069 mm = 3.07 m.
3) Scale the shaft speed based on the lower impeller: $N_2 = N_1/R$ = 588/17.24 = 34.1 rpm.
4) Since the upper impeller was able to get its job done at a slightly lower speed, we will adjust the diameter upon scale-up for equal torque: $D_2$(upper) = $D_1$(upper) * R * $(N_{upper}/N_{lower})^{1/3}$ = 152 mm * 17.24 * $(513/588)^{1/3}$ = 2504 mm = 2.5 m.
5) Typical agitator drives use AGMA (American Gear Manufacturers Association) speeds. The closest AGMA speed that is not slower than our scaled speed of 34.1 rpm is 37 rpm. Actual gear ratios may result in a somewhat different speed, but we would go for the nearest actual speed that is greater than or equal to 34.1 rpm. For now, we will calculate at 34.1 rpm to see if Reynolds number changes on scale-up would allow a lower shaft speed.
6) To calculate power draw, we must first calculate apparent viscosity, then Reynolds number, then power number. Based on the power law model described previously, we have an apparent viscosity as follows: $\mu_a$ = 1.39E6 * $N^{-0.8826}$ = 1.39E6 * $34.1^{-0.8826}$ = 61 700 cP = 61.7 Pa-s (it is best to use Pa-s for viscosity for an SI Reynolds number calculation). The Reynolds

number is calculated as follows: $N_{Re} = D^2N\rho/\mu = 3.07^2 * (34.1/60) * 1000/61.7 =$ 86.8 for the lower impeller, and $2.5^2 * (34.1/60) * 1000/61.7 = 57.6$ for the upper impeller. Both impellers are therefore in lower transition flow. Per Appendix 1, the power number of the lower impeller is about 0.75 and the power number of the upper impellers is about 0.96. Therefore, using SI units, the total power is as follows: $P = 0.75 * 1000 * (34.1/60)^3 * 3.07^5 + 2 * 0.96 * 1000 * (34.1/60)^3 * 2.5^5 = 71\,966\,W = 72\,kW$.

7) Since we are in transition flow, the pumping numbers could be somewhat higher than they were in the lab scale. For the lower lab-scale impeller, at a D/T ratio of 0.61, the pumping number is about 0.19–0.2 in laminar flow. Using the same procedure as for the scaled impeller, the lower lab impeller had a Reynolds number of about 61–62. At a Reynolds number of 86.8 in the production scale, there is no material change in the pumping number. For the upper impeller, there is actually about a 10% drop in pumping number, though it eventually increases at higher Reynolds numbers. So, we do not actually want to decrease speed. If we left the speed alone, we might want to increase the upper impeller diameter by about 3–4%. However, if we use the AGMA speed of 37 rpm, this minor adjustment is not needed.

Comment 1: If we adjust to 37 rpm, the apparent viscosity becomes 57.4 Pa-s; the Reynolds numbers become 101 and 67, respectively; and the power numbers become 0.67 and 0.85, and the total power becomes 81.8 kW, a 13.5% increase. A 100 kW motor would be sufficient.

Comment 2: Impeller placement, based on geometric similarity, would be about 0.65, 3.1, and 5.6 m off bottom. Adjustable impeller location may be useful to fine tune the process.

## Post-hydrolysis

As previously mentioned, enzymatic hydrolysis breaks down the cellular structure and frees up internal water, making it available to fluidize the remaining solids. Moreover, when cellulose is hydrolyzed, it becomes various soluble sugars. Though it is possible to test at various stages of hydrolysis and may be desirable to do so if the process involves tanks in series or other variants of a continuous flow process, it is not necessary to do so for the final condition. The author has tested numerous fully hydrolyzed slurries at up to 40% total solids (including dissolved solids), which are basically lignin slurries. In all cases, the final viscosity was less than 100 cP, and full suspension was achieved with a scale of agitation (see Chapter 5) of 3 or slightly less in some cases. Therefore, unless a viscometer indicates a viscosity greater than 100 cP, design for a blending and motion scale of agitation of 3 should be sufficient.

## Summary of Chapter

We have described the difficulty of agitating fibrous materials with yield stress, how to measure and characterize their rheology, how to conduct testing to determine minimum level of agitation needed, and how to scale-up to production equipment. We have provided a real-world example of such a test and scale-up.

## List of Symbols

| | |
|---|---|
| A | Bingham coefficient |
| B | Bingham viscosity slope above yield point |
| C | Power law exponent in DatFit; aka n |
| D | Impeller diameter |
| $D_C$ | Cavern diameter |
| M | Power law coefficient, viscosity at 1/s |
| N | Shaft speed |
| $N_P$ | Power number |
| $N_Q$ | Impeller pumping number |
| $N_{Re}$ | Reynolds number |
| R | Geometric scale ratio |
| T | Tank diameter |
| Z | Liquid level |

## Greek Letters

| | |
|---|---|
| $\rho$ | Liquid density |
| $\sigma_y$ | Yield stress |
| $\mu$ | Viscosity |

## References


**1** Arratia, P., Kukura, J., Lacombe, J., and Muzzio, F.J. Mixing of shear-thinning fluids with yield stress in stirred tanks. *AIChE Journal* 52 (7): 2310–2322. https://doi.org/10.1002/aic.10847.

**2** Elso, T., Cheesman, D., and Nienow, A. (1986). X-ray studies of cavern sizes and mixing performance with fluids possessing a yield stress. *Chemical Engineering Science* 41 (10): 2555–2562.

3 Etchells, A., Ford, W.N., and Short, D.G.R. (1987). Mixing bingham plastics on an industrial scale. *Fluid Mixing III, European Federation of Chemical Engineering Publication Series, No. 63*, University of Bradford, Bradford, England (8–10 September 1987).

4 Amanulla, A., Hjorth, S., and Nienow, A. (1998). A new mathematical model to predict cavern diameter in highly shear thinning power law liquids using axial flow impellers. *Chemical Engineering Science* 53 (3): 455–469.

5 Benz, G. (2019). What percent biomass solids can be agitated with a turbine agitator? *Blog post, Lee Enterprises Consulting blog page*, February 2019.

6 Benz, G. (2012). Determining torque split for multiple impellers in slurry mixing. *Chemical Engineering Progress* 108: 45–48.

7 Benz, G. (2018). Equipment needs to study mixing of fibrous materials. *Biofuels Digest Online* (16 April 2018).

8 Benz, G. (2010). Agitation of fibrous materials. *Chemical Engineering Progress* 106: 28–32.

9 Benz, G. (2018). Consider hydrofoil impellers for laminar flow mixing. *Chemical Engineering Progress* 114: 30–35.

10 Benz, G. (2017). Hydrofoil impellers vs. pitched blade turbines in lignocellulosic slurries. *Biofuels Digest (Online Publication)* (8 May 2017).

# 14

# Use of CFD in Biofuel Agitator Design

## Introduction

This chapter describes the use of CFD (computational fluid dynamics) in biofuel/renewable chemical agitator design, told from the viewpoint of one who is not a CFD expert. As such, theoretical discussion will be very brief. The focus of this chapter is on what CFD can do to help the designer of agitation equipment predict its performance, and the owner of processing facilities to get an idea of the possibilities.

CFD can do much more than described herein. A more complete set of examples and descriptions comes as a CD included with Ref. [1]. Our focus is on CFD simulations that are most useful for general agitator design, including the special case of yield-stress fluids, as typically found in lignocellulosic hydrolysis reactors.

The main sources of the images used in this chapter are www.Bakker.org, and private images from Tim Laska, an independent CFD consultant. All are used with permission, and will be attributed as appropriate.

## Basic Theory

In the purest form, CFD numerically solves the Navier-Stokes equations and equations of continuity. This can be done by setting up a series of grids or nodes at all surfaces, whether moving or stationary, and setting a boundary condition that a differential layer of fluid is attached, or moves, with the surface. Examples of grids are shown for a deeply concave impeller (Chemineer BT-6) and the tank wall in Figures 14.1 and 14.2.

*Agitator Design Technology for Biofuels and Renewable Chemicals*, First Edition. Gregory T. Benz.
© 2022 John Wiley & Sons, Inc. Published 2022 by John Wiley & Sons, Inc.

**Figure 14.1** BT-6 grid. *Source:* Courtesy, Bakker.org.

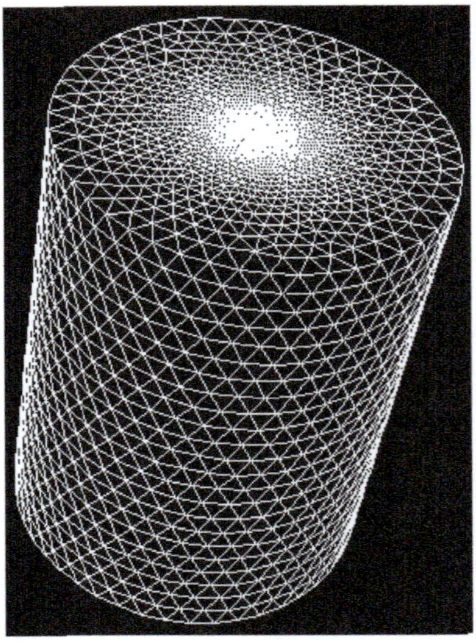

**Figure 14.2** Tank grid. *Source:* Courtesy, Bakker.org.

The solutions, of course, are approximations, as all numerical solutions are. The actual calculations at each node or gridline are arithmetic in nature, and when the grids are fine enough, the output forms a reasonably close approximation of what is happening within the tank.

## What CFD Can Do

The most basic output from these calculations is a velocity distribution. Normally, this is a time-averaged distribution. The instantaneous distribution can be very different in turbulent flow. Additional processing can be done to determine pressure distribution, power dissipation, impeller power draw, shear distribution, transient blending progress, etc. With additional functions added in, such things as solids suspension profile, bubble size distribution, holdup distribution, mass transfer coefficient distribution, dissolved species distribution, reaction progress, product distribution, local heat transfer coefficients, dead zones and cavern size, and many other things can be calculated. Not all of what can be calculated has been validated experimentally. We will explore some of the more useful calculations in subsequent sections.

Not all simulations start with an impeller grid. Some start with experimentally determined impeller boundary conditions, and then solve for velocity distributions throughout the tank. This method saves time setting up the CFD model and may be more accurate. However, it requires "wet" testing of impellers under conditions not too far from the model being simulated in order to create the boundary conditions. It is the hope of CFD modelers to come up with accurate enough CFD codes so that the wet testing can be skipped. In some cases, they may have done so well enough to skip the "wet" experimental program. However, this is not generally the case for rheologically complex macrofluids such as biomass slurries.

We will comment on whether the CFD models have been validated in subsequent sections.

## Methods of Presenting Data

Some data, such as velocity distribution, have both magnitude and direction. Thus, they can be presented as vector plots. Typically, the direction of the vector indicates the direction of flow (it can be instantaneous or time-averaged), the length of the vector is proportional to velocity, and the color of the vector is coded to velocity.

If only the magnitude of the variables is to be presented, a scalar plot is used. This is often in the form of a raster plot, in which various regions are color coded based on magnitude. For velocity raster plots, for example, the color is coded to velocity. By convention, the color red represents a high magnitude, with the magnitude reducing as the color goes through orange, yellow, and various shades of green and blue. Purple or indigo are not normally used, so we do not have the full "Roy G. Biv" rainbow color spectrum.

Simulated particle tracks can be shown, as well as stressed virtual surfaces. Results can be presented in two dimensions or simulated three dimensions. Animations or movies can be created. New methods of presenting data visually are constantly being developed.

While some of these simulations are visually amazing and fun to watch, they can take up a lot of personnel time and expense to create and a lot of computer time to execute, with no certainty that the results are valid. Just as important, many such simulations do not actually help us design some applications more effectively. Most information that is useful for common designs can be presented as a quasi-steady-state simulation in two dimensions using a raster plot. We will show a few of these in subsequent sections. The reader may think of specific needs not addressed here, and there are many CFD modelers eager to help.

We will start with the simplest and most commonplace CFD simulation.

## Velocity Distribution

The most common use of CFD for applications in the biofuels/renewable chemicals segment is velocity distribution. Many agitator vendors have the software set up so that the CFD model can be automatically loaded and solved for simple geometries and Newtonian fluids. Some have models set up for pseudoplastic fluids, and some even have models for Herschel–Bulkley fluids. Vendors that have such automated software do not usually charge money for running it as part of their equipment proposal, as it helps customers to visualize results with the quoted equipment. Samples of two-dimensional (2D) vector and raster plots are shown in Figures 14.3 and 14.4.

In general, these types of plots have been well validated by use of LDV (laser-Doppler velocimetry), DPIV (digital particle image velocimetry), and similar methods. Generally, they exhibit an error band of about ±5% when compared with experimental data.

More complicated geometries may require a three-dimensional (3D) plot, and may require the services of an application engineer or a professional CFD modeler to adequately set up the model and present the results. These services are not generally free and may cost thousands of dollars for complex applications.

Rheologically complex fluids are also an area that may require specialists to adequately simulate results. Polymer solutions, for example, may exhibit viscoelasticity. The main complex fluid we routinely deal with in this industry is high-concentration biomass slurries, which have a yield stress and are usually shear thinning also. We will deal with this in a subsequent section.

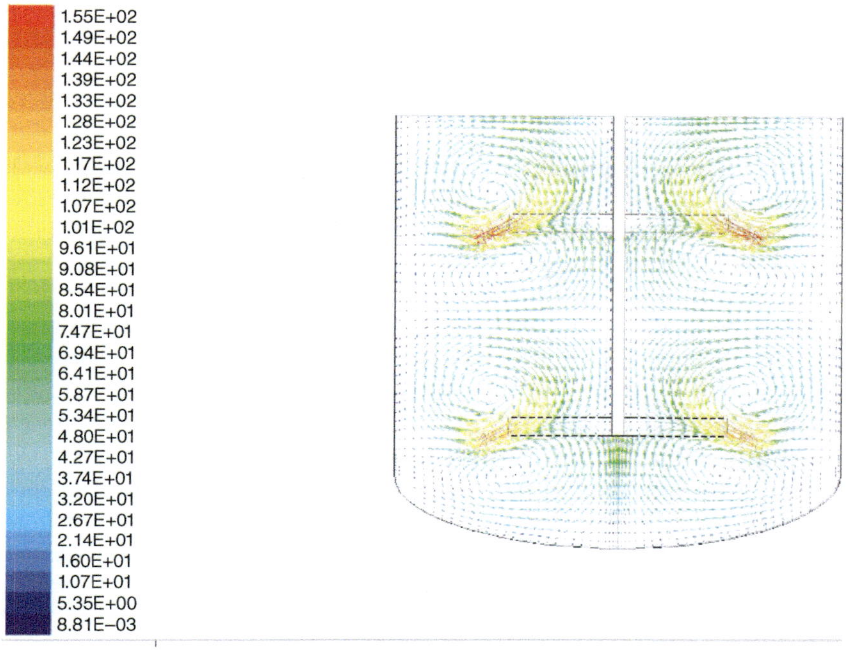


**Figure 14.3** Vector velocity plot. *Source:* Courtesy, Chemical Engineering Progress.

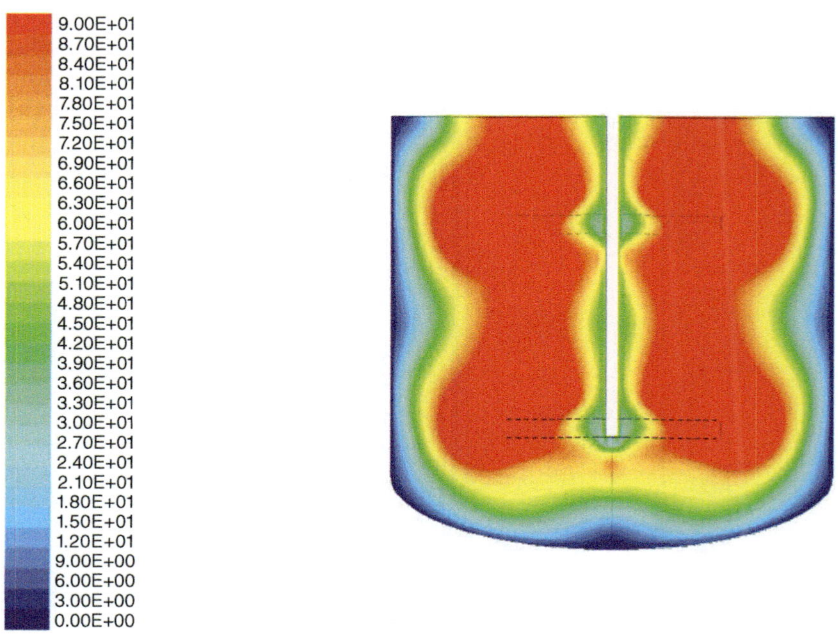


**Figure 14.4** Raster velocity plot. *Source:* Courtesy, Chemical Engineering Progress.

## Flow Around Tank Internals

Though not very common in biofuel applications (except for catabolic processes), tank internals are sometimes used for heat transfer or other purposes. Flow around and between such internals can be a concern. CFD techniques can be used to model these. The results are an extension of the general flow field, and are probably quite accurate, though not directly validated.

Figure 14.5 shows a simplified layout of a series of 12 vertical tube bundles, along with a phantom view of each impeller, in both elevation and plan views. Each bundle consists of two rows of tubes, with four tubes per row.

Figure 14.6 shows a series of cross-sectional raster plots in the plane of each impeller, going from top to bottom in the vessel and going from left to right and top to bottom on the plot. One can clearly see that the velocity distribution is not uniform, and that the distribution is quite different for the upper axial impellers from what it is for the lower radial impeller. Interestingly, the lowest velocities are found at the innermost row of tube bundles. There are no dead spots at the vessel wall.

Many other possible tank internals can be modeled.

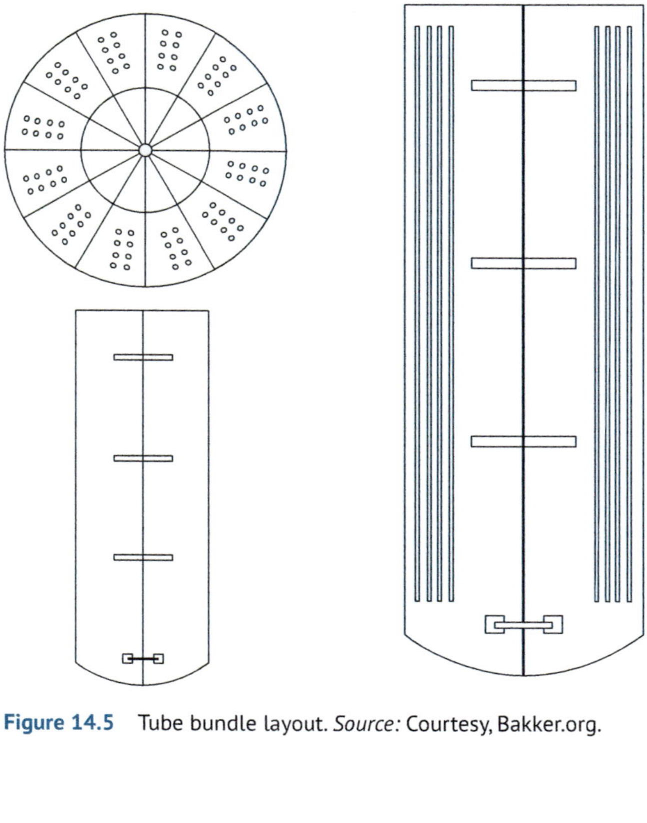

**Figure 14.5** Tube bundle layout. *Source:* Courtesy, Bakker.org.

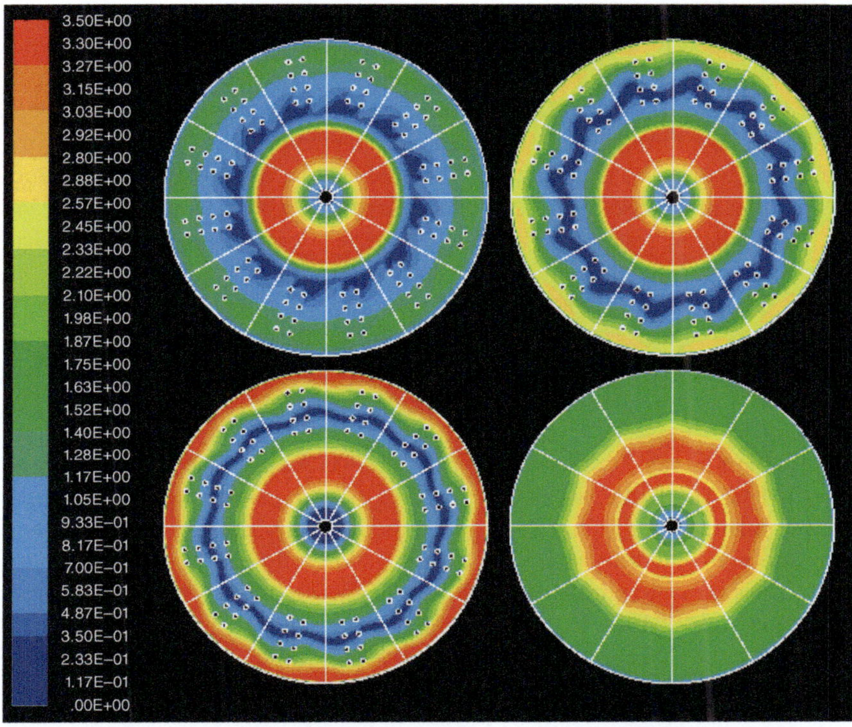


**Figure 14.6** Tube bundle velocities. *Source:* Courtesy, Bakker.org.

## Prediction of Dead Spots (Weak; Need "Wet" Test to Determine Properties)

One of the potentially useful applications of CFD is to predict dead spots or cavern formation of fluids with yield stresses, as described in Chapter 12. This would be especially useful for agitation of lignocellulosic materials, as described in Chapter 13. The results could be easily displayed using conventional vector and raster plots. Unfortunately, this turns out to be a rather weak use of CFD, not because it will not work, but because conventional viscometers cannot characterize the rheology of such fluids due to phase separation and large particles that exceed the gap in bob-and-cup rheometers. Therefore, "wet" testing is still needed to determine rheology, using an agitator as a rheometer as described in Chapter 12. The resulting rheology is normally expressed as a function of shaft speed rather than shear rate. Though in principle the data can be converted to rheology as a function of shear rate, there is uncertainty about the Metzner–Otto constant, which therefore translates into uncertainty while modeling with CFD. Moreover, since the apparatus needed to use an agitator as a rheometer is essentially the same as that needed to do a mixing test (except for possibly the number of impellers and the size of the tank if a larger mixing test scale is desired), the incremental labor to do a full mixing test is small, and the results can be scaled up with confidence. Based on this consultant's experience, the time required to conduct and interpret mixer viscosity tests is about three to five hours, including running the test, calculating results, and cleaning up the vessel, impellers, and work space. The incremental time for a mixing test based on a single concentration of biomass is about another hour. Longer time will be needed for varying solids content or observing the progress of hydrolysis and varying mixing speed or impeller systems as the sample thins out over time. In general, the time and expense of a wet test will be less than the time and cost required to model using CFD. This may change in the future.

In addition to velocity plots, there can be novel ways of visualizing caverns. CFD modeler Tim Laska has modeled caverns using apparent viscosity as a proxy for movement. Under higher shear rates, the viscosity is lower, and under low shear zones, the viscosity is higher. Thus, high viscosity may be associated with dead or slow-moving zones. Tim has illustrated this method in Figure 14.7. In this illustration, the red areas are more or less stagnant.

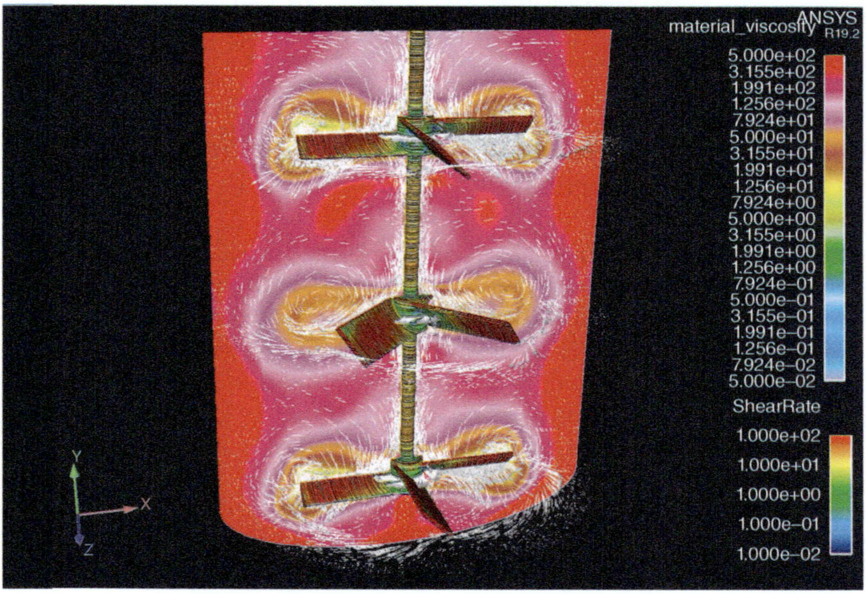


**Figure 14.7** Caverns by apparent viscosity. *Source:* Courtesy, Tim Laska.

In general, flow fields for such fluids have not been validated due to the opaque nature of the fluid, but dead areas at the wall seem to be reasonably well validated if the rheology is known.

## Blending Simulation

CFD techniques can be used to model the progress of liquid blending. In Figure 14.8a–d, we model the progress of blending in a tank that starts out with a stratified condition with roughly equal portions of two miscible species, which are

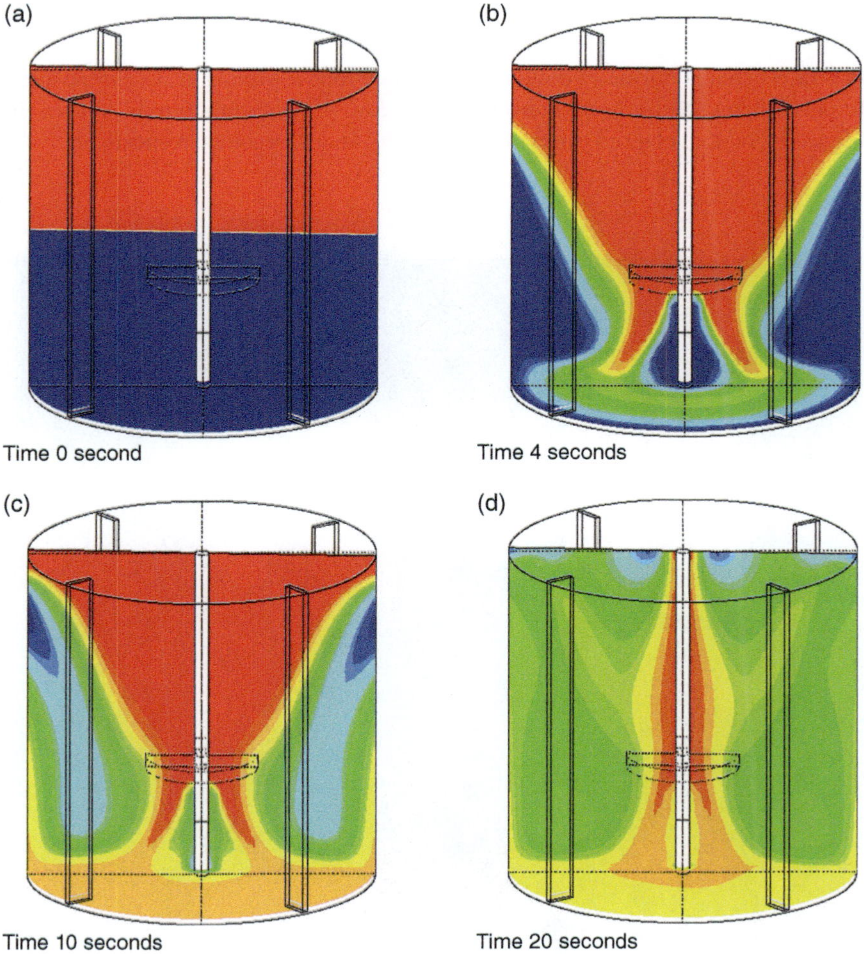


**Figure 14.8** (a–d) Blending simulation. *Source:* Courtesy, Bakker.org.

illustrated as "red" and "blue." Simulations are run at the starting condition (time = 0) and at 4, 10, and 20 seconds. The tank has been blended to a fairly high degree of uniformity after 20 seconds. Experimental test data have validated these models.

## Reaction Simulation

Though not common in this industry, competitive-consecutive reactions may be simulated by CFD. Such reactions generally are of the form:

$$A + B \rightarrow R; \quad R + B \rightarrow S \tag{14.1}$$

The goal is typically to maximize production of R by minimizing the concentration of B. This is normally done in a fed batch mode. Mixing is very important. Rapid mixing keeps local concentrations low. The simulation is shown in Figure 14.9a–f, at elapsed times of 0, 1, 2, 4, 10, and 20 seconds, respectively.

(a)

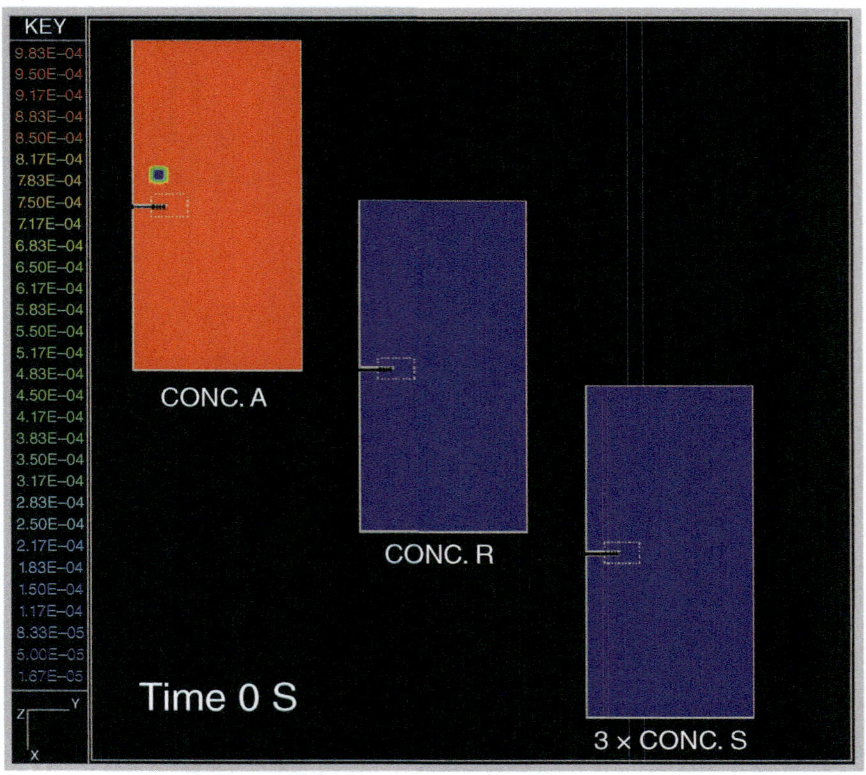


**Figure 14.9** (a–f) Reaction simulation. *Source:* Courtesy, Bakker.org.

(b)

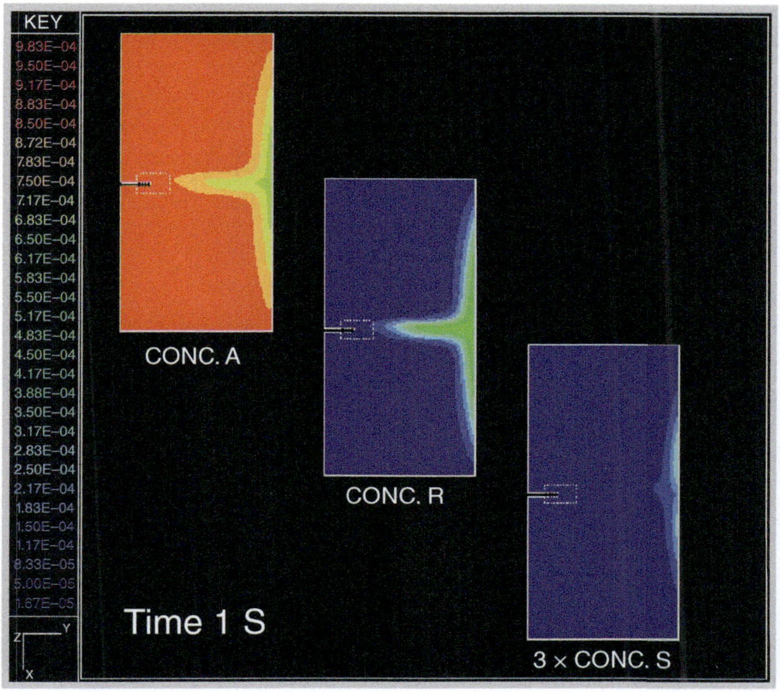


(c)

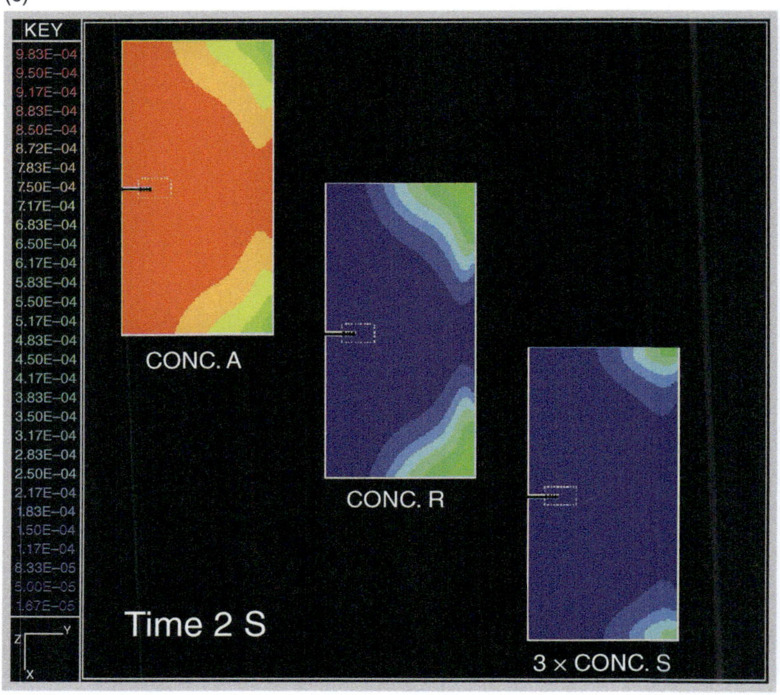


**Figure 14.9** (Continued)

(d)

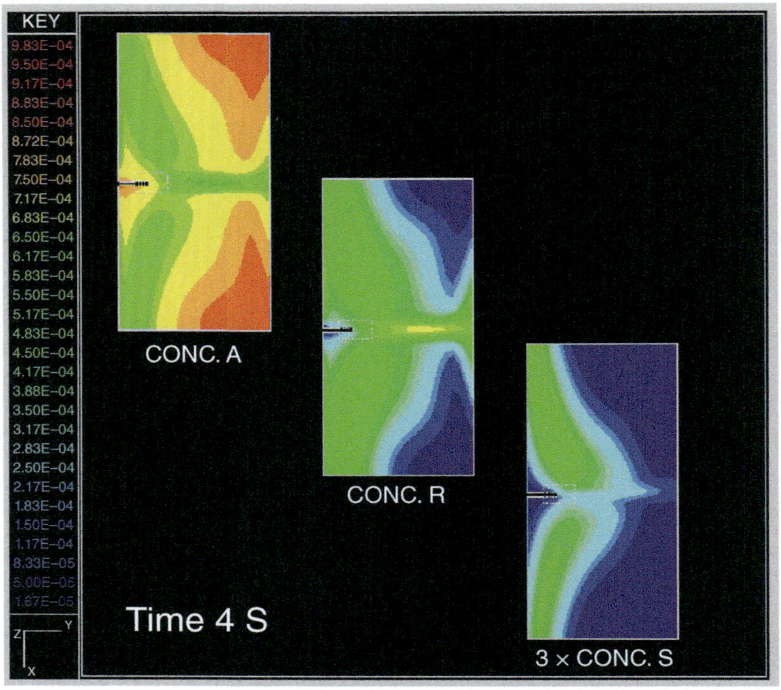


(e)

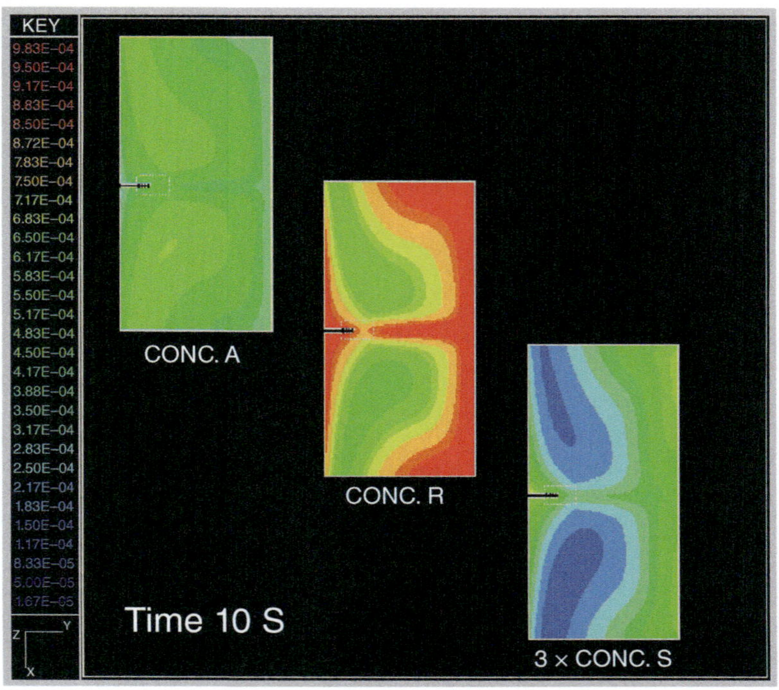


**Figure 14.9** (Continued)

(f)

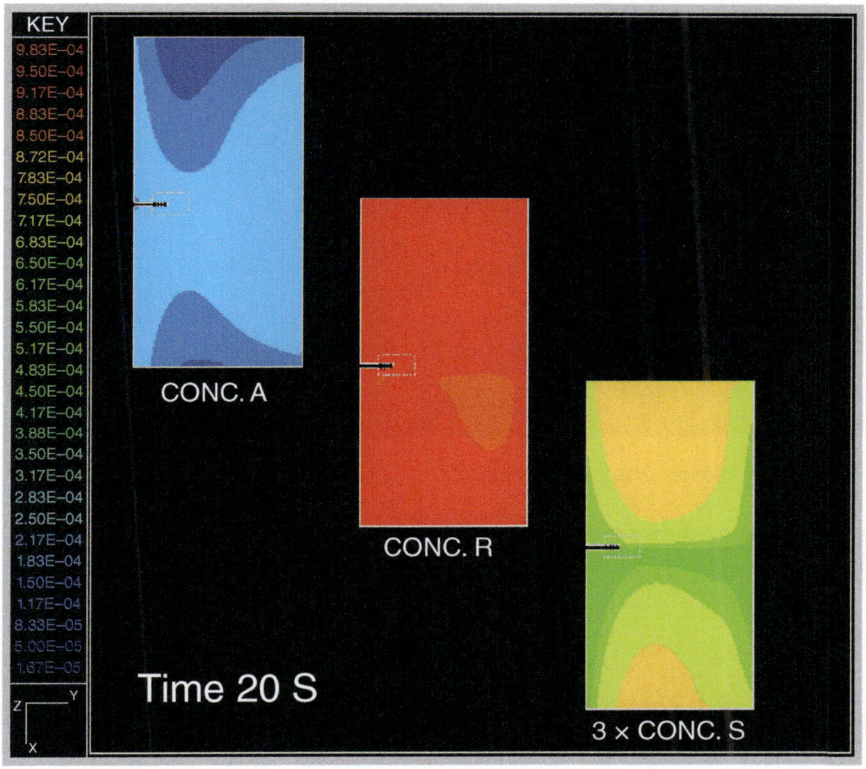


**Figure 14.9** (Continued)

The author is not sure if such results have been validated. It would be quite difficult to do so.

## Some Other Examples

Many other uses and outputs of CFD are possible, far too many to illustrate here. We show a few more mainly to emphasize the variety of what is available.

Figure 14.10 shows the velocity around a pitched blade turbine, close to the impeller, unlike the usual presentation of a cross-section of a tank. This allows more detailed analysis, and further processing can compute such things as energy dissipation and shear rate distribution.

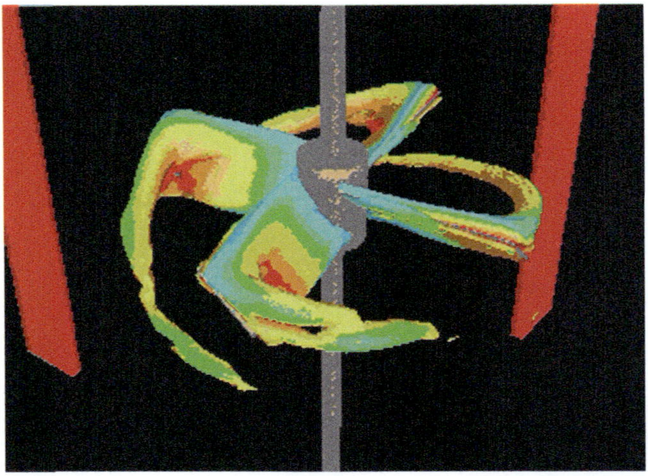

**Figure 14.10** Velocity near a pitched blade turbine. *Source:* Courtesy, Bakker.org.

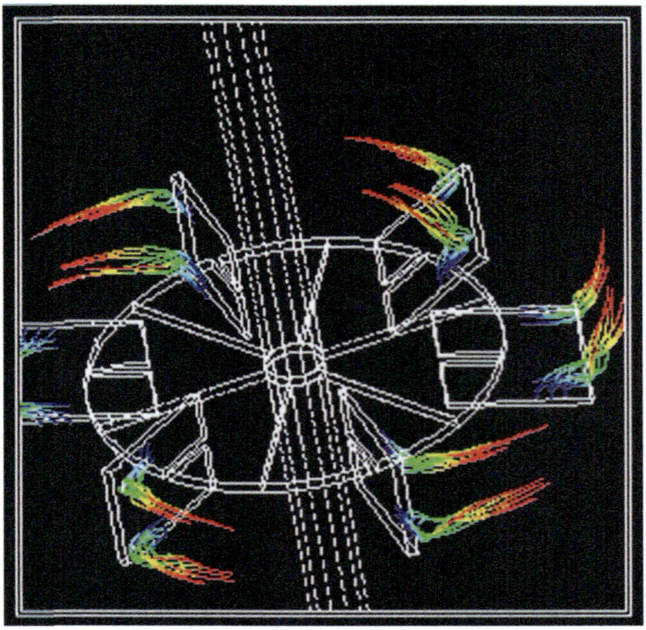

**Figure 14.11** Velocity and paths of a Rushton turbine. *Source:* Courtesy, Bakker.org.

Figure 14.11 shows the velocity near the tips of a Rushton turbine, as well as potential fluid flow paths. The lines record potential flow paths, and the color of the lines is keyed to velocity. Red is high; blue is low velocity.

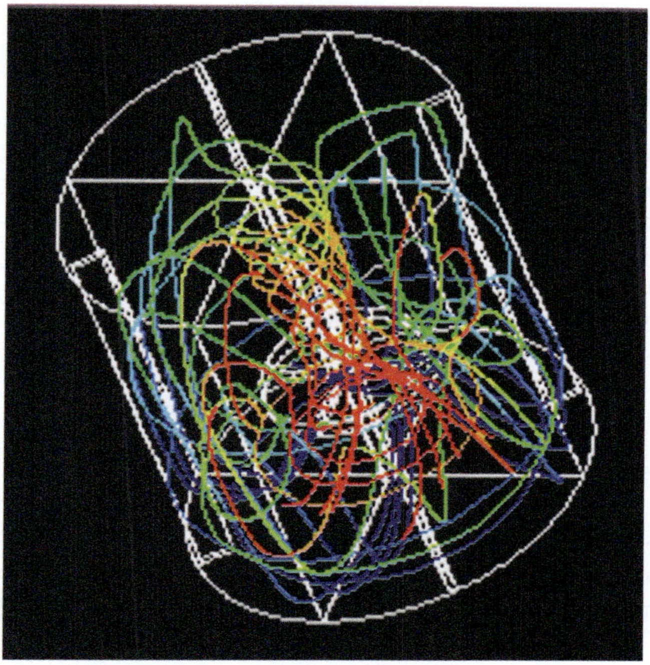

**Figure 14.12** Particle trajectories. *Source:* Courtesy, Bakker.org.

Figure 14.12 shows possible particle trajectories in a simulated 3D view of a tank. These simulations are run for a short period of time, as eventually the particle paths would fill the whole tank view. The lines represent some of the possible paths hypothetical particles might take. The color of the lines is keyed to velocity.

## A Few Words About CFD Consultants

Most of us do not have direct access to CFD simulation software or have the necessary expertise to set up and use such models. Therefore, we must depend on equipment manufacturers or consultants to do this for us. However, there are literally thousands of independent consultants offering CFD modeling services. To avoid potential frustration dealing with them, or worse, inaccurate results, I suggest limiting the field to those who also have at least several years of experience in the field of fluid agitation. Working with someone who has only modeled airplane wings or smokestack emissions may not produce the best results.

## Other Resources

Since this is a book on agitation, not specifically on CFD, we have just scratched the surface of what is possible in this chapter. Suppliers of CFD systems, such as ANSYS, Autodesk, OpenFoam, Comsol, and many others, can give a better indication of what they can do. Their websites often have very informative examples. Many independent CFD consultants are very creative, and their websites showcase what they can do.

This book does not have any animations in it, but several websites do. Reference [1] has some in the CD accompanying the book. One very good source is Dr. Bakker's website, www.bakker.org. Animations can be very helpful in understanding what is going on in a vessel.

## Summary of Chapter

We have described some of what CFD simulations can do, as well as the limitations associated with macro fluids and complex rheology. We have demonstrated some of the uses, while emphasizing that CFD cannot as of yet replace wet testing for all applications.

## List of Symbols

A Concentration of reactant "A"
B Concentration of reactant "B"
R Concentration of reactant "R"
S Concentration of reactant "S"

## Reference

1 Paul, E., Atiemo-Obeng, V., and Kresta, S. (ed.) (2004). *Handbook of Industrial Mixing*. Wiley InterScience.

# 15

# Mechanical Design of Agitators 1

Shafts and Impellers

## Introduction

For any process to work profitably, the equipment used in it must be reliable. This is certainly true of agitators. In general, agitators are reliable pieces of equipment. In normal service, such as blending of miscible liquids with a mild-to-medium degree of agitation, the products made by the leading agitator suppliers have a useful service life of more than 20 years, and have modest maintenance requirements, consisting largely of changing the oil, greasing the bearings, and changing the seals when needed. In this chapter, we will describe the mechanical design of shafts and impellers. In Chapter 16, we will delve into gear drive selection. Both topics are important for successful agitator design.

Because the biofuels/renewable chemicals industries are typically making low-value, low-margin products, there is considerable pressure to minimize capital expense (CAPEX) on all equipment, including agitators. This has sometimes led buyers of such equipment to buy based on price, and vendors of agitation equipment to quote marginal designs with little safety factor, resulting in lower reliability than what is possible and prudent. That is why the buyer should be cognizant of agitator mechanical design and be actively involved in the decision process regarding mechanical integrity and safety factors.

The intent of this chapter is not to teach the user how to do detailed mechanical design of agitators. That is still best left up to the equipment vendors, or to consultants with very specialized knowledge in this field. (That does not include me. My expertise is primarily in the chemical engineering aspects of agitator design, though I have some basic mechanical engineering knowledge about the subject. But I leave the tough stuff and detailed design to the mechanical engineering experts.) Rather, the intent of this chapter is to familiarize the reader with the basic aspects of agitator design so that he or she can understand the issues involved,

*Agitator Design Technology for Biofuels and Renewable Chemicals*, First Edition. Gregory T. Benz.
© 2022 John Wiley & Sons, Inc. Published 2022 by John Wiley & Sons, Inc.

can understand what the equipment vendor needs to know from the user, and have the ability to participate in some decisions about cost versus reliability.

Some of the inspiration for this chapter came from Ref. [1]. Other information came from a variety of sources. General mechanical engineering handbooks also supplied information, but will not be referenced by line item in the text because such information is commonplace. See Refs. [2] and [3] for examples of general mechanical engineering texts. References [4] and [5] also provide useful mechanical engineering guidance and data.

## Impeller Design Philosophy

Normally, agitator manufacturers will design each impeller for its share of the total motor power. When there are multiple impellers on one agitator shaft, this means that each impeller is unable to withstand full motor power. For many applications, this is a reasonable approach, as the motor will shut down if the total agitator power exceeds full motor load for more than a few seconds. However, if there are multiple impellers and the agitator is operated when the tank is only partially filled, such that all impellers are not fully submerged in the liquid, this can be a dangerous practice if something changes the fluid properties so as to increase the impeller power draw above normal.

One example would be if the viscosity increased. This can happen in a polymer reactor, for example. A possible scenario where this might happen in the biofuels industry would be in a lignocellulosic hydrolysis reactor, operating in fed batch mode, where the apparent viscosity at the beginning may be much higher than when the tank is full, possibly fully loading the motor with fewer impellers submerged than the total quantity installed. Depending on the number of impellers, this could result in loading the impeller to several times its prorated share of motor power.

For tanks with no possible change in viscosity, it is good and normal practice to design each impeller for its prorated share of motor power. But if there is a possibility of loading the motor with only some of the impellers submerged, the following practice is recommended.

Starting with the bottom impeller, design it for 100% of motor nameplate power at the design shaft speed. Going up the shaft, design the second impeller for its prorated power based on 100% of motor power, even if the normal total of the first two impellers is less than 100%. Repeat for each impeller until you get to the top impeller, which merely need be designed for its share of motor power. To help clarify this, we will illustrate it for a system composed of four identical impellers, to keep it simple.

- The bottom impeller should be designed for 100% motor nameplate. With a four-impeller system, that means it is designed for 4 times its normal, prorated load.

- The second impeller up from the bottom should be designed for ½ of motor nameplate, or its share of full motor load. This is double what its normal, prorated load would be.
- The third impeller from the bottom should be designed for 1/3 of motor nameplate power, which is 33% more power than its normal prorated load.
- The fourth (and top) impeller need only be designed for its normal prorated load of 25% of motor nameplate power.

If the user anticipates this kind of loading possibility, it behooves him to discuss it with the agitator vendor, as it is not their normal practice to design this way. Agitator manufacturers do not warranty against improper operation of the equipment. Operating a partially full tank with a fully loaded motor would be considered improper operation unless explicitly stated in the design basis and warranty. Therefore, this must be spelled out in the agitator data sheet and incorporated into the purchase order and warranty. If it is not, and there is a mechanical failure of one or more lower impellers due to operating at conditions beyond the stated criteria, there will be no warranty on the failed parts.

Impeller mechanical design is more complex than shaft design. Moreover, there are many kinds of impellers on the market, and many of those are proprietary. Even among well-known impellers such as the Rushton turbine, there are many variations in number of blades, blade height, and method of fastening to the disk and hub, making mechanical design complicated. Therefore, we will not get into much detail on mechanical design of impellers. We will mostly have to leave that to the equipment vendors. However, we will illustrate the design concepts in some detail for a generic low-solidity hydrofoil impeller, which is the most common type in use for most liquid motion and solids suspension problems. We will do this after the next section, which discusses hydraulic force, as hydraulic force is involved in both impeller design and shaft design.

## Discussion on Hydraulic Force

As we will see later, the two major loads determining shaft design for stress are torque and bending moment. The origin of torque is obvious: it is the result of power being transmitted at a certain shaft speed. But where does the bending moment come from? It is the result of random unbalanced hydraulic forces acting on the impeller(s), and the moment arms those forces act upon. But where do the hydraulic forces come from? Over the years, I have often been asked why the manufacturers do not just make the impellers highly precise and fully balanced in order to avoid such forces. And, indeed, there are forces associated with mechanical imbalance and lack of perfect symmetry in impeller construction. But these are minor forces compared to the *hydraulic* imbalanced forces produced. And those forces do not originate from any imperfection in impeller construction, but from the nature of turbulence.

To understand this better, imagine riding along in a car at 100 km/h, with your hand sticking out of the window. You will note there is a drag force on your hand. However, it is far from constant. In fact, your hand will be buffeted about as the forces fluctuate wildly. This is due to turbulence. It is the same for agitator blades in a tank.

Each blade, just as your hand does, experiences varying loads. Importantly, those variations are not synchronized or symmetric among each of the blades. Therefore, there is also a fluctuating net side load on the impeller. This load is highest under highly turbulent conditions, and theoretically falls to zero in laminar flow. Having more blades per impeller reduces the peak magnitude of this load. However, there are practical limits to the number of blades that can be used, including expense and the fact that, for producing flow, fewer blades are better for a given power input. The most common hydrofoil impellers have three blades as a reasonable compromise between pumping performance and mechanical stability. As noted in Chapter 4, a one-blade impeller of a given blade geometry would pump more for the power input than a two- or three-bladed impeller, but the hydraulic forces would be unmanageable for a one-blade impeller, and still quite high for a two-bladed impeller.

The magnitude of this load can be empirically correlated by means of a concept called dimensionless hydraulic force:

$$F_h = F_D * T_Q / D \tag{15.1}$$

where:

$$T_Q = P / (2\pi N) \tag{15.2}$$

The value of $F_D$ depends on the number of blades, the agitation intensity, the mounting arrangement (it is higher for off-center mount, for example), the flow direction, and whether it is single phase or gas–liquid.

Table 15.1 summarizes dimensionless hydraulic factors for typical generic impellers and should be adequate for high-solidity impellers also. The values have not been previously published to my knowledge and come from several sources and my own experience. They are intended for design purposes and may be slightly conservative. However, irregular geometries could have much higher forces.

For example, let us suppose that we have a 1.5 m diameter axial flow impeller with three blades operating at 100 rpm (1.667/second), and it draws 10.55 kW. The resultant torque is (10 550 W/(1.667/second * 2 * π) = 1007 N-m. (1 W = 1 N-m/s)). We assume for biofuel applications that it is in a large enough tank that it produces

**Table 15.1** Dimensionless hydraulic forces.

| Impeller number of blades | Direction of pumping | Problem class | Dimensionless hydraulic force |
|---|---|---|---|
| 1 | Does it matter? | Any | 2.0 |
| 2 | Any | Mild blending | 0.6 |
| 2 | Any | Vigorous blending | 0.75 |
| 2 | Radial | Blending | 0.6 |
| 2 | Radial | Gas/liquid | 0.75 |
| 2 | Down | Gas/liquid | 0.9 |
| 2 | Up | Gas/liquid | 0.75 |
| 3 | Any | Mild blending | 0.45 |
| 3 | Any | Vigorous blending | 0.5 |
| 3 | Radial | Blending | 0.45 |
| 3 | Radial | Gas/liquid | 0.7 |
| 3 | Down | Gas/liquid | 0.9 |
| 3 | Up | Gas/liquid | 0.65 |
| 4 | Any | Mild blending | 0.3 |
| 4 | Any | Vigorous blending | 0.4 |
| 4 | Radial | Blending | 0.3 |
| 4 | Radial | Gas/liquid | 0.7 |
| 4 | Down | Gas/liquid | 0.9 |
| 4 | Up | Gas/liquid | 0.6 |
| 5 | Radial | Gas/liquid | 0.75 |
| 6 | Radial | Gas/liquid | 0.6 |

The above applies to center mounted agitators in cylindrical vessels only.

normal agitation, rather than violent agitation. From Table 15.1, our dimensionless hydraulic force design value is 0.45, so our hydraulic force is 0.45 * 1007 N-m/1.5 m = 301 N. For those who think in English units, this equates to 67.9 lb force. Some would use 30.9 kg-f, but that is not an Systeme Internationale (SI) unit.

High hydraulic forces also increase the stress on individual blades due to variation about the mean force per blade. Some vendors do not take this into account. While we do not delve into detailed impeller calculations here except for the conceptual example below, it may be useful to ask vendors to show their blade stress calculations and show how they take hydraulic force into account.

## Example Mechanical Design of Generic Hydrofoil Impellers

Normally, hydrofoil impellers (and most others) are constructed by attaching blades to a central hub, which is mounted to the shaft. A typical example is shown in Figure 15.1. In most cases, vendors will have a series of standard hubs, rather than design them for each application. This is especially true when the shaft diameter is less than 150 mm, as most such hubs will be investment castings, and it would be prohibitive to custom make an investment casting mold for each application. The hub characteristics must be known before impeller blade thickness can be calculated and total impeller weight determined, which will be needed later for shaft critical speed calculations. The hub is selected based on three attributes: shaft size, design torque, and maximum blade width. We will use impeller diameter rather than blade width, using the simplifying assumption that our hydrofoils will have a blade width of D/6. (Sometimes vendors will have certain standard blade widths, so the w/D ratio may vary somewhat.)

Table 15.2 lists a series of standard hydrofoil hub designs, generated using software owned by the author. It does not necessarily match the hubs used by any actual agitator manufacturer, but it should be reasonably close. The proper hub would be selected from such a table based on the three criteria just mentioned. Our table is based on the strength and density of 316SS, at a temperature of up to 100 °C. It is also based on a dimensionless hydraulic force value of 0.45, which is typical for such impellers mounted in a vertical cylindrical tank, on center, with a scale of agitation

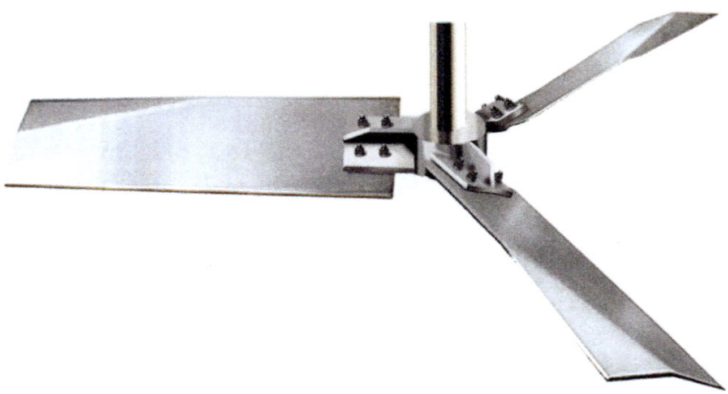

**Figure 15.1** Chemineer HE-3 impeller. *Source:* Courtesy, NOV.

**Table 15.2** Hydrofoil hub selection.

| Shaft diameter (in./mm) | Maximum impeller torque (in.-lb/N-m) | | Maximum impeller diameter (in./m) | | Hub diameter (in./mm) | | Hub mass (lb/kg) | |
|---|---|---|---|---|---|---|---|---|
| 2/51 | 10 000 | 1 130 | 60 | 1.52 | 12 | 305 | 57 | 26 |
| 2/51 | 35 000 | 3 955 | 90 | 2.29 | 16 | 406 | 96 | 44 |
| 2.5/64 | 10 000 | 1 130 | 60 | 1.52 | 12 | 305 | 54 | 25 |
| 2.5/64 | 35 000 | 3 955 | 90 | 2.29 | 16 | 406 | 91 | 41 |
| 3/75 | 10 000 | 1 130 | 60 | 1.52 | 12 | 305 | 50 | 23 |
| 3/75 | 35 000 | 3 955 | 90 | 2.29 | 16 | 406 | 88 | 40 |
| 3/75 | 61 000 | 6 893 | 108 | 2.74 | 20 | 508 | 176 | 80 |
| 3.5/90 | 35 000 | 3 955 | 90 | 2.29 | 16 | 406 | 85 | 39 |
| 3.5/90 | 61 000 | 6 893 | 108 | 2.74 | 20 | 508 | 170 | 77 |
| 4/100 | 35 000 | 3 955 | 90 | 2.29 | 16 | 406 | 79 | 36 |
| 4/100 | 61 000 | 6 893 | 108 | 2.74 | 20 | 508 | 163 | 74 |
| 4/100 | 125 000 | 14 125 | 144 | 3.66 | 24 | 610 | 283 | 129 |
| 4.5/114 | 35 000 | 3 955 | 90 | 2.29 | 16 | 406 | 72 | 33 |
| 4.5/114 | 61 000 | 6 893 | 108 | 2.74 | 20 | 508 | 154 | 70 |
| 4.5/114 | 125 000 | 14 125 | 144 | 3.66 | 24 | 610 | 276 | 125 |
| 5/127 | 61 000 | 6 893 | 108 | 2.74 | 20 | 508 | 145 | 66 |
| 5/127 | 125 000 | 14 125 | 144 | 3.66 | 24 | 610 | 267 | 121 |
| 5/127 | 200 000 | 22 600 | 144 | 3.66 | 26 | 660 | 313 | 142 |
| 5.5/140 | 125 000 | 14 125 | 144 | 3.66 | 24 | 610 | 258 | 117 |
| 5.5/140 | 200 000 | 22 600 | 144 | 3.66 | 26 | 660 | 303 | 138 |
| 5.5/140 | 420 000 | 47 460 | 180 | 4.57 | 30 | 762 | 591 | 269 |
| 6/152 | 125 000 | 14 125 | 144 | 3.66 | 24 | 610 | 247 | 112 |
| 6/152 | 200 000 | 22 600 | 144 | 3.66 | 26 | 660 | 292 | 133 |
| 6/152 | 420 000 | 47 460 | 180 | 4.57 | 30 | 762 | 579 | 263 |

of 6 or less, in turbulent flow. The hub allowable torque ratings would possibly need to be decreased for higher hydraulic factors. We will not get into that detail here.

## Hub Selection Example

Suppose we have a hydrofoil impeller of 1.75 m diameter, rotating at a shaft speed of 68 rpm in a water-like fluid with a density of 1200 kg/m$^3$. Its power draw would be $0.3 * 1200 * 1.75^5 * (68/60)^3 = 8601$ W, or 8.6 kW. If this is the only impeller on the shaft, we would select a standard motor size of 11 kW. Our design torque, based on the philosophy of designing for full motor power, would therefore be $11\,000/(2\pi(68/60)) = 1545$ N-m. The minimum shaft diameter at the impeller location, based on a design shear stress of 40 000 kPa, would be 0.0581 m, or 58.1 mm. (The equation we used for this is number 15.10 in the shaft design section). Based on a standard shaft size of 64 mm, the nearest size that is at least 58.1 mm, and a design torque of 1545 N-m, and using the 1.75 m impeller diameter, we would choose the hub rated for 3955 N-m torque, tentatively mounted on a 64 mm shaft. The upper shaft may need to be larger for critical speed. Based on a 64 mm shaft, the hub weight is 41 kg. If it is later determined that a larger shaft is needed, the same hub can be used with a larger bore, up to 114 mm diameter, and the hub weight would be somewhat reduced because of the larger bore. For example, the same hub with a 114 mm bore would weigh 33 kg.

## Blade Stress Calculation Example

Though some sources and vendors do not account for hydraulic force in impeller design, it really should be taken into account. For a 30° three-bladed hydrofoil only, we can use the following equation, which was communicated privately to the author:

$$\text{Force on blade}\left(F_b\right) = \left(1.378 + 1.333F_D\right) * T_Q / D \tag{15.3}$$

For our above example, assuming a normal dimensionless hydraulic force of 0.45, we would thus get:

$$\text{Force on blade} = \left(1.378 + 1.333 * 0.45\right) * 1545\,\text{N} - \text{m} / 1.75\,\text{m} = 1746\,\text{N}$$

The blade stress can be calculated as follows:

$$\text{Blade tensile stress} = \left(F_b * \left(0.75D - D_h\right)\right) / 2Z_m \tag{15.4}$$

where $Z_m$ is the section modulus of an extension blade, defined next:

$$Z_m = wt^2 / 6 \tag{15.5}$$

The last several formulae are for down-pumping in fluid. Up-pumping would have to account for blade weight on stress. The factor of 0.75 on D in Eq. (15.4) is based on the effective load radius of a constant angle blade. It is slightly conservative for a blade that has an angle of attack that decreases toward the blade tip, which most hydrofoil impellers do.

For our sample blade calculation, assuming our 1.75 m diameter impeller would have a blade width, w, of 0.292 m, and also starting with an estimate of blade thickness of 16 mm (0.016 m), we would have the following results: $Z_m = 0.292 * (0.016)^2/6 = 1.25\text{E-}5\,\text{m}^3$.

The corresponding blade stress is 1746 N(0.75 * 1.75 m − 0.406 m)/(2 * 1.25E-5 $m^3$) = 6.33E7 N/$m^2$ = 63 500 kPa. The recommended allowable blade stress for 316SS used as extension blades is 80 000 kPa, or about 12 000 psi. Our initial estimated thickness resulted in about 79% of the allowable value, so this would be an acceptable blade thickness.

The total blade weight would be (based on a density of 8045 kg/$m^3$) 3(1.75/2–0.07) * 0.406 * 0.016 * 8045 = 126.2 kg. (The value of 0.07 is the estimated hub body radius). The total impeller weight is the blade weight plus the hub weight, or 126.2 + 41 = 167.2 kg. This would be used in shaft critical speed calculations.

For other impeller styles, blade angle and thrust would be involved. These factors are embedded in the formulae used for a 30° impeller. We will not delve into other impellers in this chapter, as there are too many vendor-specific practices. We just wished to illustrate the concept.

## Shaft Design Philosophy

Similar to the impeller design philosophy, the shaft should be designed for 100% of motor power and peak bending moment, at least at the upper part of the shaft. However, many large tanks in the biofuels industry have long enough shafts that they must be made in several sections. With multiple impellers, it is often the case that some impellers are mounted above some shaft sections. In such a case, it is permissible to design the lower shaft sections for the maximum power they could possibly see, which is generally the sum of the actual power of all impellers below the coupling at the top of the respective shaft section, up to a maximum of motor nameplate power. If the motor is not fully loaded, it is advisable to design each section for its actual maximum power divided by the total motor loading factor. For example, if the motor is loaded to 80% with all impeller submerged, each shaft section should be designed for its actual power divided by 0.8. This allows future operation at 100% motor nameplate. As an illustration of this principle, if a shaft is built in three sections and has four impellers, but the bottom section has only two impellers mounted on it, that section need only be designed for the sum of the power of those two impellers divided by the total motor loading factor.

This philosophy does not apply if it is possible to draw more than prorated impeller power, due to changes in viscosity, startup in settled solids, etc., whereby full motor power might be invested with fewer than all impellers submerged. In such cases, all shaft sections must be designed for full motor power. Startup in settled solids will be covered in more detail later in this chapter, under its own heading.

## Shaft Design Based on Stress

Most mechanical system designs for stress start out with a load diagram. For an agitator shaft, the two principal kinds of loads are torque and bending moment. The torque arises from the power draw of the impellers, delivered at a certain shaft speed. The maximum torque for design is generally based on 100% motor nameplate power, as in Eq. (15.2).

The bending moment arises from the hydraulic forces described earlier, applied to a moment arm. For a cantilevered shaft (no steady bearing), the bending moment is the sum of the hydraulic force of each impeller times its distance from the lower gear drive bearing, or the mechanical seal bearing if there is one. (This is slightly conservative, as the hydraulic forces of each impeller are not necessarily aligned at any given moment. However, they can be on occasion, so we design as though they were.) For a unit with a steady bearing (which is commonly the case for tall vessels), the analysis is more complicated, as there are at least two bearing points: one at each end of the shaft. Though computer programs are the best way to design for such a case, a simplified way of doing it manually is feasible and conservative. Refer to Figures 15.2 and 15.3 for the dimensional variables used in the next equations.

For both design concepts, we would normally design for torque calculated based on motor power, using Eq. (15.2). However, for multiple shaft sections, we may not have all impellers contributing power, so we could look at that section summing the torque of impellers that actually contribute torque to that section, as described earlier. For the bending moment, we use Eq. (15.1) to calculate the hydraulic force for each impeller, and then calculate bending moment as follows. We may also calculate bending moment for each shaft section using the moment arm to the nearest shaft coupling at the drive end of that section.

For a cantilevered unit, bending moment is the sum of each impeller's contribution, or:

$$M = \Sigma F_h * L = F_{h1} * L_1 + F_{h2} * L_2 \ldots \tag{15.6}$$

For steady bearing units, each impeller contributes a bending moment in proportion to its distance from each bearing, that is:

$$M_i = F_{hi} * B_i * C_i / L \tag{15.7}$$

Though the bending moments are not totally additive, and occur at different places along the shaft, we can treat them as though they were by summing up the Eq. (15.7) for all impellers:

$$M = \Sigma F_{hi} * B_i * C_i / L = F_{h1} * B_1 * C_1 / L + F_{h2} * B_2 * C_2 / L \ldots \tag{15.8}$$

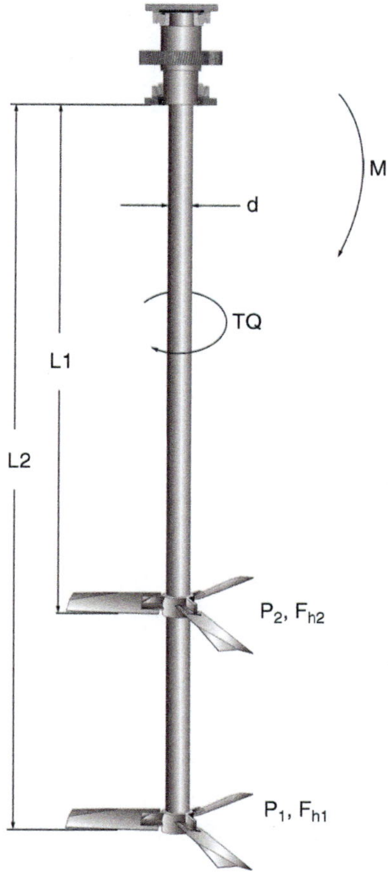


**Figure 15.2** Cantilevered load diagram.

Once we have the torque and bending moment, we can calculate shaft tensile and shear stresses. First, we calculate the combined moment resulting from torque and bending moment:

$$M_C = \left(M^2 + T_Q{}^2\right)^{0.5} \tag{15.9}$$

Then we can calculate shear stress as follows:

$$\sigma_s = M_C * 16/\left(\pi d^3\right) \tag{15.10}$$

Tensile stress is similar, but adds the bending moment to the combined moment:

$$\sigma_t = \left(M + M_C\right) * 16/\left(\pi d^3\right) \tag{15.11}$$

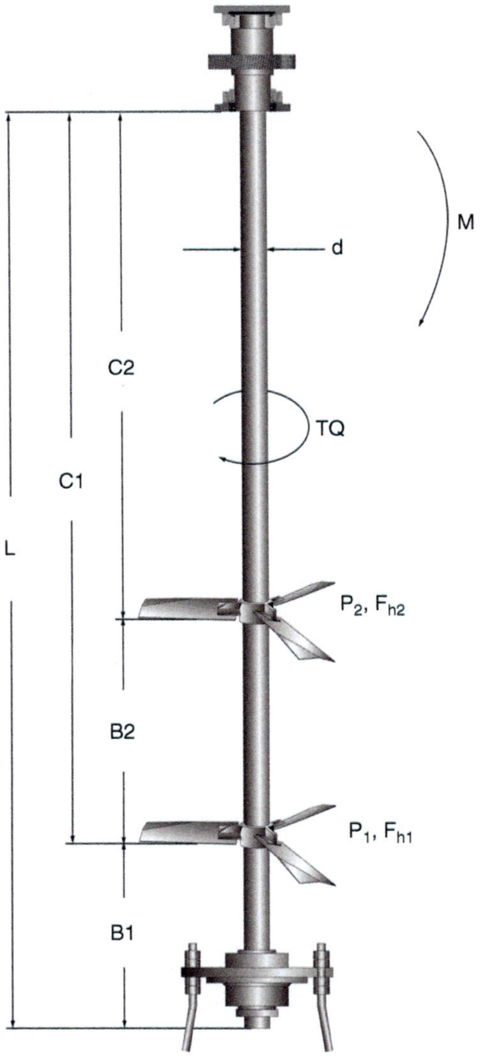


**Figure 15.3** Steady bearing load diagram.

In the case of hollow shafts, which will be discussed later, instead of $d^3$, use $(d_o^3 - d_i^3)$. Though other cross-sections are possible, such as square shafts, I have never personally seen them used in biofuel applications, so they will not be discussed here.

To choose or evaluate shaft size using these equations, it is necessary to know what the allowable shear and tensile stresses are. These vary with material of

**Table 15.3** Allowable stresses.

| | Allowable shear stress | | Allowable tensile stress | |
|---|---|---|---|---|
| **Material** | **psi** | **Pa/1000 or kPa** | **psi** | **Pa/1000 or kPa** |
| 304SS | 5500 | 37923 | 9150 | 63089 |
| 316SS | 5935 | 40922 | 9890 | 68192 |
| 304LSS | 5081 | 35033 | 8470 | 58401 |
| 316LSS | 5240 | 36 130 | 8730 | 60193 |
| 316LSS dual certified | 5935 | 40922 | 9890 | 68192 |
| Alloy 20 | 6340 | 43714 | 10500 | 72398 |
| Alloy C-276 | 7960 | 54884 | 13300 | 91704 |
| Alloy C-22 | 7900 | 54471 | 13200 | 91014 |
| AL-6XN | 7950 | 54815 | 13250 | 91359 |
| Titanium Grade 2 | 3150 | 21719 | 5250 | 36199 |

The above is based on a process temperature of 125 °C.

construction and temperature. To allow for stress concentration and safety factors, it is customary to use yield stress/4 as a maximum tensile stress and about a 5 divider for shear stress. Table 15.3 lists typical values for some alloys. Most commonly, 304 or 316 stainless steel is used in the biofuel industry, though I have included some higher alloys for reference. Consult a corrosion specialist for detailed evaluation and material selection.

A very commonly used material these days is dual certified 316/316L stainless. It has the low-carbon feature of an L-grade material but maintains the strength of the standard 316SS. It is included in the table.

## Sample Example Problem: Cantilevered

Suppose we have an agitator assembly consisting of a 93.1 kW motor, gear reduced to a shaft speed of 100 rpm. It has three impellers, each of which is a 1.8 m diameter low-solidity hydrofoil impeller. Though they are sized to draw a total of about 85% of motor nameplate, we will design for 100% of the motor power, prorated for each impeller. On that basis, we will design for 31.04 kW for each impeller. The shaft extension from the lower gearbox bearing is 14 m. The impeller locations are 0, 3.5, and 7.5 m from the bottom end of the shaft, which is cantilevered. The proposed shaft diameter is 0.127 m, solid. The shaft material is 316SS. Agitation is normal, and the unit is center mounted in a cylindrical tank. Our problem is to determine if this is strong enough.

First, we look at the total torque, based on motor power, using Eq. (15.2):

$$T_Q = P/2\pi N = 93.1\,kW * 1000\,W/kW/(2\pi * 100\,rpm/60\,s/min) = 8890\,N-m$$

Next, to determine hydraulic force, we must first calculate the torque of each impeller, using the same equation but with the prorated power:

$$\text{Impeller } T_Q = 2963\text{ N}-\text{m each}$$

Now we can calculate the hydraulic force of each impeller, using Eq. (15.1) and dimensionless hydraulic force values from Table 15.1. For the kind of impellers used here, Table 15.1 suggests a dimensionless hydraulic force of 0.45. Therefore, for each impeller:

$$F_h = F_D * T_Q/D = 0.45 * 2963\,N-m/1.8\,m = 741\,N$$

Now we can calculate the bending moment at the bearing at the top of the shaft. We must first calculate the impeller moment arm from that bearing, which equals the shaft length minus the impeller spacing from the bottom end of the shaft. Thus, for impeller 1 we have:

$$L_1 = 14\,m - 0 = 14\,m$$

Similarly, $L_2 = 14 - 3.5 = 10.5\,m$ and $L_3 = 14 - 7.5 = 6.5\,m$.

Now we can calculate the total bending moment using Eq. (15.6):

$$M = \sum F_{hi} * L_i = 741\,N * 14\,m + 741\,N * 10.5\,m + 741\,N * 6.5\,m = 22\,971\,N-m$$

Now we have the design torque and bending moment. To determine shaft shear stress and tensile stress, we calculate the combined bending moment using Eq. (15.9):

$$M_C = \left(M^2 + T_Q{}^2\right)^{0.5} = \left(22\,971^2 + 8\,890^2\right)^{0.5} = 24\,631\,N-m$$

We can calculate the shear stress from Eq. (15.10):

$$\sigma_s = M_C * 16/\left(\pi d^3\right) = 24\,631\,N-m * 16/\left(\pi * (0.127\,m)^3\right)$$
$$= 6.124E7\,N/m^2 = 61\,241\,kPa$$

Tensile stress can be calculated by using Eq. (15.11):

$$\sigma_t = (M + M_C) * 16/\left(\pi d^3\right) = (22\,971 + 24\,631) * 16/\left(\pi * 0.127^3\right)$$
$$= 1.18E8\,N/m^2 = 118\,400\,kPa$$

From Table 15.3, the maximum allowable shear stress for 316SS is 40 922 kPa, and the maximum allowable tensile stress is 68 192. So, our 0.127 m proposed shaft diameter is insufficient. The reader is encouraged to calculate the minimum shaft diameter required, which is about 0.153 m, or nominally 6″. This would potentially be an expensive design for stress, as it might require a larger gear drive to support this shaft than would be needed for torque alone, and it probably would not work for critical speed considerations. A steady bearing would most likely be required.

## Sample Problem with Steady Bearing

Using the same problem, we will look at adding a steady bearing. Let us assume that we add it 0.5 m below the lower impeller. This changes L from 14 to 14.5 m. Noting that for each impeller, B + C = L, we have $C_1 = 0.5$ m, $B_1 = (14 - 0.5) = 14$ m; $C_2 = (3.5 + 0.5) = 4$ m, $B_2 = (14.5 - 4) = 10.5$m; $C_3 = (7.5 + 0.5) = 8$ m; and $B_3 = (14.5 - 8) = 6.5$ m.

Our torque remains the same, at 8890 N-m. The bending moment is obtained by summing the hydraulic forces times BC/L:

$$M = 741\,N*\left(14*0.5\right)/14.5 + 741\,N*\left(10.5*4\right)/14.5 + 741\,N*\left(6.5*8\right)/14.5 = 5161\,N-m$$

Our combined moment, from Eq. (15.9), is:

$$M_C = \left(8\,890^2 + 5\,161^2\right)^{0.5} = 10\,279\,N-m$$

Our shear stress, using Eq. (15.10), is:

$$\sigma_s = M_C * 16/\left(\pi d^3\right) = 10\,279*16/\pi\left(0.127\right)^3 = 2.556E7\,Pa = 25\,560\,kPa$$

Our tensile stress, using Eq. (15.11), is:

$$\sigma_t = \left(M + M_C\right)*16/\left(\pi d^3\right) = \left(5\,161 + 10\,279\right)*16/\pi\left(0.127\right)^3 = 3.839E7\,Pa = 38\,390\,kPa$$

As before, from Table 15.3, the maximum allowable shear stress for 316SS is 40 922 kPa, and the maximum allowable tensile stress is 68 192. So, our 0.127 m proposed shaft diameter is sufficient from a stress standpoint. In fact, as the reader can verify, the minimum shaft diameter for strength using a steady bearing is 0.109 m, or to the nearest ½″ rounding up, 4.5″. The savings in shaft and hub costs normally would more than make up for the cost of adding a steady bearing, though maintenance costs are higher.

The next section discusses designing with steady bearings as part of a discussion on critical speed. We will illustrate a steady bearing design with this same example problem.

## Shaft Design Based on Critical Speed

All physical systems have finite stiffness, inertia, and damping, making them possible to model as a combination of springs, masses, and inertial dampers. For simplicity, most dynamic models of agitator shaft systems assume zero damping, so the resultant deflections are "worst case." The primary mode of agitator shaft deflection is laterally, i.e. normal to the shaft axis. The resistance to deflection is the stiffness of the shaft and the resistance of the bearings supporting the shaft, as well as the structure supporting the agitator. The main inertial masses are the impellers, which are treated as concentrated loads, the in-tank couplings, and the distributed mass of the shaft itself. The shaft and impeller combined system has a series of natural frequencies at which they will tend to resonate if an oscillatory load is applied. The lowest such frequency is called the first lateral natural frequency, or, more commonly (and arguably, incorrectly), the shaft critical speed. However, the use of the term "critical speed" is so widespread and accepted that we will use it here for brevity.

There are many natural frequencies above the critical speed. However, they would not normally be a concern unless the operating speed of the agitator is above critical speed, which is often the case for small propeller agitators but almost never the case for large production agitators. So, we will concern ourselves only with the first lateral natural frequency, or critical speed.

There are also torsional natural frequencies. These tend to be much higher than the lateral natural frequencies, and I have yet to see a single case where they affected shaft design. However, they can affect phenomena within the gear drive. For example, if the torsional natural frequency matches a gear tooth meshing frequency, it can cause chatter and destroy the gears. I have actually known of cases where this happened. We will come back to this when discussing gear drive selection in Chapter 16.

The key to avoiding critical speed problems is to avoid cyclical loads that have frequencies that get too close to the critical speed of the shaft. If a load has an oscillatory frequency (also known as an exciting frequency) that matches the natural frequency of the shaft, and if there is no damping, the shaft will resonate and increase in the amplitude of deflection until shaft failure. (The theoretical amplitude increase would be infinity, but the shaft will fail first.) There is always some damping. Some of it comes from the internal friction within the shaft material, but most of it comes from the fluid the agitator is immersed in. So, the deflection will not reach infinity in the real world, but could reach the point of shaft failure. If the load frequency is sufficiently far from the critical speed, the amplification factor of the loads will be small enough to allow safe and stable operation. Also, if the system has more intrinsic damping it can be operated closer to critical speed. We will provide guidelines on these two phenomena.

But first, we need to determine where the exciting frequencies come from. The most obvious one is the rotation of the shaft itself. If the shaft is not perfectly straight (and it never is), has a mechanical or hydraulic imbalance, is slightly eccentric due to coupling eccentricity or some other factor, it will create an oscillating load at an applied exciting frequency equal to the shaft speed. It is recommended that the shaft be straightened to a total indicator runout (TIR) of 0.003″/ft. of length, or 0.25 mm/m, or a slope of 2.5E-4. The shaft speed must not be too close to the critical speed. But this is not the only source of oscillatory loads. Impellers create oscillatory pulses at a frequency of the number of blades times the shaft speed, and smaller pulses at any even dividend of the blade number times the shaft speed. For example, a four-bladed impeller creates an exciting frequency of 4 times the shaft speed and a smaller magnitude oscillation at 2 times the shaft speed. Three-bladed impellers create an exciting frequency of 3 times the shaft speed. Six-bladed impellers create frequencies of 2, 3, and 6 times the shaft speed, with the largest magnitude at 6 times the shaft speed. It is important that none of these frequencies match the shaft critical speed. Based on large-scale lab measurements, there is no measurable interaction between the number of blades and the number of tank baffles. This may come as a surprise to some. But what it means is that there is no problem using four baffles with a three-bladed impeller (or vice-versa) as long as the total baffle width is equivalent to about 0.33–0.4 times the tank diameter.

How close can we come to critical speed? It depends on the amount of damping in the system. Impeller type and design affect this, as does whether the unit is cantilevered or has a steady bearing. We call the difference "stabilized" versus "unstabilized." Units with steady bearings are considered to be stabilized. Cantilevered units are considered to be intrinsically stabilized with some impellers, and for other impellers, stabilizing fins (no illustration shown) can be added to stabilize the system. These act somewhat as rudders, resisting lateral movement. This concept principally applies to pitched blade turbines, which are not recommended for most biofuel/removable chemical applications, as they require much more torque for equivalent performance. This is discussed in Chapters 4 and 13 in some detail. Stabilizing fins are not needed for any unit with a steady bearing. Table 15.4 presents guidelines for critical speed operating ranges.

Now we can discuss how to calculate critical speeds. Broadly speaking, there are two cases: cantilevered design and steady bearing design. Of course, there are

**Table 15.4** Critical speed guidelines.

| Shaft style | Impeller style | Number of blades | Stabilized? | $N/N_C$ range | Blade passing frequency range |
|---|---|---|---|---|---|
| Cantilever | Pitched or radial | 4 | Yes | $0.80 \leq N/N_C$; $N/N_C \geq 1.25$ | Avoid $N/N_C$ between 1.67 and 2.4 as well as between 3.33 and 4.8 |
| Cantilever | Pitched or radial | 4 | No | $0.65 \leq N/N_C$; $N/N_C \geq 1.54$ | Avoid $N/N_C$ between 1.67 and 2.4 as well as between 3.33 and 4.8 |
| Steady bearing | Pitched or radial | 4 | No | $0.80 \leq N/N_C$; $N/N_C \geq 1.25$ | Avoid $N/N_C$ between 1.67 and 2.4 as well as between 3.33 and 4.8 |
| Cantilever | Hydrofoil | 3 | — | $0.80 \leq N/N_C$; $N/N_C \geq 1.25$ | Avoid $N/N_C$ between 2.5 and 3.6 |
| Cantilever | Hydrofoil | 4 | — | $0.80 \leq N/N_C$; $N/N_C \geq 1.25$ | Avoid $N/N_C$ between 1.67 and 2.4 as well as between 3.33 and 4.8 |
| Steady bearing | Hydrofoil | 3 | — | $0.80 \leq N/N_C$; $N/N_C \geq 1.25$ | Avoid $N/N_C$ between 2.5 and 3.6 |
| Steady bearing | Hydrofoil | 4 | — | $0.80 \leq N/N_C$; $N/N_C \geq 1.25$ | Avoid $N/N_C$ between 1.67 and 2.4 as well as between 3.33 and 4.8 |
| Cantilever | Radial | 6 | — | $0.80 \leq N/N_C$; $N/N_C \geq 1.25$ | Avoid $N/N_C$ between 2.5 and 3.6 as well as between 3.33 and 4.8 and 5 and 7.2 |
| Steady bearing | Radial | 6 | — | $0.80 \leq N/N_C$; $N/N_C \geq 1.25$ | Avoid $N/N_C$ between 2.5 and 3.6 as well as between 3.33 and 4.8 and 5 and 7.2 |

other variants, such as stepped shafts, multiple steady bearings, etc. These more complicated cases are best handled by computer programs written for the purpose. But we will first discuss simplified hand calculations for basic cases.

## Cantilevered Designs

For the case of a cantilevered design with constant shaft diameter, the critical speed may be calculated using Eq. (15.12) for stainless steel:

$$N_C = 27\,082 * d^2 / \left(L * W_e^{0.5} * (L+S)^{0.5}\right), \text{ where units are kg, m, and s.} \quad (15.12)$$

This equation is straightforward except for calculating the equivalent mass, $W_e$. This has two components: impeller weight and shaft weight. For a constant diameter shaft, the shaft equivalent weight is the actual shaft weight/4, or:

$$W_e(\text{shaft}) = \text{shaft weight}/4 = 1579 d^2\ L \text{ for stainless steel.} \quad (15.13)$$

For the impeller, it is the sum of individual impeller weights ratioed to the cube of their proportion of their extensional position relative to the total shaft length, or:

$$W_e(\text{impeller}) = W_1 + W_2 (L_2/L)^3 + W_3 (L_3/L)^3 + \cdots \text{ or } W_e(\text{impeller}) = \Sigma W_i (L_i/L)3 \quad (15.14)$$

Total equivalent weight is the sum of the shaft and impeller equivalent weights:

$$W_e = W_e(\text{shaft}) + W_e(\text{impeller}) \quad (15.15)$$

The obvious question now is: Where do we get the impeller weights? Given the vast array of proprietary impellers on the market, and the inherently greater complexity of designing them compared to shafts, the best place to get them is from your selected agitator vendor. I spent some time trying to come up with a simplified procedure for calculating impeller weights for a multitude of impeller styles for this chapter and was unable to do so. What I decided to do instead is to use software I own or have access to in order to create Table 15.2 of some possible hub weights for the most commonly recommended impeller type for biofuel applications: the low-solidity, three-blade hydrofoil. The extension blade weights can then be calculated based on Eqs. (15.4) and (15.5) to calculate blade stresses, and then calculating blade weights based on the blade length, thickness, and material density. Allowable blade stress for stainless is 80 000 kPa (~12 000 psi). For convenience, Eqs. (15.4) and (15.5) are repeated here:

$$\text{Blade tensile stress} = \left(F_b * (0.75D - D_h)\right)/2Z_m \quad (15.4)$$

where $Z_m$ is the section modulus of an extension blade, defined next:

$$Z_m = wt^2/6 \tag{15.5}$$

Note that the stress in Eq. (15.4) should be in Pa for consistent units, not kPa.

As stated earlier in the chapter, the density of stainless steel is taken as 8045 kg/$m^3$. Because stainless plate comes in discreet thicknesses, one must round up to the nearest standard thickness. It is also true that some manufacturers make the blade width in discreet increments, which would affect power draw as well as weight. For purposes of this chapter, we will assume that the blade width is 1/6 of the impeller diameter, the hub body is 1.5 times the shaft diameter, and the blade thickness comes in 1/8″ (3.175 mm) increments. But local suppliers may have different standards. In other words:

$$w = D/6 \tag{15.16}$$

$$l = (D - 1.5d)/2 \tag{15.17}$$

$$v = 3 * l * w * t \tag{15.18}$$

$$W_b = \rho_m * v \tag{15.19}$$

$$W = W_b + W_h \tag{15.20}$$

## Example Problem: Determining Impeller Weight

Let us use our original simple example of a cantilevered design for stress, but change the shaft to 0.153 m (6″) diameter, as we know we will need this much for stress. The impeller diameter is still 1.8 m, and the design torque is still 2963 N-m per impeller. The blade force is still 741 N. The smallest hub that will fit that shaft size, per Table 15.2, has a hub diameter of 610 mm and a mass of 112 kg. Our blade width, per Eq. (15.16), equals 1.8 m/6 = 0.3 m. Our blade length, per Eq. (15.17), equals (1.8 − 1.5 * 0.153)/2 = 0.785 m. Combining Eqs. (15.4) and (15.5), we get that 80 000 000 Pa = 741 N * (0.75 * 1.8 m − 0.61 m)/(2 * 0.3 m * $t^2$/6), or t = 8.28E-3 m, or about 0.326″. Since we assume only 1/8″ increments are available, the nearest standard thickness equal or greater than the minimum requirement is 3/8″, or 9.53E-3 m. Therefore, per Eq. (15.18), v = 3 * 0.785 m * 0.3 m * 9.53E-3 m = 6.73E-3 $m^3$. Therefore, per Eq. (15.19), the blade weight is 8045 kg/$m^3$ * 6.73E-3 $m^3$ = 54.1 kg. Therefore, the impeller weight, per Eq. (15.2), is the sum of hub weight plus blade weight, or 112 + 54.1 = 166.1 kg. Note that the bottom-most impeller could be mounted on a turned-down section of shaft. This would save shaft weight and hub weight. For example, if we used a 76 mm shaft through the bottom hub, which would carry the prorated torque of the impeller, the net impeller weight would be about 98 kg, including the credit for shaft weight savings. A longer step would require dynamic analysis.

## Example Problem: Cantilevered Critical Speed Calculation

To illustrate a simple cantilevered shaft critical speed calculation, we will use the same example that we used for a cantilevered stress calculation, but adjusted for the shaft size we needed for stress. In that problem, we did not have the shaft support bearing span, so, for this problem, we will assume a span of 0.4 m. We also did not have impeller weights, but in the earlier example we determined an estimated impeller weight of 166.1 kg, though the bottom impeller weight could be reduced to 98 kg if the shaft were stepped to 76 mm diameter.

We know that the impellers are 0, 3.5, and 7.5 m from the lower end of the shaft, and that the shaft length, L, is 14 m. To use Eq. (15.14), we need to know the extension from the lower support bearing. This is just (L-C), where C is the distance from the end of the shaft to the impeller. So, we have $L_1 = 14 - 0 = 14$, $L_2 = 14 - 3.5 = 10.5$, and $L_3 = 14 - 7.5 = 6.5$ m. Now we can solve Eq. (15.14):

$$W_e(\text{impeller}) = 166.1 + 166.1(10.5/14)^3 + 166.1(6.5/14)^3 = 252.8\,\text{kg}$$

The equivalent weight of the shaft, which has a diameter of 0.152 m, is calculated using Eq. (15.13):

$$W_e(\text{shaft}) = 1579 * (0.152\,\text{m})^2 * 14\,\text{m} = 510.7\,\text{kg}$$

The total equivalent weight is the sum of these two:

$$W_e = W_e(\text{shaft}) + W_e(\text{impeller}) = 510.7 + 252.8 = 763.5\,\text{kg}$$

Now we can calculate the critical speed using Eq. (15.12):

$$\begin{aligned} N_C &= 27\ 082 * d^2 / \left(L * W_e^{0.5} * (L+S)^{0.5}\right) \\ &= 27\ 082 * (0.152)^2 / \left(14 * 763.5^{0.5} * (14+0.4)^{0.5}\right) \\ &= 0.426/\text{second} \quad \text{or} \quad 25.6\ \text{rpm} \end{aligned}$$

Our shaft speed is 100 rpm, therefore $N/N_C = 100/25.6 = 3.91$. That means the unit is operating well above critical speed. If we had used the stepped shaft to save weight, the critical speed would be increased to 26.8 rpm, which would give an $N/N_C$ value of 3.73. It is not good practice to operate a large agitator above the first critical speed. Therefore, this design, which is adequate from a shaft stress standpoint, is inadequate from a critical speed standpoint. Addition of a bottom steady bearing will help for both stress and critical speed. We will see in the next section whether a solid 0.152 m shaft will be adequate with a steady bearing.

This was a fairly simple design. More complicated designs, such as those involving stepped shafts, hollow shafts, etc., are best handled with a computer program or finite element analysis (FEA). A number of manual calculation methods are included in Ref. [4].

## Units with Steady Bearings

Shaft design with steady bearings is rather complex. Though Ref. [4] has a number of cases presented, they are not that easy to use. It is best to use a computer program that has a simplified FEA routine built in. We will show such a simulation here. But first, let us discuss a quick and reasonably accurate way to do it for the case that the shaft does not have too many impellers. There is an equation for the critical speed of the shaft alone that is based on the shaft being simply supported (imagine pivot points) at each end. Adding impeller weights would tend to lower the critical speed. However, the actual way shafts are generally supported is with two bearings at the top of the shaft plus the steady bearing below. In the case of mechanical seal units, they may have another bearing in the seal cartridge. This is a much more rigid system than a simple support at both ends. In my experience, in most cases, that additional rigidity more than compensates for the impeller weight. So, for a quick estimate of the critical speed, we can use that equation for a constant diameter stainless steel shaft. It is presented as Eq. (15.20) next:

$$N_C = 2180d / L^2; \text{ MKS units. (MKS is a measurement system of units, standing for meters, kilograms, and seconds)} \tag{15.21}$$

For the hypothetical sample problem just described, we would calculate:

$$N_C = 2180 * 0.152 / 14.5^2 = 1.576 / \text{second} = 94.56 \text{rpm}.$$

That would put $N/N_C$ at 1.06, which would not be acceptable. However, we will check this again with our simplified FEA/Dynamic analysis model.

To model this, we must conceptually divide the shaft into sections. We can use as many sections as we want, but we must have at least a new section for each discontinuity of any type, including lumped masses such as impellers and in-tank couplings; lateral forces; power draw; changes in shaft section, bearings and any other discontinuity. We generally number the spans from the gear drive end to the bottom end. Each section or span begins with a station number and ends with the next higher station number. For boundary conditions, we add a fictitious short span at each end of the shaft. (This makes the algorithm stable and more likely to converge.)

The dynamic analysis program used for this example was developed exclusively for Benz Technology International, Inc. It uses English units only at this point. So far in this chapter, we have used SI units, but will illustrate the problem in English units. As an aid to understanding the program inputs, we have created Table 15.5.

The basic model is conceptualized in Figure 15.4. To keep it simple, we did not show any gears that might be between the upper two bearings as stations, or any in-tank couplings. A gear would add some weight and a side load, but the effect on

**Table 15.5** Dynamic analysis inputs.

| Station number | Span Length (m/in.) | Shaft outside diameter (m/in.) | Shaft inside diameter (m/in.) | Spring constant to ground [(kN/m)/(lb/in.)] | Lumped mass (kg/lb) | Power (kW/Hp) | Hydraulic force (N/lb) |
|---|---|---|---|---|---|---|---|
| 0 | | 0.152/6 | 0/0 | | | | |
| 1 | 0.0254/1 | 0.152/6 | 0/0 | 1.75E5/1E6 | | | |
| 2 | 0.5/19.7 | 0.152/6 | 0/0 | 1.75E5/1E6 | | | |
| 3 | 6.6/259.8 | 0.152/6 | 0/0 | | 166.1/365.4 | 93.1/41.67 | 741/166.6 |
| 4 | 4/157.5 | 0.152/6 | 0/0 | | 166.1/365.4 | 93.1/41.67 | 741/166.6 |
| 5 | 3.5/137.8 | 0.152/6 | 0/0 | | 166.1/365.4 | 93.1/41.67 | 741/166.6 |
| 6 | 0.5/19.7 | 0.152/6 | 0/0 | 1.75E4/1E5 | | | |
| 7 | 0.025/1 | 0.152/6 | 0/0 | | | | |

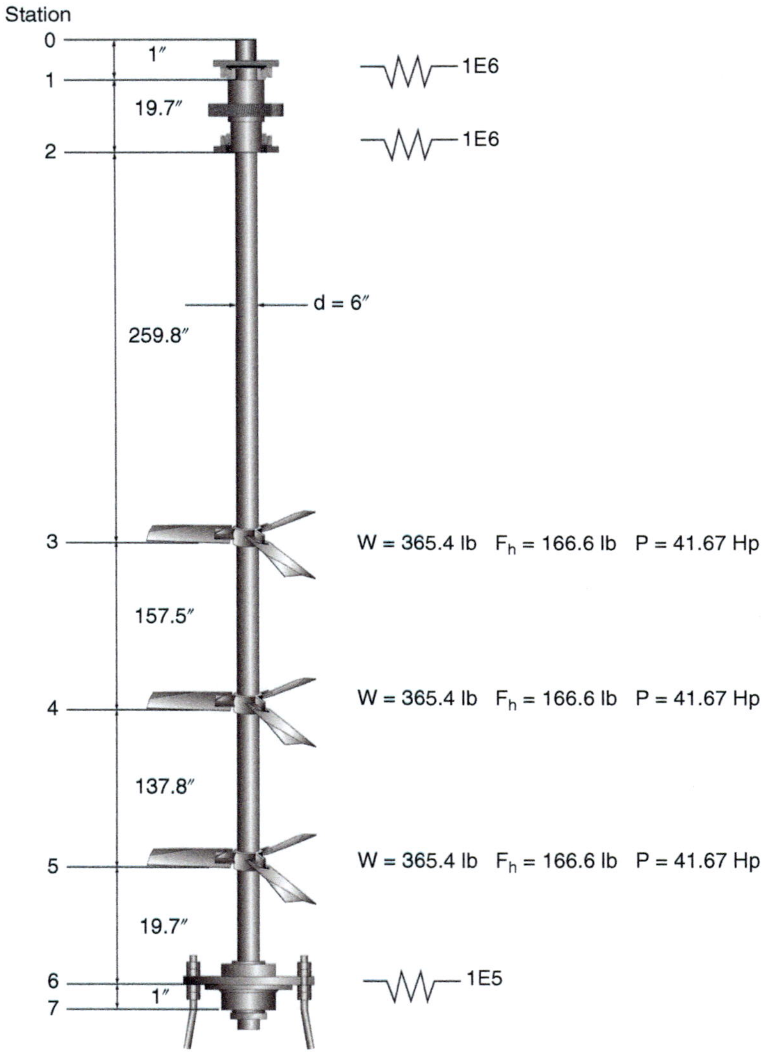


**Figure 15.4** Critical speed model.

shaft design would be minimal due to the proximity to the shaft support bearings. A shaft of 14.5 m total length would need an in-tank coupling for manufacturing purposes, and there would need to be a coupling between the gear drive and the extension shaft. Again, we did not include these for simplicity. The coupling weight would be small compared to the impeller weights, likely about 40 kg for this example. So, adding in the coupling weights to the model would possibly decrease the critical speed by a couple of percent.

The shaft is supported by bearings, which are modeled as linear springs to ground. As long as the bearing stiffness is high, the actual value has little impact on the calculations. It cannot be infinite, however, as the program would not converge on a result. The values shown are reasonable, in terms of order of magnitude. The steady bearing is shown as 1/10 as stiff as the gear drive bearings, as it is not mounted in as rigid a structure. This type of modeling has been validated by many field measurements, not only for agitators, but also for many kinds of rotating equipment. A 1″ long fictitious span has been added at each end of the shaft to allow boundary conditions of zero stress, making for better convergence of the algorithm.

The program calculates stresses and deflections, as well as critical speeds. If a station had been added for a mechanical seal, it could predict deflection through the seal, which could be important in some cases. The program output is shown as Figure 15.5.

Several key results are worth mentioning. The first critical speed is calculated as 110.3 rpm, which would have our agitator operating at 90.7% of critical speed, as compared to 106% using our simple hand calculations, so the hand calculation, in this case, was conservative by about 17%. As noted earlier, the simple calculation is usually conservative, except when there are many impellers or unusually heavy ones. Though the shaft stresses are okay, we wish to operate at or below 80% of critical speed. Therefore, this is still an unacceptable design. Adding in-tank couplings could possibly push it over 95%, so for an actual agitator design, this would need to be checked. Later in this section we will look into what we can do to make the design work.

The maximum shear stress is calculated as 2449 psi, and the maximum tensile stress is 2890 psi. These are well within the allowable numbers. Converting to kPa, the corresponding values are 16 886 and 19 927.

On this printout, the maximum shaft deflection shown is about 0.3″, or about 8 mm. This is based on all of the impeller hydraulic forces being at peak magnitude and all in the same direction. That does not happen very often. On the other hand, it does not account for amplification due to the $N/N_C$ ratio, which would increase it, nor the effects of fluid damping, which would reduce the effect of the critical speed amplification ratio. On balance, the deflection numbers are probably close to a worst case scenario. However, they do not include deflection due to shaft straightness issues, errors in coupling squareness, or any other shaft manufacturing issue. Though we did not put in a station for a mechanical seal, we could easily do so, and thereby check for deflection in the seal area.

Shear force is also shown. It is listed as Shear (lbs). Its main use is to determine bearing reaction loads. For this simulation, the shear load for the span terminating in the steady bearing is 306 lb, or 1361 N. Typical design practice is to design the steady bearing support structure for 3 times the actual load, so we would design for roughly 4100 N side force.

| | | | |
|---|---|---|---|
| Carbon Steel<br>Stainless Steel<br>Titanium, Grade 2<br>Alloy C276 | | Top Entering<br>Side Entering | Calculate |
| | | 1st Critical Speed | 2nd Critical Speed (RPM) |
| Customer | Steady bearing<br>1 | | |
| Shaft Speed (RPM) | 100 | *110.3 (90.7%)* | 365.5 (27.4%) |
| Temperature (F) | 223 | | |

| Shaft Length (in) | Shaft O.D. (in) | Shaft I.D. (in) | Support Stiffness (lb/in) | Lumped Mass (lb) | Impeller Force (lb) | Impeller Power (HP) | Shaft Deflection (in) | Bending Moment (in-lbs) | Shear (lbs) | Tensile Stress (psi) | Shear Stress (psi) |
|---|---|---|---|---|---|---|---|---|---|---|---|
| 1 | 6 | 0 | 1.00E+06 | 0.000 | 0.000 | 0.000 | −1.501E-03 | 0.000E+00 | −1.501E+03 | 2.348E+03 | 2.348E+03 |
| 19.7 | 6 | 0 | 1.00E+06 | 0.000 | 0.000 | 0.000 | 1.696E-03 | −2.958E+04 | 1.941E+02 | 1.752E+03 | 2.449E+03 |
| 259.8 | 6 | 0 | 0.00E+00 | 365.400 | 166.600 | 41.670 | 3.048E-01 | 2.085E+04 | 2.752E+01 | 2.289E+03 | 1.797E+03 |
| 157.5 | 6 | 0 | 0.00E+00 | 365.400 | 166.600 | 41.670 | 2.938E-01 | 2.519E+04 | −1.391E+02 | 1.852E+03 | 1.259E+03 |
| 137.8 | 6 | 0 | 0 | 365.400 | 166.600 | 41.670 | 4.778E-02 | 6.022E+03 | −3.057E+02 | 6.525E+02 | 5.105E+02 |
| 19.7 | 6 | 0 | 1.00E+05 | 0.000 | 0.000 | 33.000 | 3.057E-03 | −8.014E-09 | 1.899E-09 | 1.180E-16 | 1.890E-10 |
| 1 | 6 | 0 | 0 | 0.000 | 0.000 | 0.000 | 7.759E-04 | −1.277E-08 | 1.899E-09 | 7.406E-17 | 3.011E-10 |

The maximum tensile stress of 2.890E+03 psi occurs at 2.804E+02 inches.
The maximum shear stress of 2.449E+03 psi occurs at 2.070E+01 inches.

**Figure 15.5** Solid shaft steady bearing dynamic analysis.

Since this design is unacceptable from a critical speed standpoint, we have to alter it in some way. Possibilities include a larger shaft, stepped shaft, hollow shaft, and maybe several other ways. In the next section we will discuss the use of hollow shafts.

## Solid Shaft Versus Hollow Shaft

The earlier discussion and examples were based on using solid shafts, which is the most common practice. But there are some cases where a hollow shaft is sometimes desirable or necessary. We will list some of the reasons. But first, allow me to address one of my pet peeves about shaft terminology. Hollow shafts are often referred to in the industry as "pipe shafts." While hollow agitator shafting may be thought of as a pipe, commercial pipe is not a suitable material for manufacturing hollow shafts. This is because it is not round enough, not straight enough, and the common use of welded seams with filler metal that may not match the metallurgy of the pipe can increase corrosion problems. It is also rather rough, so if the shaft must be polished, a considerable material removal allowance must be part of the design calculation. Proper hollow shaft material should be ordered as "seamless" pipe, with careful controls on roundness, straightness, and actual wall thickness, as well as appropriate surface smoothness. In some cases, it is better to use mechanical tubing rather than tightly controlled pipe, as tubing is manufactured to tighter tolerances as standard.

With that out of the way, we can move on to when it is appropriate to consider using a hollow shaft. Here are a few reasons:

- When the solid shaft diameter needed to stay within critical speed limits would force the use of a larger gear drive than needed for the rest of the loads. (This may not apply if the shaft has its own support bearings and is flexibly coupled to the gear drive. This configuration will be discussed in our section on gear drives.) For the earlier example problem, that would definitely be the case.
- When the material cost per unit mass is high enough that hollow shafting will save enough money to offset the increased labor costs associated with hollow shafting. This will vary according to labor costs and the experience of the vendor making hollow shafts. As a rough guide, for major US vendors, it may be more economical to consider hollow shafts when the material of construction is 316SS and the required solid shaft diameter is 150 mm or larger. For higher alloys, such as Titanium, C-276 and the like, the transition may be less, possibly as low as 100 mm.
- When the weight of the wetted assembly would cause an unacceptably low bearing life in the shaft support bearing system. (This is rare.)

The use of hollow shafting can make a big difference in shaft critical speed. Usually, only part of the shaft is hollow, as it is difficult to use hollow shafts through seals and steady bearings. The hollow shaft is coupled to the solid shaft by use of welded couplings resembling forged weld-neck flanges, though they are specially made and machined to assure concentricity and perpendicularity. We will illustrate this critical speed effect by converting the main span on our sample problem from solid to hollow, ignoring the top coupling, and adding some weight to the upper and lower impellers to simulate coupling weight. The shaft through the gear drive and the steady bearing is solid. This change and the resulting changes to stress and critical speed are shown in Figure 15.6.

We replaced most of the shaft with a hollow section having an outside diameter (OD) of 8.625″ and an inside diameter (ID) of 7.625″, or a nominal wall thickness of 0.5″. This matches 8″ Sch 80 pipe dimensions. The steady bearing shaft is 4.5″. The impellers are lighter, as a massive hub is not needed. All the stresses are still okay, and the critical speed has increased from 110.3 to 198.1 rpm, almost an 80% increase, changing $N/N_C$ from 0.907 to 0.505. The design is stable and acceptable. Shaft deflection has decreased to 0.2″, or 5 mm. Note that this model assumes that there is no shaft seal. It is not possible to use a hollow shaft through a shaft seal, unless it is heavy wall and machined to an accurate, round external diameter. This is usually not practical. So, the normal design approach for sealed units with hollow shafts is to use a solid shaft through the seal, coupled to a hollow shaft in the tank.

Hollow shafting has some drawbacks. Among them are:

- The larger shaft diameter means the shaft couplings are larger also, which may affect the required agitator mounting flange size. This could reduce space available for other nozzles on the top head, unless a bottom-entering agitator is used.
- Impeller hubs in most cases must be welded in place. Adjustment of impeller location is not possible if this is the case. With heavy-wall pipe, clamped-on hubs may sometimes be possible if the hub location is machined accurately. In rare cases, a keyed design may be possible, but the wall thickness required for this may render the design uneconomical. External welded keys (welded to the shaft) have sometimes been used, with the hub slipped over the key.
- If impeller hubs are welded on, the required opening for installation is larger still. If installed through the agitator mounting flange, the flange diameter may need to be very large.
- Shaft couplings are more expensive than on solid shafts.
- Large-diameter hollow shafts may need to be machined outside agitator vendor facilities due to lathe swing limitations, making quality control more difficult. That would probably not be the case for this example problem, however.
- A separately coupled solid stub shaft is generally required through the steady bearing, though this could be a plus for maintenance.

| | | | | |
|---|---|---|---|---|
| Carbon Steel<br>Stainless Steel<br>Titanium, Grade 2<br>Alloy C276 | | Top Entering<br>Side Entering | | Calculate |
| | | 1st Critical Speed | 2nd Critical Speed (RPM) | |
| | Steady bearing | | | |
| Customer | 2 | | | |
| Shaft Speed (RPM) | 100 | 198.1 (50.5%) | 663.6 (15.1%) | |
| Temperature (F) | 223 | | | |

| Shaft Length (in) | Shaft O.D. (in) | Shaft I.D. (in) | Support Stiffness (lb/in) | Lumped Mass (lb) | Impeller Force (lb) | Impeller Power (HP) | Shaft Deflection (in) | Bending Moment (in-lbs) | Shear (lbs) | Tensile Stress (psi) | Shear Stress (psi) |
|---|---|---|---|---|---|---|---|---|---|---|---|
| 1 | 6 | 0 | 1.00E+06 | 0.000 | 0.000 | 0.000 | −1.427E-03 | 0.000E+00 | −1.427E+03 | 2.348E+03 | 2.348E+03 |
| 19.7 | 6 | 0 | 1.00E+06 | 0.000 | 0.000 | 0.000 | 1.619E-03 | −2.812E+04 | 1.916E+02 | 1.777E+03 | 2.440E+03 |
| 259.8 | 8.625 | 7.625 | 0.00E+00 | 240.000 | 166.600 | 41.670 | 1.948E-01 | 2.165E+04 | 2.498E+01 | 2.001E+03 | 1.559E+03 |
| 157.5 | 8.625 | 7.625 | 0.00E+00 | 200.000 | 166.600 | 41.670 | 1.853E-01 | 2.559E+04 | −1.416E+02 | 1.614E+03 | 1.093E+03 |
| 137.8 | 8.625 | 7.625 | 0 | 240.000 | 166.600 | 41.670 | 3.191E-02 | 6.072E+03 | −3.082E+02 | 5.658E+02 | 4.419E+02 |
| 19.7 | 4.5 | 0 | 1.00E+05 | 0.000 | 0.000 | 33.000 | 3.082E-03 | 2.983E-09 | −3.878E-09 | 3.335E-10 | 1.667E-10 |
| 1 | 4.5 | 0 | 0 | 0.000 | 0.000 | 0.000 | 1.585E-03 | −6.990E-09 | −3.878E-09 | 3.207E-16 | 3.907E-10 |

The maximum tensile stress of 2.520E+03 psi occurs at 2.804E+02 inches.

The maximum shear stress of 2.440E+03 psi occurs at 2.070E+01 inches.

**Figure 15.6** Steady bearing dynamic analysis example output with hollow shaft.

- Except for small shafting or repetitive designs, FEA is almost always required to assure localized stresses in the hollow shaft are not too high. This is especially true if thin-wall pipe is used. Thin-wall pipe may have a high enough section modulus for the bulk loads, but there can be high-stress concentration factors where hubs are welded on. For example, a 20″ OD pipe with a 0.085″ thick wall is theoretically as strong as an 8.625″ OD pipe with a 0.5″ thick wall. But who would dare to attach anything to a 0.085″ thick wall? The safest attachment is to use a cylindrical hub with stub blades integral to the hub. This hub would be welded to the hollow shaft. With heavy enough pipe wall thickness, it is possible to weld stub blades for axial turbines directly to the hollow shaft, but FEA must be used to check this, at considerable engineering expense. But if the stress concentrations are too high, one might witness (as I have) a phenomenon I call the can opener effect. (You can use your imagination.)
- Though not commonly a problem for most biofuel designs, external pressure can be an issue for some hollow shaft applications, since the shaft is sealed and acts as a pressure vessel. This is more likely to be an issue for hydrogenators and similar reactors, as they run at much higher pressures. There are some creative ways to get around this, which the author would be happy to discuss.

In short, hollow shafting can be a viable shaft design option for cost savings or critical speed considerations, but it is not without a few drawbacks.

## Role of FEA in Overall Shaft Design: Simplified Discussion

Finite element analysis, or FEA, has been used in industry in general and agitator design in particular for decades. It can pinpoint design problems that manual calculations might not. Though the shaft design spreadsheet illustrated in the last several figures is a sort of FEA program, most people associate FEA with color coded stresses and deflections. Though I do not have a gallery of illustrations here, I make the following suggestions for where FEA should be used in agitator design to supplement manual methods:

- For most hollow shaft designs.
- For unusual impeller hub designs, or cases where the stub blade is less than 40% of the extension blade width.
- For deflection of concave blades, to make sure they do not "open up" too much under load, which would increase power draw and may even lead to catastrophic failure.
- For slots and bolt hole areas in disk turbine designs, especially those with adjustable diameter.

- To determine dynamic deflection and natural frequency of the impeller disk.
- To evaluate vessel head deflection and stress.
- To determine deflection at the shaft end for "slap ring" design, as described later in this chapter.

## Steady Bearing Materials

If the process fluid were solids-free, use of steady bearings would not present much of a maintenance issue. With a lubricating fluid, the useful life of a steady bearing may be comparable to the useful life of the entire agitator. Indeed, I know of a case where a Nylon polymer reactor showed no measurable wear on any steady bearing components after years of operation. (The Nylon polymer melt has similar flow characteristics and lubricity to standard temperature and pressure (STP) oil treatment, Hastings Motor Honey, and similar viscous lubricants.)

Alas, most fluids in the biofuels/renewable chemicals area are aqueous, and hence are not great lubricants. (The exceptions are biodiesel and renewable diesel, where at least the solids-free oil-based fluids are reasonably good lubricants.) Moreover, most aqueous fluids except some of the final products are actually slurries, and some can be quite abrasive. For example, starch is moderately abrasive. Most plant biomass is more abrasive than starch, especially if it still has appreciable amounts of plain old dirt carried over from harvesting. Therefore, steady bearing components often wear out quickly, sometimes as quickly as just several months. One possible solution is to flush the steady bearing with clean liquid. This works well mechanically, but for long-batch cycles, it can result in significant dilution of the tank contents. Typical flush rates for a steady bearing are in the ballpark of 250 ml to 1 l/min, depending on the size and design details of the steady bearing. If the tank contents are in the tank for several days, which can easily be the case in some processes, thousands of liters of flush liquid will be added to the tank. This may be unacceptable.

The only other approach is to choose steady bearing materials that will last longer and are compatible with the process fluids. In general, the vendor standard default materials may not be the best. Most vendors will supply a wear sleeve or other sacrificial metal member made of the same material as the shaft. They will typically provide a bushing or other stationary member made of glass-filled Teflon (GFT; glass filled polytetrafluoroethylene [GFPTFE]). (Virgin polytetrafluoroethylene [PTFE] is not suitable, as it will creep or extrude, resulting in opening up the clearance between shaft and bushing.) The main reasons vendors default to this material are twofold:

1) It is easily machined to the desired size and shape.
2) It is generally compatible with the great majority of process fluids, from a corrosion standpoint.

However, GFT and any other material that has a waxy surface will have a rather short system wear life in abrasive service. This is not due to the bushing wearing out. Instead, abrasive particles get embedded in the surface of the waxy material, slowly turning it into a pretty good imitation of sandpaper. The abrasive-embedded bushing then proceeds to wear out the wear sleeve. If not caught before the wear sleeve wears through, it will proceed to start wearing out the actual shaft, ultimately requiring a shaft replacement.

Longer life can be achieved by using a rigid, fairly hard material for the bushing, or at least one that does not allow particles to become embedded in it. Such a material must also be compatible with process fluids. Some bushing materials the author has seen successfully being used include lead-free bronze, metallized graphite, graphite with a polymeric binder (typically food grade), and various grades of ceramic. Such materials can change the steady bearing wear life from months to years.

Longer life can also be achieved by hard-surfacing the sacrificial metal part. Common choices include Stellite and Colmonoy products. Generally, these will be compatible with most process fluids. One notable exception is high-temperature acid pretreatment of biomass. Corrosion can be a big problem under those conditions, and material compatibility must be carefully checked.

## Startup in Settled Solids

Whenever slurries are handled, there is a chance that they will settle out, whether due to mechanical failure of the agitator drive, or a power failure. For this chapter, we will concern ourselves with only the latter scenario, as the first scenario will require shutdown and repair, and likely, the tank will need to be emptied and cleaned out.

When solids have settled out, they will need to be re-suspended. There are both process and mechanical issues with this requirement. Though this chapter is about mechanical design, we will cover some of the process aspects first for the sake of continuity, rather than discuss this requirement in other chapters.

### Process Issue #1: Are the Solids Intrinsically Re-suspendable?

This is not always straightforward to answer. It generally depends on particle shape, particle size distribution, and whether there is any sort of adhesion between particles. In general, the solids easiest to re-suspend are those consisting of spherical particles of uniform size, with no adhesion. These do not exist in the target industries of this book. Small particles, though they have a lower settling rate than large particles, are often more difficult to re-suspend. Flat plate-shaped particles of a wide size distribution can be difficult to re-suspend, especially if there are any adhesive forces. More extreme shapes may be impossible. For example, though we have not tried it, it is likely that a settled bed of Christmas ornament hangers or fishhooks would not be re-suspendable. Of more practical importance for this book are grain and biomass slurries. For example, starch slurries, if allowed to settle completely, generally cannot be re-suspended. They will pack down hard and seem to have some degree of adhesion. They have to be broken up before starting the agitator. We will describe how to do this later. Sometimes they need to be cut up with a pressure washer and physically lifted out of the tank. I have heard stories of people trying to break up a settled starch bed with explosives but bursting the tank instead.

Spent grains in a starch-based ethanol plant, various biomass fibers, etc., generally can be re-suspended. They may require more agitation to re-suspend than to maintain suspension, which is not the case with most slurries with a narrow particle size distribution.

### Process Issue #2: If the Slurry Is Re-suspendable, Does It Take More Agitation Than to Maintain Suspension?

Conceptually, this can be determined in the pilot scale by operating at the just-suspended condition, shutting the agitator off, and letting it settle for a while (this could be minutes or a day or more, depending on settling rate), and then restarting at the same shaft speed. If the lower impeller is to be buried in the settled bed, protective measures may have to be taken, as described later. Ideally, the test would be done with the lower impeller above the settled bed; if the slurry will re-suspend, then, it has a decent chance of re-suspending in the production scale. However, this is not always the case. Various factors such as bed compression, adhesion, or mechanical particle interaction may require more agitation to re-suspend than to maintain suspension.

### Process Issue #3: Is the Settled Bed Depth Low Enough That the Lowest Impeller Can Be Located Above It?

This would be the best situation. If the lower impeller is above the settled bed, it will not require more than the normal operating power to start up, and will not be subjected to high hydraulic loading. But, if low-level agitation is required, it may be difficult to avoid a buried impeller.

### Process Issue #4: If an Impeller Must Be Buried in the Settled Bed, How Much Torque Does It Take to Start Rotation?

Unfortunately, that may depend on the settled bed depth as well as vibration effects as described under the next heading. But, as a minimum, settled beds generally behave as Bingham plastic fluids. That is, they have a yield stress. What that means in practice is that the starting torque for a given settled bed condition and impeller type is directly proportional to the cube of the impeller diameter. If the bed properties do not change, this can be used to predict the minimum starting torque required in the production scale by doing a smaller-scale test and measuring the breakaway torque. If that is less than the operating torque, then the unit should start. If it is more than 100% of normal operating torque but less than 200%, the unit may still start, but it will not be able to run more than a minute under such an overload without tripping circuit breakers, so that may not be enough. We will describe measures to reduce starting torque later.

For solids that essentially cannot settle because there is almost no free liquid, usually there is enough torque to start up, because the settled bed is in reality a thick yield-stress fluid. Specifically, high-solids fibrous biomass slurries usually have no great difficulty starting up. Starch slurries, on the other hand, may be impossible to start up in without breaking up the bed. In general, fine slurries are more difficult for an impeller to start up in than coarse slurries.

To test for minimum starting torque in the lab, it is best to have a torque meter that either logs data or holds peak values. The agitator shaft is best turned by hand, though startup at very low Hz with a variable frequency drive (VFD) drive may also be considered. Figure 12.16 illustrates a possible way to measure torque in the lab.

## Process Issue #5: Are There Sources of Vibration Nearby That Can Affect Settled Bed Properties?

Another factor is what kind of equipment may be operating near the slurry tank. If the tank is operating by itself with no other equipment nearby, the bed will not compact as much as when there is a vibration source in the vicinity. This may not matter as much if the whole plant loses power, but if only the agitator must shut down, other equipment can affect the settled bed density by exposing it to slight vibrations that can cause smaller particles to settle between larger ones, greatly increasing the difficulty of re-suspension. Examples of such equipment may include reciprocating air compressors, hammer mills, and vibratory screeners. In some cases, if a road is nearby, passing traffic can transmit enough micro-vibrations to affect bed properties. If such is the case, scale-up is difficult. It may be necessary to break up the bed before starting the agitator, which will be described later.

## Process Issue #6: Are the Settled Bed Properties Affected by the Absolute Height of the Settled Bed?

As the scale of operation increases, the absolute height of the settled bed increases. For beds that can tend to compress, re-suspension may become more difficult with increasing bed depth. Generally, organic solids, especially those with a cellular nature, become worse upon scale-up, whereas rigid solids (such as minerals or glassy materials), especially with a narrow particle size distribution, exhibit little or no sensitivity to bed depth.

## Mechanical Solution #1: Control Starting Torque

There are two conflicting goals when it comes for designing for startup in settled solids: the desire to get the impeller up to speed so it can re-suspend solids, and the desire to avoid breaking things and burning out motors. Controlling starting torque is one aspect that may reduce the chance of damage to the equipment, but it may result in preventing the re-suspension of solids when the impeller is buried in the settled bed. The solution to this is in the section "Mechanical Solution #2."

In this subheading, we will describe various devices that have been used to limit starting torque, and some of the pros and cons. We have opted not to show any illustrations, as pictures are readily available on the Internet. We will include the base case of no specific torque limiting device. The choice of what device to use depends on bed properties and how long a torque increase may be required.

- No device. In this case, the motor stalling torque is the torque limit, and the motor control center limits the duration of overload. This is simple, but it requires the buried impeller and shaft to be designed for about 300% of the normal running torque. This could greatly increase the cost of the agitator. Moreover, the motor is likely to "kick out" within a few seconds, leaving the settled bed unsuspended. This is not a recommended practice for any but easy applications, such as uniform particle size or beds that do not pack; it may work okay for high-solids biomass slurries.
- Electrical soft start. This spreads the starting time out, reducing the maximum current and generally limiting torque to about 200% of running torque. This may still not be sufficient for some settled beds. The shaft and impeller must still be designed for more than running torque.
- Fluid coupling. This works similarly to the torque convertor in an automatic transmission. Technically, this does not limit torque, but allows gradual acceleration of the shaft. If this is not enough, the torque will eventually build up, and given enough time, the working fluid will get very hot, causing a fusible plug to melt, spraying hot oil all over the place, and failing to suspend the solids as the rpm drops back to zero. If the startup is successful, there will be a 3–5% slip of the coupling under normal conditions, resulting in a 3–5% power loss, making the system somewhat energy inefficient. A variation of this type of coupling uses metal shot inside the coupling rather than a liquid. These usually lock up to zero slip at full speed, which is more energy efficient. However, with each startup, the shot becomes distorted, making the unit slip less each time.

- Slip clutch. These are often found on farm equipment such as rotary tillers and some brush cutters. Basically, they are set to slip when the torque reaches a certain value, usually 125–150% of normal torque. Under normal running conditions, they do not slip, so they are energy efficient. Under overload, they will slip, limiting torque but perhaps preventing successful re-suspension. Moreover, if the slippage lasts for more than a few seconds, they get hot enough to start fires. Their use is often prohibited in Division 1 and 2 hazard areas, such as near ethanol fermenters and storage tanks. They are really only suited to shock loads of short duration, which makes them ideal for farm equipment, but maybe less than ideal for starting up an agitator in settled solids.
- There are couplings that use ball bearings resting in detents between two spring-loaded plates. When the torque exceeds a pre-set value, the plates spread apart a little and the balls move along a spiral path to a circumferential groove, allowing the coupling to freewheel at almost zero torque. However, when this happens, the output from the coupling is basically 100% slippage, that is, 0 rpm. So, this type of coupling protects the equipment from torque damage but is useless to get the agitation job done.
- Since VFD drives can save energy in operation, many facilities have them. As it turns out, VFD drives are also one of the best torque limiters. They can have the maximum current set to anywhere up to 150% of full load. Current is basically proportional to torque. Moreover, sophisticated models can be set to ramp up to speed slowly, and to reduce the output frequency, and hence, shaft speed, to limit the torque. This is normally the best solution. A caveat is that, if the bed requires more torque than the VFD is set to deliver, the agitator will never fully start up, and the solids will not be re-suspended.
- Hydrostatic drives. Though not common, these use pressure as an analog for torque, and can be set up so that speed is automatically reduced to limit system pressure. In that sense, they work a lot like VFDs. They will also stop an overload at a pre-set torque level by means of a pressure relief valve, but this generates a lot of heat if operated in this mode for very long. One drawback is that they are only about 60–80% efficient, which is why they are rare in agitator use.

There are probably other torque limiting technologies available. But torque limitation is only one of the steps to a successful design for re-suspending solids.

## Mechanical Solution #2: Break Up the Settled Bed

Because settled beds of solids can have wildly variable properties, which can result in significant torque requirements and can cause massive bending loads if they shift unevenly upon startup, it is often desirable to alter the bed properties before starting. In essence, this means breaking up the settled bed. In practice, this usually means introducing either a gas (typically air) or a high-pressure liquid in the area below the impeller. This author favors air (or an inert gas if oxidation is a problem), because it does not dilute the product and because it does not require high pressure to get the job done. Air can be introduced on a temporary basis by inserting a pipe (also called a lance) of appropriate length below the impeller and turning on the air supply, with the agitator turned off. This usually works in a few seconds up to a minute or so. If the required pipe length is too much for a person to handle, it is possible to build in a permanent sparge pipe to break up the solids. A potential problem with this approach is that the pipe may become plugged in normal operation. A spring-loaded check valve at the tip of the pipe may help prevent this.

If the solids do not settle quickly, the lance may be removed before starting the agitator.

## Mechanical Solution #3: Limit Shaft Deflection

Limiting torque and designing the bottom impeller to withstand the maximum available torque and a dimensionless hydraulic force of 2 will generally prevent failure of the bottom impeller. However, it may not be sufficient to prevent shaft failure, even if the shaft is designed for the maximum available torque. This is because the torque may not be evenly applied, resulting in large side loads on the bottom impeller, which results in a large bending moment in the case of a cantilevered shaft. If the bottom impeller sees a solid bed on one side and not the other, it can try to pivot about the tip of one blade. This creates a dimensionless hydraulic force of 2, which creates a huge bending load. Generally, it is not practical to design for this. The shaft deflection must be limited to prevent shaft failure.

One way to do this is to use a steady bearing. That solves the problem, but introduces a wearing part inside the tank, as has been discussed previously. A cantilevered design would be preferred for most slurry applications. Since startup in a settled bed is not normal operation, there is a way to limit deflection on an intermittent basis that does not have normally wearing parts. That method is called by some a limit ring. This author prefers the term "slap ring" for its more graphic appeal.

A slap ring is quite similar to a steady bearing, but it is designed to be non-contacting in normal operation. Therefore, it has no bushing or wear sleeve, and it operates at a substantial clearance from the shaft. It is good practice to turn down the shaft through the slap ring area, so that if the shaft ever does hit the slap ring, the small burrs that will be created will not interfere with impeller hub removal if that ever becomes necessary. A schematic representation is shown in Figure 15.7.

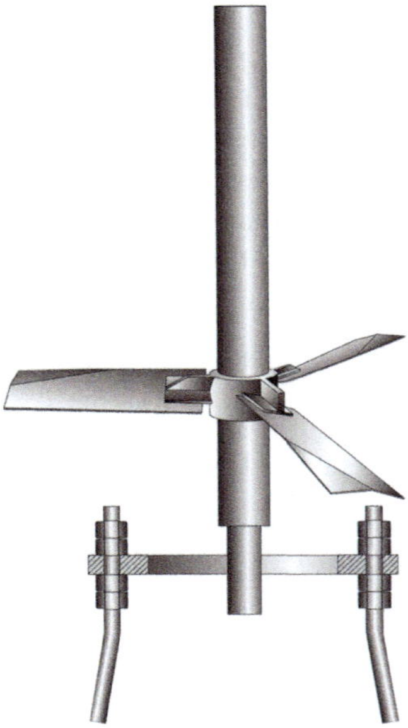

Figure 15.7 Slap ring.

The question naturally arises: How do we determine the amount of clearance to design for in the slap ring? There are different opinions on this. Almost all agree, however, that the slap ring must not contact the shaft during normal operation. That means the shaft must work acceptably as a cantilever design under normal operating conditions. The procedure suggested by this author is as follows:

1) Design the shaft to have acceptable stresses and critical speed under normal conditions.
2) Design the lower impeller to withstand the maximum available torque from the machine, using whatever torque limiting device is chosen. Also use a dimensionless hydraulic force of 2 for the impeller, but not for the shaft. This may make the impeller quite heavy, which will affect shaft critical speed.
3) For dynamic shaft analysis, put all of the available power into the bottom impeller.
4) Vary the hydraulic force until the shaft stresses are about 50% higher than the normal allowable stress limit. There is enough safety factor in the allowable limits to handle *occasional* excursions into higher stresses.
5) Use a dynamic shaft analysis to determine the shaft deflection under the above condition.
6) Choose a slap ring inside diameter equal to 2 times the deflection plus the shaft diameter through the slap ring.

An example of the results of this philosophy is presented in Figure 15.8. This problem is similar to the previous steady bearing problem, except that we changed the shaft speed to get the $N/N_c$ ratio below 0.8. We put 156 Hp into the bottom impeller, representing 125 Hp with a 25% over torque limitation, such as by a VFD set point. We did not do detailed analysis of impeller weights, as we mainly wanted to illustrate deflection. We varied the hydraulic force until we got stresses about 50% higher than allowable, which resulted in our dynamic analysis program flagging the stresses in red. The resulting deflection where the slap ring would be (second station from the bottom) is 11.5″. With a 4.5″ shaft diameter at this point, the suggested slap ring ID would be $2 \times 11.5 + 4.5$, or 27.5″ (0.7 m). Some might be tempted to make it smaller. But if it is too small, it is likely that the shaft may contact the slap ring more frequently, perhaps even in normal operation. The goal of the slap ring is to prevent failure in the event of startup in a settled bed, not to

| | | | | |
|---|---|---|---|---|
| Carbon Steel<br>Stainless Steel<br>Titanium, Grade 2<br>Alloy C276 | | Top Entering<br>Side Entering | Calculate | |
| | | 1st Critical Speed | 2nd Critical Speed (RPM) | |
| Customer | Steady bearing<br>2 | | | |
| Shaft Speed (RPM) | 37 | 54.4 (68.0%) | 351.2 (10.5%) | |
| Temperature (F) | 223 | | | |

| Shaft Length (in) | Shaft O.D. (in) | Shaft I.D. (in) | Support Stiffness (lb/in) | Lumped Mass (lb) | Impeller Force (lb) | Impeller Power (HP) | Shaft Deflection (in) | Bending Moment (in-lbs) | Shear (lbs) | Tensile Stress (psi) | Shear Stress (psi) |
|---|---|---|---|---|---|---|---|---|---|---|---|
| 1 | 7.5 | 0 | 1.00E+06 | 0.000 | 0.000 | 0.000 | −3.381E-02 | 0.000E+00 | −3.381E+04 | 3.208E+03 | 3.208E+03 |
| 19.7 | 7.5 | 0 | 1.00E+06 | 0.000 | 0.000 | 0.000 | 3.501E-02 | −6.661E+05 | 1.200E+03 | 6.162E+02 | ***8.658E+03*** |
| 259.8 | 12.75 | 12 | 0.00E+00 | 500.000 | 0.000 | 0.000 | 3.539E+00 | −3.544E+05 | 1.200E+03 | 1.011E+03 | ***5.054E+03*** |
| 157.5 | 12.75 | 12 | 0.00E+00 | 300.000 | 0.000 | 0.000 | 7.267E+00 | −1.654E+05 | 1.200E+03 | 1.684E+03 | 3.571E+03 |
| 137.8 | 12.75 | 12 | 0 | 400.000 | 1200.000 | 156.000 | 1.096E+01 | −8.540E-07 | 1.077E-10 | 0.000E+00 | 9.745E-09 |
| 19.7 | 4.5 | 0 | 1.00E+00 | 0.000 | 0.000 | 0.000 | 1.150E+01 | −3.143E-09 | 1.077E-10 | 0.000E+00 | 1.757E-10 |
| 1 | 4.5 | 0 | 0 | 0.000 | 0.000 | 0.000 | 1.153E+01 | 4.476E-08 | 1.077E-10 | 5.003E-09 | 2.502E-09 |

***The maximum shear stress of 8.658E+03 psi occurs at 2.070E+01 inches.***

**Figure 15.8** Slap ring dynamic analysis.

contact the shaft in normal operation. It is still important to try to avoid such start-ups by having a backup power supply and using an air lance or similar method to break up the settled bed.

### Mechanical Solution #4: Lift the Impeller Up Out of the Settled Bed Before Starting

This is a rare and unusual approach, but it is worth mentioning because it is innovative, it works, and, to the author's knowledge, it has been successfully applied (by Philadelphia Mixers) to the largest agitators ever built, in terms of torque and total unit weight, though not power. The particulars are that this design was used on (three) eight-million-gallon coal slurry tanks, with a working volume of six million gallons (almost 23 000 $m^3$). The particle size and % solids are unknown to this author. The unit had a motor power of about 373 kW, operating at a shaft speed of 6 rpm. That works out to a maximum possible torque of over 593 000 N-m. Each agitator weighs over 100 tons. Though the coal slurry probably re-suspends easily, requiring no more agitation than required to maintain suspension, starting up in a settled bed might require more torque and carries the risk of shaft or impeller failure unless an extremely robust design is used. Given the massive design required to even operate at normal torque, it might not have been possible to fabricate equipment that could withstand such a startup. The impellers alone are more than 10 m in diameter. A steady bearing was ruled out due to the abrasive nature of coal particles.

The solution employed for the above coal slurry mixer was to use a large hollow shaft on which the upper impeller was mounted in a fixed elevation off bottom, and a smaller, concentric shaft going through the outer shaft and gear drive, on which the lower impeller was mounted. This shaft was coupled to a raising and lowering mechanism, so that the lower impeller could be raised above the settled bed level before starting. That way, there is no torque overload and no abnormal bending moment. After the solids have been re-suspended, the lower impeller can be lowered to its normal operating position. The author does not know if this could be done while the agitator is running. However, the lower impeller can be disengaged from the upper one, allowing the upper impeller only to be used for re-suspension if needed.

A conceptual arrangement of this design, drawn by a technical illustrator and inspired by the old literature of Philadelphia Mixers, Inc., appears in Figure 15.9. Images of people are there to give a sense of scale. The "ghost" impeller outline indicates the position of the lower impeller when raised.

Though this design is expensive and mechanically complex, it is a viable approach to both preventing damage to the equipment and enabling solids to be re-suspended. It may prevent the need to get in the tank and dig the contents out after a power failure. However, its use should be considered only when it has been verified that the solids can be re-suspended without an impeller buried in the solid bed. Generally, this means relatively hard solids (incompressible) with a narrow size distribution, no adhesion, and particle shapes ranging from roughly spherical to roughly cubic.

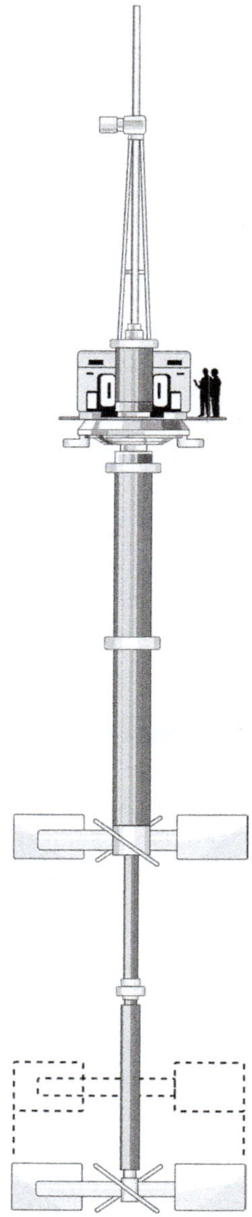

**Figure 15.9** Agitator with impeller lift mechanism.

## Summary of Chapter

In this chapter we have covered all of the major aspects of shaft and impeller design, including the special topic of startup in settled solids, which has little published about it. We have covered the pros and cons of solid versus hollow shafts, steady bearings versus cantilevered design, etc. We have discussed the degree of conservatism that may be needed for various applications, and how to protect the equipment during upsets.

## List of Symbols

B Distance of impeller from steady bearing
C Distance of impeller from lower gearbox support bearing
d Shaft diameter
D Impeller diameter
$D_h$ Hub diameter
$F_b$ Force on blade
$F_D$ Dimensionless hydraulic force
$F_h$ Peak value of hydraulic force
l Blade length
L Shaft length, or impeller distance from support bearing
M Bending moment
$M_C$ Combined bending moment and torque
N Shaft speed
$N_C$ Critical speed; *aka* first natural lateral frequency
P Impeller power

$S$ Bearing span
$t$ Blade thickness
$T_Q$ Impeller torque; shown as TQ in figures
$T$ Tank diameter
$v$ Blade volume
$w$ Blade width
$W$ Impeller weight
$W_b$ Blade weight
$W_e$ Equivalent mass
$W_h$ Hub weight
$Z_m$ Section modulus

## Greek Letters

$\delta$ Shaft deflection
$\rho_m$ Material density
$\sigma_s$ Shear stress
$\sigma_t$ Tensile stress

## References


**1** Ramsey, W. and Zoller, G. (1976). How the design of shafts, seals and impellers affects agitator performance. *Chemical Engineering Magazine* (30 August): 52–59.
**2** Young, W.C. and Budynas, R.G. (2002). *Roark's Formulas for Stress and Strain*, 7e. General reference. McGraw-Hill.
**3** Glover, T.J. (1998). *Pocket Ref*, 2e. Littleton, CO: Sequoia Publishing.
**4** ENGINEERSEDGE. Critical speeds of rotating shafts with single loads equations and calculators. https://www.engineersedge.com/bearing/critical-speed-single-load.htm (accessed 11 October 2020).
**5** LINEARMOTIONTIPS. Beam deflection equations for linear systems. https://www.linearmotiontips.com/beam-deflection-equations-for-linear-systems/ (accessed 11 October 2020).

# 16

# Mechanical Design of Agitators 2

Gear Reducers, AKA Agitator Drives

## Introduction

Chapter 15 explored how the design of shafts and impellers impacts equipment reliability. Equally important is the reliability of the mechanisms that support and drive the shaft and impeller system.

Most agitators operate at a shaft speed slower than motor speed. There are a number of reasons for this, including shaft design considerations and the fact that, for motion-controlled applications, lower shaft speeds with larger impellers will require less power to achieve the desired process result than higher speeds with smaller impellers. Thus, some means must be provided to reduce speed and increase torque. We have opted to only cover gear reducers for this purpose in this chapter, for reasons explained in the next heading.

A second function of the drive system is to support the shaft, unless the loads are isolated from the drive. We will use this fact as a major principle in agitator gear drive selection.

Some of the inspiration for this chapter came from Ref. [1]. Other information came from a variety of sources, including internal communications from Mr. Bill Fryers, who is a mechanical engineering expert who has a great deal of agitator design experience, specifically in the design of purpose-built agitator drives. General mechanical engineering handbooks also supplied information, but will not be referenced by line item in the text because such information is commonplace.

## Agitator Gear Drive Selection Principles and Concepts

As stated in the introduction, except for the case of very small agitators, most agitator shafts and impellers do not operate at motor speeds. They need to be slowed down to allow large enough impellers to be used to control the full liquid volume, and to address such issues as shaft critical speed and seal life.

*Agitator Design Technology for Biofuels and Renewable Chemicals*, First Edition. Gregory T. Benz.
© 2022 John Wiley & Sons, Inc. Published 2022 by John Wiley & Sons, Inc.

There are multiple ways of reducing speed and increasing torque. The most commonly used ones in agitator design are belt drive and gear drive, though in principle other mechanisms can be used, such as capstan drive and direct hydrostatic drive, among others. Though belt drives are simple and inexpensive, they are relatively inefficient (typically 92% per reduction stage) under normal maintenance conditions, require belt tension adjustment frequently, and require belt replacement after a year or two; sometimes more frequently. Gear drives using helical or spiral bevel gears have a typical efficiency of 98.5% per reduction stage, and gear life is about 5–25 years in biofuel service, depending on design. Routine maintenance consists mainly of changing the oil and greasing the bearings. Moreover, there is little to differentiate one manufacturer's belt drive design from another. For these reasons, we have opted to discuss gear drives only in this chapter. We also will not discuss worm gear drives, as they are less efficient and wear out more quickly (typically one to five years). We will confine our gear drive discussion to those using helical or spiral bevel gears. Planetary gearing would also be acceptable, but we are unaware of any agitator manufacturer offering this.

What we will do is go over the several ways a successful agitator drive system can be configured, offer opinions on what might be the best way, and then get into some specific details needed to specify a design.

But first, we will delve into a brief history of agitator design and construction, reliability problems encountered, the loads involved, and why agitator drives are a specialized product, distinctly different from a commercial gear drive.

## Early History

The first manufacturers of agitators used whatever was commercially available to reduce speed and increase torque. But, because the magnitude of the loads was not well understood and the commercially available gear drives were not specifically designed for agitator service, most units had a short lifespan. Some failed catastrophically within days or weeks. Larger service factor helped, but people did not always know how to be sure of a selection that would be successful. The key to successful design turned out to be developing an understanding of the loads imposed on an agitator drive, and how to accommodate them. Specialized mechanical testing equipment was needed to develop load correlations, and most manufacturers did not have such equipment until the 1960s. Some still do not have them today. So, what are those loads and how do they differ from those a commercial gear drive is designed to handle?

## Loads Imposed

Figure 16.1 depicts the loads a typical commercial gear drive is designed to handle. Such gear drives are generally suitable for close-coupled loads, such as a gear

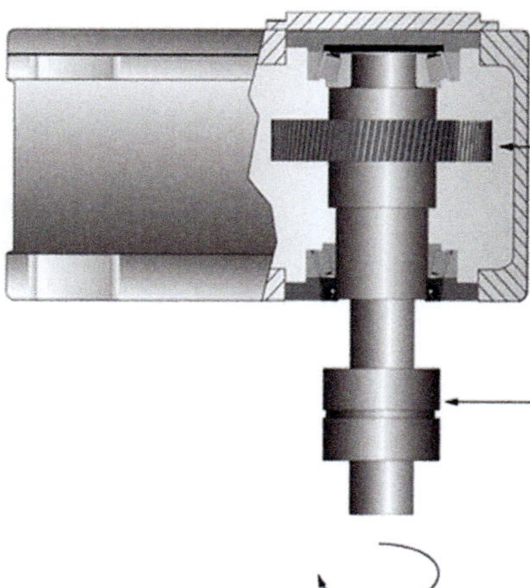


**Figure 16.1** Commercial gear drive design load diagram.

or chain-and-sprocket on the output shaft. These close-coupled loads create a small bending moment, well within the range of the gearbox output shaft bearing capacity. There is also a bending moment created by the gear reaction force. These two moments create a deflection of the output shaft, but it is small enough to create no problem for the gear alignment. Typically, the internal shafting is not much larger than that required to transmit torque. In fact, torque is the primary load a commercial gear drive is designed for. Axial thrust, or weight, imposed on the output shaft is limited to very small values compared to agitator loads.

An agitator experiences different operating conditions from most commercial gear drives. For example, it often runs 24 hours per day for at least 90% of the time in many plant applications, and the torque transmitted is relatively steady. An agitator also has a very different load diagram, as shown in Figure 16.2. In addition to torque, agitator service imposes such loads as the weight of the shaft and impellers, impeller thrust, impeller side loads and the attendant bending moment they create due to being far from the gear drive bearings, and pressure thrust if there is a seal and the vessel operates under pressure or vacuum. The bending moment not only affects bearing lives, but greatly increases gear tooth misalignment due to drive shaft deflection. If these loads are not taken into account, bearing and gear failures can and will occur. That is what happened frequently in the early days of the agitator industry. There is an additional factor. The vast majority of commercial gearboxes are designed so that the output shaft bearings are a slide fit in the housing. This is fine for gear loads and side loads on the output shaft such as belt or sprocket loads but it is not so good if the gearbox must support an agitator shaft.

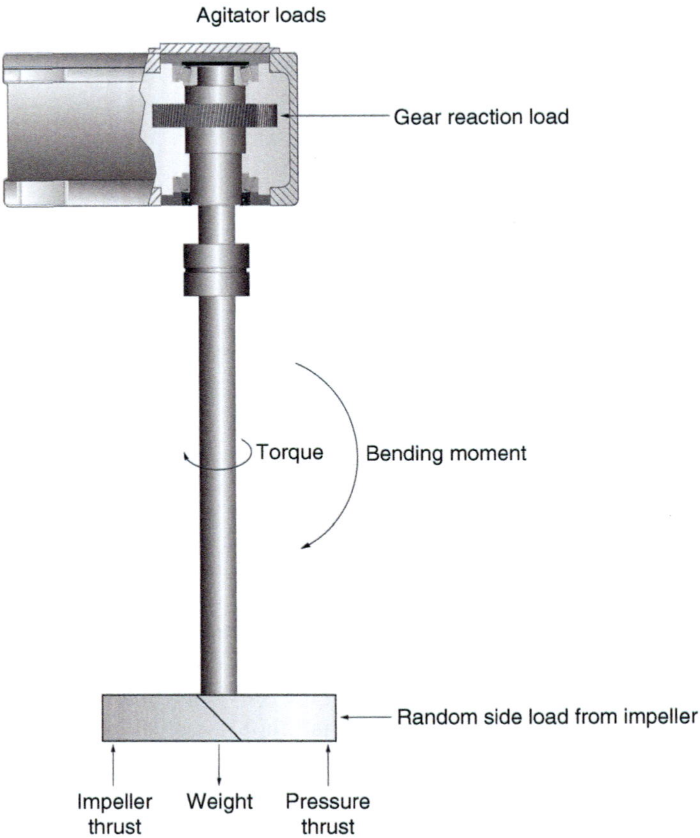


**Figure 16.2** Agitator load diagram.

The changing load directions can result in the bearing fretting in the bearing housing leading to a failure of the gearbox. If the bearings on the output shaft are very large for their duty, then the rate at which the casing is damaged may not be enough to cause early failure. Also, some smaller commercial gearboxes have bearing retention techniques that are not suited to a shaft which sees high levels of rotating bending moments. So how do we design to accommodate these loads?

## Handle or Isolate Loads?

Broadly speaking, there are two major design approaches that can be successfully employed: designing the gear drive to handle the loads, or trying to isolate the loads from the gear drive by mounting the impeller drive shaft on a separate set of bearings. We will look at several approaches in each category.

## Handle Loads Option 1: Oversized Commercial Gear Drive

It is sometimes possible to get satisfactory life in agitator service by using a larger commercial gear drive to get the necessary shaft diameter and robust bearings needed. Figure 16.3 conceptually compares a commercial gear drive with a service factor (defined as the ratio of American Gear Manufacturers' Association [AGMA] rating/motor nameplate power) of 1.25 with one that has a service factor of 3.

Figure 16.3a shows the smaller gear drive; Figure 16.3b shows the larger one. If the smaller gear drive has a shaft diameter of d, the one with a 3.0 service factor will have a shaft diameter of about 1.3d. It will also be taller and heavier, and its gears will theoretically last longer. However, due to the bending load, the gear deflection will be higher than what was intended for the product, so its life will be reduced somewhat. It has been this author's experience that a service factor of 3.0 is generally the minimum required when applying a commercial gear drive to agitator service. We will get into more detail on service factors later.

The oversized gear drive approach has one advantage: it is usually the cheapest option. But it is still not usually as reliable as a purpose-built drive, and its longer internal shafting results in lower shaft critical speeds Ref. [2], reducing allowable shaft extensions, and its increased deflection will allow more wear on mechanical seals. It is also slightly less efficient, as there are more fixed losses in a larger gear drive.

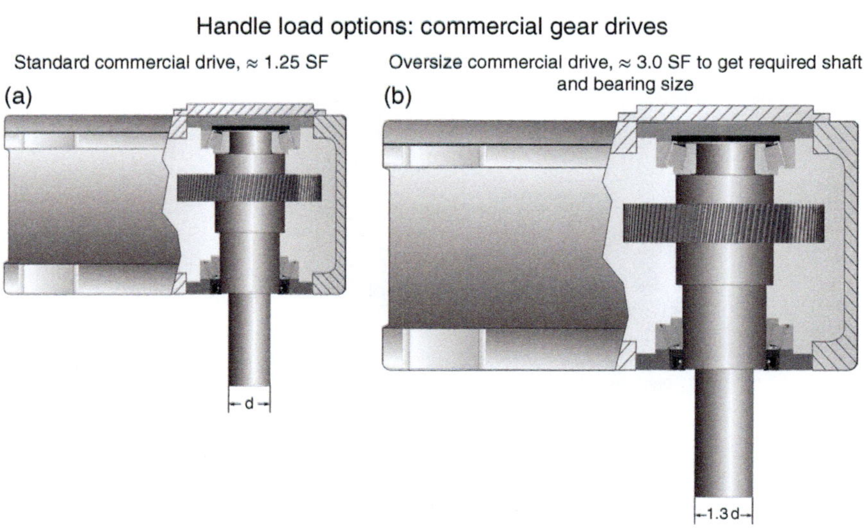


**Figure 16.3** (a, b) Commercial gear drive size comparison.

## Handle Loads Option 2: Purpose-Built Agitator Drive

The concept of this drive design, depicted in Figure 16.4, is to use an output shaft that is much larger than that of a commercial gear drive, combined with massive bearings that are mounted in such a way as to prevent rocking or fretting in the housing; a very rigid housing; and a short bearing span internal to the drive. The reason to use a short bearing span is to minimize deflection, both inside the gear drive and outside, at such locations as the mechanical seal and along the entire extension shaft. This logic can be understood by referring to Eq. (16.1), which can be derived from Ref. [3]:

$$\delta \propto S^3 / d^4, \quad \text{where } \delta \text{ is the deflection} \tag{16.1}$$

One can readily see that combination of a short bearing span and a big shaft diameter dramatically reduces deflection. This makes the gears last longer as well as improves seal life. It also raises the shaft critical speed, which can allow slightly longer shaft extensions. Compared to a commercial reducer, the shaft diameter is generally about 1.8d. It is usually a solid shaft going through the gearbox.

Such gear drives have been manufactured by major agitator manufacturers since the 1970s. In normal biofuel/renewable chemicals service, such a gear drive, when applied with a service factor of 1.25, will last for 15–25 years before a rebuild is needed. It will have output bearing lives of several hundred thousand hours.

Such drives have also been manufactured by companies that manufacture commercial gear drives. They are clearly advertised as agitator drives. In many services, they may give satisfactory results. However, in a study I did for a client, I found that they all had smaller output shafts and lower output bearing lives

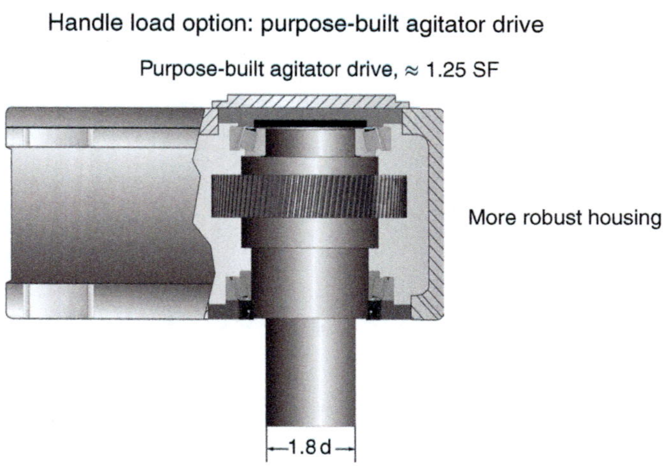


**Figure 16.4** Purpose-built solid shaft agitator drive.

than those made by major agitator manufacturers; at least compared to those using tapered roller bearings on the output shaft.

The only disadvantage of the purpose-built drive is higher initial cost than even a commercial drive with a 3.0 service factor. However, lower downtime probably makes the total cost of ownership much less.

A variation of this design used by some manufacturers uses a hollow output shaft, with the extension shaft rigidly locked into position. This is slightly less rigid, leading to lower shaft critical speeds and higher seal deflection and wear. If the hollow reducer shaft is made of steel and the extension shaft is made of an alloy such as stainless steel, plant moisture has sometimes been known to cause galvanic corrosion between the two materials, making it very hard to remove the gear drive for servicing, or to change the mechanical seal in some designs.

## Isolate Loads Option 1: Hollow Quill Integrated Drive with Flexibly Coupled Extension Shaft

This design concept has been in use since at least 1960, and is offered today by several agitator manufacturers as well as some gear drive manufacturers. A schematic of the concept may be found in Figure 16.5. The basic concept is that the output gear is mounted on a hollow quill shaft on its own set of bearings. The agitator drive shaft is concentric with the hollow quill, and it is mounted on its own separate set of

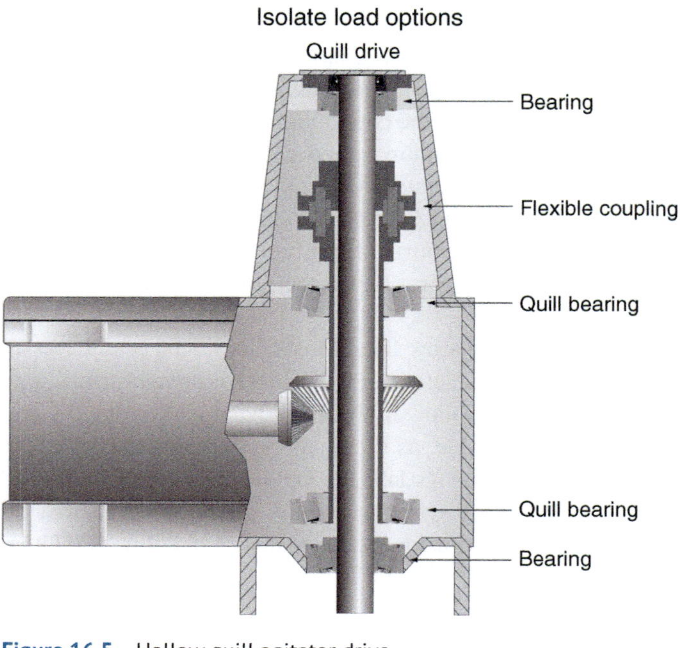


**Figure 16.5** Hollow quill agitator drive.

bearings, attached to the same overall housing, above and below the quill shaft bearings. In principle, these bearings take all of the external loads described in the load diagram in Figure 16.2, whereas the quill shaft and its bearings take only the reaction loads from the low-speed gear set. The two shafts are connected by means of a flexible coupling located above the upper quill shaft bearing. It is claimed that such construction isolates the external loads from the gear quill shaft, reducing gear shaft deflection and promoting long life. This design has proven much more reliable than commercial gear drives for most applications. However, it does not actually work the way it is described in manufacturers' literature.

To begin with, the degree of isolation of the extension shaft loads from the gear shaft is questionable. To understand this, one must understand a little about how flexible couplings are selected. Basically, such couplings have two selection criteria: the mechanical rating and the bore size. For a coupling to have long life and yet provide reasonable torsional and lateral flexibility, it should typically have a mechanical rating around 125–150% of the applied torque. However, when the coupling must be sized to fit a very large shaft, such as the hollow quill shaft, it will be a much larger coupling than needed for the torque rating. In many cases, the actual coupling rating will be 5–15 times the applied torque. When applied at such a high service factor, the coupling will be almost rigid. It will provide no appreciable attenuation of torsional shock loads (although this is not a problem if the high-speed motor coupling is properly selected to do just that). It will also transmit much or most of the side loads directly to the quill shaft if the drive shaft starts to deflect. That means the quill shaft bearings must be designed to handle such loads. Fortunately, they usually are.

Secondly, there is the matter of gear tooth deflection due to flexing of the quill shaft itself. In a Master's Thesis report (copy not available), done by R. Stephen Hill in the 1980s, it was determined that the common hollow quill designs of the day actually had greater gear tooth deflection from the gear reaction forces alone than the most common purpose-built agitator drives did from all loads combined. This is because the quill shaft was much less stiff (had a lower moment of inertia) than the massive solid shafts used in the purpose-built gear drives, and the hollow quill also had a much longer bearing span. (Remember, the deflection is proportional to the cube of the bearing span.)

Thirdly, we come to the matter of bearing lives. Most commercial gear drives as well as the solid-shaft agitator drive depicted in Figure 16.4 have two bearings supporting all of the output loads. The hollow quill design has four bearings. Some might think this makes it more reliable, as the loads are spread out over more bearings. However, if any of the bearings fail, the agitator fails. And if all four bearings have the same life, the overall system reliability will be less. This may be easily derived from the common equation for mean time between failures (MTBF):

$$\mathrm{MTBF} = 1/\left[\Sigma(1/\mathrm{L})^2\right]^{0.5} \tag{16.2}$$

If each of the bearings has the same life, the Eq. (16.2) leads to an overall life of a four-bearing design that is 70.7% as long as for a two-bearing design. However, most hollow quill designs use ball bearings for the drive shaft support instead of the more rugged tapered roller bearings, so their life may well be even less.

Finally, we get to shaft deflection and critical speed. The quill shaft design has a substantially longer drive shaft than does the purpose-built solid shaft design, and its drive shaft is smaller in diameter as well. This leads to more shaft deflection and a lower critical speed. These two effects combine to produce lower seal life and a shorter allowable shaft extension.

The main positive points of this design are:

- Design proven over at least 50 years of production.
- Much more reliable than commercial gear drives.

The main disadvantages are:

- Much more expensive than other purpose-built agitator drives as well as commercial gear drives.
- Rather tall. Overhead clearance could be an issue.
- Shorter maximum shaft extension, compared to solid shaft purpose-built drives.
- More deflection through the seal, reducing seal life.
- If built to the same gear rating and bearing life standards as solid-shaft purpose-built drives, they probably are less reliable overall.

## Isolate Loads Option 2: Outboard Support Bearing Module

Shown in Figure 16.6, this concept can use either a commercial gear drive or a purpose-built agitator drive. However, there is little reason to use a purpose-built agitator drive as long as the commercial gear drive in question has ample gear and bearing life ratings.

The basic idea is that the extension shaft loads are supported in a bearing module independent of the gear drive. That module often also contains the mechanical seal as well. The shaft exiting the top of the bearing module is flexibly coupled to the gear drive above. The gear drive and the bearing module are normally connected together by means of a spacer pedestal. It is also possible to mount the gear drive independently, with a universal-joint shaft or similar shaft connecting the two, as shown in Figure 16.7.

The bearing module approach allows a shorter bearing span for the shaft support than the hollow quill design. This allows standard seals to be used and permits longer shaft extensions. The bearing module shaft diameter can be as large as needed, independent of the gear drive selection. With this design concept, it is

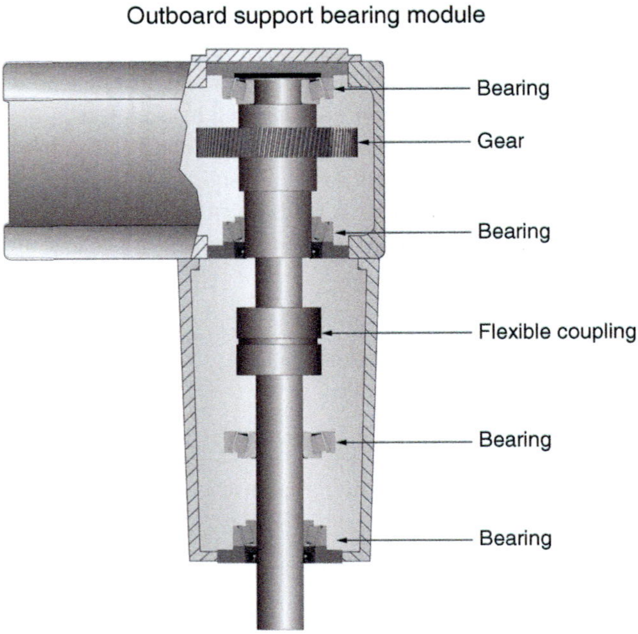


Figure 16.6 Outboard support module.

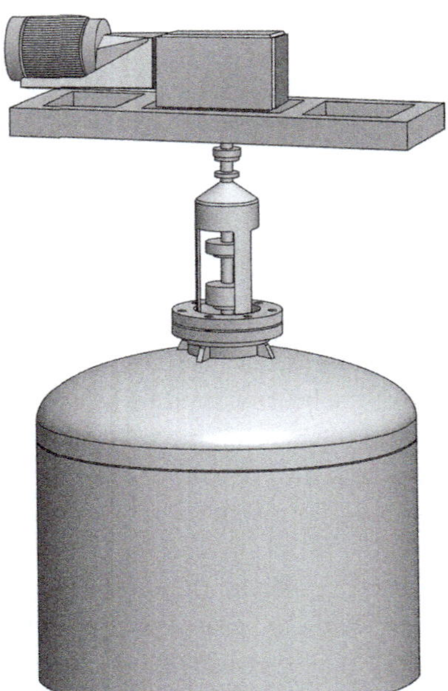

Figure 16.7 Auxiliary mechanical seal with beams to building.

never necessary to oversize the gear drive to get the required shaft diameter. The shaft out of the top of the module is normally turned down so that a flexible coupling with a service factor of 1–1.5 can be used, better isolating the loads and possibly attenuating shock somewhat. Though this design approach is not common, it finds use when the size of the equipment is such that no standard agitator drive is available, and when vessel pressure creates thrust far beyond what standard agitator drives are designed for. Though high pressure is not normally a feature of biofuel and renewable chemical processes, it is possible that future applications may require high pressure, making this a possibility. Also, if the scale of operation becomes larger than any purpose-built drive on the market today can handle, this is a viable approach. For example, many of the world's largest fermenter agitators have been built this way. In most cases, for agitators of greater than 500 kW motor size, it is the only option available. Some pros and cons are as follows:

**Pros:**
- Cheaper than purpose-built agitator drives in large sizes.
- Available in larger sizes than purpose-built drives.
- Independent bearing module allows for high pressure or high bending applications without affecting gearbox life or selection.
- Can remove gearbox for service without having to support the shaft in the tank.

**Cons:**
- Even taller than quill design.
- Expensive in small sizes.
- More wearing parts; likely less reliable than a purpose-built agitator drive.

## Bearing Life Considerations

To achieve a target gear reducer life of 15 years or more, the weakest bearing in the gearbox will usually need to have an $L_{10}$ life (time to 10% failure) in the range of 30 000 hours minimum. It may need to be more than 50 000 hours if all the bearings have similar lives. Table 16.1 lists the requirements for bearing lives imposed by AGMA. In a competitive market, many commercial reducers just meet the bare minimum. One would logically conclude that we would want a minimum service factor of 1.75 when using a commercial gear drive. This would only apply if the agitator shaft loads were isolated from the gear drive, as in the bearing isolation module design. Much higher service factors would be needed to get the required shaft diameters and bearing sizes on the output shaft. Purpose-built gear drives, in general, have much higher bearing lives than AGMA requires, especially on the output shaft bearings. Typical figures are a minimum $L_{10}$ life in normal service of 30 000 hours at a 1.25 service factor and more than 100 000 hours on the output shaft. Such drives also have more rigid housings to accommodate the bending and thrust loads while keeping gear deflection well below AGMA allowances.

For AGMA-rated gearboxes the shortest calculated bearing lives are shown in Table 16.1.

Based on the Table 16.1, and a good bit of field history experience, I have compiled a rough guide to recommended gear drive service factors as a function of drive arrangement and service conditions. This is found in Table 16.2.

I emphasize that this is just a guide, and is intended to be conservative. For specific applications or designs, the vendor should calculate all loads and bearing lives for the particular design in question. The vendor should be required to provide a summary of gear ratings and bearing lives for the particular service. Nonetheless, if a vendor quote seems to have substantially lower service factors than recommended in Table 16.2, buyer beware!

To emphasize the comparison, I have included Table 16.3, which lists gear drive weight of a mid-range agitator drive, comparing a top-tier, purpose-built gear drive versus two standard commercial gear drives and a commercial gear drive manufacturer's "agitator" drive. This study was done for a client. Clearly, the top-tier product is more ruggedly built and can handle larger shaft loads.

**Table 16.1** AGMA bearing life requirement.

| | Bearing life to 10% failing ($L_{10}$) | |
|---|---|---|
| **Gearbox service factor** | **Ball bearing (h)** | **Roller bearing (h)** |
| 0.8 | 2 560 | 2 376 |
| 0.9 | 3 645 | 3 519 |
| 1 | 5 000 | 5 000 |
| 1.25 | 9 766 | 10 520 |
| 1.5 | 16 875 | 19 317 |
| 1.75 | 26 797 | 32 292 |
| 2 | 40 000 | 50 397 |
| 2.25 | 56 953 | 74 630 |
| 2.5 | 78 125 | 106 032 |
| 3 | 135 000 | 194 704 |

**Table 16.2** Recommended gear drive service factors.

| Service | Gear drive type/arrangement | | | | |
|---|---|---|---|---|---|
| | Commercial | Purpose-built to handle loads by agitator manufacturer | Purpose-built by commercial gear drive manufacturer to handle loads | Hollow quill drive built by agitator manufacturer | Isolation module with commercial drive |
| Water-like, moderate agitation | 3.0 | 1.25 | 1.5 | 1.25 | 2.0 |
| Water-like, intense agitation | 4.0 | 1.5 | 1.75 | 1.5 | 2.5 |
| Make down of hard-to-wet solids | 4.0 | 1.5 | 1.75 | 1.5 | 2.5 |
| Viscous agitation | 2.5 | 1.25 | 1.25 | 1.25 | 1.75 |
| Gas–liquid contacting, e.g. fermenters | 5.0 | 1.75 | 2.0 | 2.0 | 2.75 |

**Table 16.3** Gear drive comparison.

| Mid-sized gear drive, ~24 000 in.-lb torque rating | | |
|---|---|---|
| **Description** | **Weight, lb (kg)** | **Shaft size, in. (mm)** |
| Top tier purpose-built | 919 (418) | 3.5 (89) |
| Commercial "A" | 552 (251) | 2.75 (70) |
| Commercial "B" | 234 (106) | 2.375 (60) |
| Commercial "agitator" drive | 485 (220) | 3 (76) |

## Noise Considerations

Large agitator drives tend to be noisy. Using AGMA quality 10 or higher will reduce surface roughness and mitigate noise. At a given pitch line velocity, helical gearing is quieter than spiral bevel gearing. Therefore, for right angle drives, it is preferable to make the first reduction stage helical rather than spiral bevel in a 2 or 3 reduction stage gear drive. That having been said, most agitators operating with a motor size of greater than 100 kW will require hearing protection in the area. Some will need area noise abatement measures. However, most biofuel applications require less than a 100 kW motor.

## Torsional Natural Frequency

Torsional natural frequency is excitation in a rotary direction, as opposed to shaft critical speed, which is excitation in the lateral direction.

Though torsional frequency analysis is beyond the scope of this book and the skill of this author, it can be an important consideration. Generally speaking, the torsional natural frequency is quite high compared to any oscillatory frequencies in the system. However, as the size of the equipment becomes greater, there is a greater likelihood that it can become a problem. I recommend that it be checked for all units with a motor size of 200 kW or greater. I have personally seen gear tooth destruction on a 300 kW unit. A potential resonance problem was detected in the design phase of multiple 750 kW units, and it was dealt with in the design phase.

What kinds of problems can torsional resonance cause? The main one is gear tooth chatter. If the torsional natural frequency is too close to the gear tooth meshing frequency of any of the gear sets, there can be a problem. Bearing roller frequencies are theoretically a possible problem area as well. For best attenuation, the torsional natural frequency should be well below any gear tooth meshing frequency. So, you might ask, how does one adjust the torsional natural frequency? Though the answer requires sophisticated finite element analysis, the concept is quite simple: select the correct torsional stiffness in the high-speed coupling, i.e. the motor coupling.

Most designers of high-power equipment prefer to use metal grid couplings, as they are very compact and relatively inexpensive for the torque transmitted. But they are also very stiff in both the torsional and lateral direction. To lower the torsional natural frequency, it is desirable to use a softer coupling. Some couplings use a tire-like rubber element. These may do the job, but they try to pull the motor shaft and gear drive input shaft together as the rubber element winds up, possibly reducing bearing life of both the motor and the gear drive. A better design, in my opinion, uses a floating rubber or polyurethane element. Some designs use rubber blocks inside a "spider" metal housing, placing the rubber blocks under compression, but allowing them to float axially. With either kind of elastomeric element, there can be several choices of elastomer, resulting in different degrees of stiffness. Coupling service factor will also affect the stiffness. In short, coupling stiffness is the key to changing the torsional natural frequency, and there are many options available, covering a wide range of stiffnesses.

Most major agitator manufacturers should be able to calculate the torsional natural frequency. Alternatively, there are a number of independent mechanical engineering consultants who can do so.

## Important or Useful Mechanical Design Features

- *Dry well seal around low speed shaft.* This keeps oil in the gearbox in the event of a slow leak in the lip seal or grease seal at the lower output shaft bearing. Overall, for top-entering units, this is a *must have.* For bottom entry units, it is not necessary, as the lower bearing cap is sealed. To keep the images simple and easy to understand, we did not show a dry well in our gear drive illustrations. Agitator manufacturers will normally show these in their literature.
- *Shaft coupling below gear drive.* If it ever becomes necessary to remove the gear drive for servicing back at the maintenance shop, a rigid shaft coupling below the gear drive is a *must have.* Taper lock or compression fit shaft connections within a hollow gear drive have proven very difficult to remove, especially when the hollow gear drive shaft is carbon steel and the extension shaft is stainless or high alloy. Galvanic corrosion will almost certainly occur in a typically humid plant environment.
- *Change gear feature.* This can allow for future process changes, as well as allowing for some commonality of parts when the same gear drive size is used for several applications. A *nice-to-have* feature.

- *Magnetic drain plug*. Helps keep wear particles away from gears and bearings in between oil changes. *Nice to have.*
- *Extended oil drain pipe to allow access away from the center of the vessel.* Could prevent a dangerous slippery oil mess on the top head during oil change. *Nice to have.* In some locations, it might be a *must have.* This can easily be a field modification.
- Extended grease lube fitting pipes for ease of access. *Nice to have.*
- On units equipped with gear drive cooling, a filter and a temperature indication, as well as an interlock to prevent operating the agitator when no oil or coolant is flowing, and a temperature cooling water shutoff to keep the gear drive from getting so cold as to have internal condensation. A *must have.*

## Summary of Chapter

We covered the various concepts used for gear drive selection and shaft support in agitator drives, including bearing life and gear rating considerations. We believe this will prove invaluable to engineers in the biofuels industry, as they have often been forced to make equipment choices that resulted in more downtime than necessary, based on capital cost considerations. This chapter will enable a better analysis of reliability versus cost.

## List of Symbols

| | |
|---|---|
| d | Shaft diameter |
| D | Impeller diameter |
| $F_D$ | Dimensionless hydraulic force |
| $F_h$ | Peak value of hydraulic force |
| L | Shaft length, or impeller distance from support bearing |
| L′ | Mean life of a piece of equipment or component |
| M | Bending moment |
| $M_C$ | Combined bending moment and torque |
| MTBF | Mean time between failures |
| N | Shaft speed |
| $N_C$ | Critical speed; AKA first natural lateral frequency |
| P | Impeller power |
| S | Bearing span |
| $T_Q$ | Impeller torque. Shown as TQ in figures |

## Greek Letters

$\delta$ Shaft deflection
$\sigma_s$ Shear stress
$\sigma_t$ Tensile stress

## References

1 Hill, R. and Kime, D. (1976). How to specify drive trains for turbine agitators. *Chemical Engineering Magazine* (August 1976): 46–51.
2 ENGINEERSEDGE. Critical speeds of rotating shafts with single loads equations and calculators. https://www.engineersedge.com/bearing/critical-speed-single-load.htm (accessed 11 October 2020).
3 LINEARMOTIONTIPS. Beam deflection equations for linear systems. https://www.linearmotiontips.com/beam-deflection-equations-for-linear-systems/ (accessed 11 October 2020).

# 17

# Vendor Evaluation

## Introduction

In this book, we have endeavored to educate the user of agitation equipment on all the factors needed to deliver the desired process result, at minimum energy consumption, and with a high degree of mechanical reliability. But, unless you intend to build your own agitators, you will need to buy them from a manufacturer of agitation equipment. This chapter is about how to select the vendor of such equipment, acknowledging that there may be several compromises made in the process. We will go over several factors in the evaluation process, but it is up to you, the user, to weight those factors according to what is most relevant for your needs.

## Buying Practice in Ethanol Industry

Though this book is about agitator applications for both biofuels and renewable chemicals, in this chapter, we will focus on the buying habits of the ethanol industry. We do so for two reasons: by volume of product and by capital expenditures, the ethanol industry is by far the largest in this category, and it probably operates on the slimmest profit margins. This has a direct impact on the buying practices in this industry. Getting project funding is often difficult, motivating buyers and project managers to try to minimize capital cost, even if the long-term profitability might suffer due to the compromises made to meet a capital budget. Many ethanol plants are built using second-tier suppliers, though some such suppliers do a good job of taking care of their customers. A goal of this chapter is to illuminate the factors that may enhance value within the constraints of a low capital budget.

*Agitator Design Technology for Biofuels and Renewable Chemicals*, First Edition. Gregory T. Benz.
© 2022 John Wiley & Sons, Inc. Published 2022 by John Wiley & Sons, Inc.

## Integrity, Reliability of Product

Down time is expensive! In addition to repair costs and possible loss of in-process inventory, there is lost opportunity cost when product orders are placed with others if the plant cannot deliver product (e.g. ethanol) due to downtime. These costs dwarf any possible savings in capital cost.

As a minimum, vendors should make shafts and impellers that do not break in normal service, or even upset service when specifically called for in the specifications.

Most of the wearing parts are found in the gear drives, as described in Chapter 16. In general, purpose-built agitator drives manufactured by the agitator supplier have the greatest reliability, but are also the most expensive. If capital constraints force one to buy from a supplier that does not offer such a product, it is a good idea to have spares in the maintenance shop, and a system in place to effect rapid replacement with minimal downtime.

## Energy Efficiency

As mentioned in Chapter 5, for motion-controlled applications, it is possible to get the same process result with various combinations of power, shaft speed, and impeller type and diameter. It is often the case that, for 10–20% higher capital cost, the power requirements can be cut 30–50%. The payout period for such savings is usually a matter of months, yet it is rare, in my experience, to see customers pay a premium for energy savings. Vendors should be encouraged to quote both their cheapest option and their minimum energy option, so that a present worth of power cost analysis can be presented to management. Vendors should be prepared to justify why their best design creates such energy savings, as not all vendors' "energy efficient" impellers are equal.

## Service

Given that all mechanical equipment will fail eventually, the ability of a vendor to service the equipment is important. This means they should be able to rapidly repair or replace any worn out or broken components. Some vendors have a dedicated team of professionals who work on repairs only or at least on a priority basis. Others just put it in the queue along with new equipment sales. It behooves the buyer to ask about repair programs.

Field service engineers should be available on short notice. Training the customer on how to maintain the equipment is a valuable function of such engineers.

## Tech Support, Transparency

Agitators are not, in general, pre-packaged, off the shelf products. Rather, they are engineered products, configured on a job-by-job basis for the specific customer application. Though the primary responsibility for equipment sizing lies with the vendor, best results are usually obtained by the use of teamwork. The team can consist of vendor application engineers, user process engineers, and quite often, a consultant. The degree to which this team is successful can depend on how much information the vendor is willing to share about the performance attributes of its products, especially impellers.

Vendors vary widely in their willingness to share such information. Some publish frequently, attend mixing conferences, and share a great deal of information with the customer. Others keep all such information closely guarded; some do not even share such basic information as pumping and power numbers with their own sales engineers! This makes it impossible for the end user to participate in the design discussion or to quantitatively evaluate the bids. It removes the "verify" from the old adage "Trust but verify."

Ideally, not only will a vendor share data, but will be willing to educate its customers by presenting technical seminars. These should be actual technical presentations, of sufficient quality to earn Continuing Education Units (CEUs) or Professional Development Hours (PDH), even if the courses are not registered as such. Some vendor seminars are nothing more than an extended infomercial. Don't waste your time on these.

## R&D, Customer Test Lab

As in any other field of endeavor, agitation technology is constantly changing. This includes process technology, impeller technology, mechanical technology, and manufacturing technology. For a vendor to keep up with it, a significant portion of sales revenue should be devoted to R&D. My own opinion is that at least 2% of sales should be devoted to R&D. Some industry-leading vendors approach 10%. Since the absolute amount of money devoted to R&D is potentially greater in larger organizations, most new technology is developed by them. This can be a very strong advantage.

Ideally, the R&D function should be organized and carried out by people who are recognized experts in their field. One way to evaluate the relative expertise of a company's R&D staff is to check for their publications. Most major agitator manufacturers publish frequently, in such publications as the AIChE Journal, Chemical Engineering Magazine, Chemical Engineering Progress, International Journal of Chemical Engineering and many others, and they attend scientific and engineering conferences around the world, such as the North American Mixing Forum (NAMF), various AIChE meetings, the Fuel Ethanol Workshop, and many other venues. Smaller companies are less visible in their publications. Strangely, I know of one company that claims to be one of the largest and most technically capable agitator manufacturers in the world, with the largest staff of PhD engineers, but I have yet to see any articles published by them. They must be so great that they keep everything secret!

Obviously, for a vendor to have a good R&D program, it must have a good testing lab (though some vendors farm out a lot of their R&D to universities and their labs). Such labs should be able to test performance and power draw characteristics in a variety of vessel sizes, ranging from benchtop scale to at least several thousand liters. They should be able to measure stress and strain on impellers and shafts. They should have the ability to measure velocity profiles using such technologies as Laser-Doppler Velocimetry (LDV), Particle Image Velocimetry (PIV) or similar technologies, and appropriate software to correlate the results.

Of similar importance is the availability of a customer test lab for application testing. Some applications are unusual or complex enough that lab testing is necessary to assure a valid design. Some examples include:

- Rheologically complex fluids, such as high-solids biomass slurries, water soluble polymers, polymer melts, consumer paints and coatings, etc. These may include shear thinning or thickening behavior, a yield stress, viscoelasticity, etc.
- Slurries having solid particle shapes that differ markedly from spherical or cubic, such as metal turnings, punch press drops, and needle-like crystals (Examples include the magnesium turnings often used for Grignard reactions, wood splinters as a biomass source prior to hydration, and iron oxide production as a pigment from sheet steel punch-press drops).
- Reaction selectivity with multiple reaction pathways, such as series/parallel reactions.
- Mass transfer between phases (liquid/liquid, liquid/gas, and liquid/solid).
- Unusual vessel geometries, such as large amounts of heat transfer surface inside, very shallow aspect ratio, irregular shapes, etc.

The vendor should have a customer test lab with the capability to do at least the following tests:

- Power draw tests in unusual tank geometries. This has often been done for large quantities of vertical tube bundles or panel coil baffles. The power draw can easily be 70% or less of the power draw using standard baffling.
- Gassed and ungassed blend time.
- Power draw and holdup at equal superficial velocity to proposed production equipment. This can involve really high gas flow rates when measured as

volume of gas per volume of liquid per minute (VVM). Many tests I have done over the years required a VVM of 30 or more.

- $k_la$ measurements. These can be used for some comparative results. However, broth-specific $k_la$ correlations must be developed by the user with real fermenter broths, preferably during a "live" fermentation series. This is not something any agitator manufacturer I am aware of can do.
- Viscosity measurement. Simple viscometers are found in most mixing labs. A few have advanced rheometers, which are preferred for highly non-Newtonian liquids, such as soluble gums or polymers, as well as high-solids slurries. A properly instrumented agitator is the best choice for high solids biomass slurries, as described in Chapter 12.
- Lab vessels of at least 0.3 m diameter for testing high-solids biomass slurries, with adequate torque and speed as described in Chapter 13.
- The ability to handle hazardous or flammable materials. Most vendors are very cautious about doing this, as they usually do not have explosion proof lab equipment, and typical lab mixing equipment needed for a good mixing test is too large to fit into a standard laboratory fume hood. Some vendors have a custom built, walk-in hood large enough to wheel in or out sufficiently large apparatus as needed.
- Having a bevy of almost all of the vendor's standard impeller styles in a variety of sizes to meet the test needs.
- Having the ability to make custom impellers as needed.
- Having a variety of vessel sizes available, with a variety of tank bottom styles.
- Most tests should be conducted in vessels made of clear materials so that mixing and process results can be observed visually.
- Proper disposal or return of materials subject to material safety data sheet (MSDS) requirements.
- Written safety protocols that are rigorously followed.
- The ability to record and edit lab tests for a condensed and relevant presentation of results. (Your management does not want to sit through an eight hour long recording of a lab test!)
- Vendor may own the lab test data, or you may own it, depending on whether you pay for the test or not, and the details of any such lab test purchase orders.

The main takeaway of the above is that, if you need a lab test, work with a vendor that can develop the data you need and present it to you in a neat summary.

If the vendor does not have such a lab, this is a minus. The customer may need to invest in his own lab to do the testing. Some consultants have limited in-house testing capability as well. Another option might be some university test labs or contract research facilities. Some vendors rely heavily on computational fluid dynamics (CFD) simulations instead of lab tests. But, in some cases, a "wet" lab test is required to develop the rheological model that must be input to the CFD simulation, so it may make more sense to just do the wet test and scale up from it.

## ISO Certification (Necessary vs Sufficient)

In the long run, for a company to deliver a good product and good service, it must do things consistently in a well-documented way. An ISO 9001 or similar certification is a good way to assure that. However, it should be considered a necessary condition, but not necessarily sufficient. As some have said, it is possible under ISO 9001 to deliver a consistently bad product. So, more is needed than such certification to assure a good product is being produced.

### Quality Control Program (Not Lot Sample; 100%)

Companies that make vast quantities of consumer goods can generally assure good quality by sample inspections or lot inspections. This is because they make repetitive products, and need only check to see if anything is drifting off the product specifications over time. The very nature of their manufacturing processes is geared toward repeatability. They do not need to inspect every product specimen.

Agitators are a different case. While there may be a number of standardized components, there are no truly standard agitators. Each one is custom designed for a customer's particular process, tank dimensions, and other specifications. In essence, it is an engineered product. As such, lot inspection is unacceptable. The agitator manufacturer should practice 100% inspection to all quality control standards applicable to each part of every agitator it produces, except for commodity items it purchases outside, such as motors, seals, fasteners, and bearings. For those purchased items, it must do vendor quality audits from time to time. A non-exhaustive list of inspections that should be carried out would include, at a minimum:

- Dimensional check on all finished parts.
- Weld check by visual inspection, dye penetrant. or X-ray as appropriate.
- Shaft straightness/runout check.
- Pressure test of mechanical seal assemblies.
- Impeller balance check.
- Gear drive noise check.
- Gear tooth contact pattern check (if the agitator vendor makes its own gear drive).

- Material certifications as required.
- Segregation of all raw materials by alloy designation.
- Welding and machining of alloy components to be done in an area isolated from carbon steel fabrication and machining.
- Paint job quality.
- All bearings are greased.
- Vapor phase rust inhibitor installed as needed.
- Crating must be substantial to assure delivery undamaged. Special attention is needed for polished components. For international shipments, the crating wood shall be heat treated and certified insect-free.
- If wetted parts are polished, a profilometer test plus visual inspection are both required, and dye penetrant testing is mandatory to uncover cracks and pits.

## Typical Delivery Times and Performance

All capital projects are planned based on a schedule of receipt of capital goods. It is easier to meet the planned schedule if the vendor has a short delivery cycle. However, the project manager should try to develop his plan based on the realistic deliveries typical of the vendors in a given category. It is unfortunate to have to choose a second-tier vendor just to meet a delivery schedule. That will be costly in the long run.

Perhaps even more important than a short delivery cycle is a predictable delivery cycle. Vendor performance and credibility on delivery promises are very important. A good vendor will deliver within $\pm 20\%$ of the promised delivery date; sometimes even better. But I have seen some vendors flat out lie about deliveries. I have seen cases where the vendor sales people have claimed that the order was released to manufacturing and all materials ordered, when the drafting had not even started. I have seen cases when the drawings were not completed by the promised equipment delivery date, and the actual delivery date was 6–9 months after the promised date, when the promised delivery time was already 12 weeks. Once the order is placed, you are committed. Research the vendor's delivery credibility, and be aware that it can change over the years, whether due to changes in personnel, owners, new computerized business systems or a myriad of other potential reasons.

It also helps if the customer project manager only releases the order to manufacturing after everything is checked and no change orders are anticipated. Every change order after release to manufacturing incurs delays and has potential cost impact. In my earlier career at an equipment vendor, I have seen the problems change orders cause. I have seen some customers issue 10–15 change orders during the manufacturing cycle, seemingly unaware of the headaches they cause.

I have even seen some customers try to submit a change order after the equipment had shipped. As the song says, "It's too late, baby!" Of course, the vendor will charge for each of these change orders. But, in my experience, they almost never recoup the true cost of such changes. They just take the losses to maintain customer goodwill. You might not care about that. But you should care about what it will do to your project schedule.

## Parts Inventory

A good vendor will try to keep all normally wearing parts in stock. Custom parts, such as shafts and impellers, must be built-to-order. However, some manufacturers have a special rapid manufacturing program for custom-built parts that need to be replaced, either due to failure or customer-requested modifications, such as impeller retrofits or changes in impeller locations. Due to low quantities sold per year, large gear drives and their internals are not always kept in stock, as government taxes penalize companies for having a large parts inventory. This unfortunate tax structure discourages keeping low volume parts on hand. In any case, it is wise for an end user to have a reasonable number of spares on hand in its maintenance shop.

## Location(s) of Parts and Service

The location of the parts inventory is also important. If the manufacturer is located far from your plant, or even in a foreign country, do they have a local parts warehouse? If not, this may not disqualify a vendor, but may require the user to keep critical spares in stock. In my opinion, this should be done anyway. The cost of downtime dwarfs any possible savings by not keeping critical parts in inventory at the plant site. Forgive me for being repetitively redundant about this.

## Reputation and References of Vendor

Ultimately, for a vendor to stay in business and maintain or increase market share, it must satisfy its customers. Therefore, what long-term users have to say about a vendor's quality, reliability, and service is very important. However, companies do change with time, turnover among key personnel and ownership. A balance between long-term reputation and recent history is needed, especially if there has been a major management or ownership change.

## Price (Least Important Factor)

Many buyers of process equipment, when racking competitive bids, will focus on the price. I guess their company reward system encourages this illusory way of “saving money.” And, indeed, if two or more vendors are completely equal in terms of process performance, mechanical reliability, customer service and delivery, then a decision based on price makes sense. But are they ever really truly equal? In this book, we have tried to help the buyer get all vendors to an equal footing, but actual plant history can be a more powerful guide than what is presented in this book. Two or more vendors are rarely truly equal. If it can be proven, for example, that one vendor’s offering is likely to have 5% less downtime than another’s over, say, a 15-year period, then it would be better to buy that vendor’s equipment at full list price or above than to accept the second vendor’s equipment even if they gave it away for free. The difficulty lies in proving the differences. And, admittedly, it is a good idea to try alternate suppliers once in a while, as companies and products do change. But, at a minimum, if there are differences in gear drive quality and bearing lives, it is always a good idea to buy the best quality offering. Though purchasing agents may not like this idea, as they often get rewarded for “saving money,” in many cases it would be best to have the key decision makers choose the successful bidder without even seeing the price, and then, after the price is revealed, negotiate with the chosen vendor to see if a better deal can be attained. Getting all vendors to an “apples-to-apples” basis in the bidding cycle generally causes them to try to be as competitive as possible.

## Use of a Consultant

Though most vendors are proud of their internal expertise, most are also willing to work with the customer’s consultants, just as they would with a customer’s own in-house mixing “guru.” After all, the consultant acts as a member of the customer’s team. The consultant not only aids in writing specifications and bid evaluation, he may need to inform the customer of installation problems he is aware of with a particular vendor, and suggest that the customer check newer installations to see if the problems have been solved.

Unfortunately, not all vendors are cooperative with consultants. I have personally been threatened with lawsuits for pointing out deviations in a vendor’s bid from the specification, for analyzing process and mechanical performance problems in comparison with how I would do it, and for pointing out problem installations. (They even threatened to sue my customer in one case!) Yet these functions are legitimate and reasonable activities of an engineering consultant. If such suits were really actionable, Consumer Reports would be unable to stay in business.

In my opinion, the real purpose of such lawsuit threats is intimidation. But who wants to work with such a vendor?

## Why Use an Outside Consultant?

Over the years, the methods used by CPI companies to specify, evaluate, and purchase process equipment have changed considerably. Since no one engineer can be expert in the design of all kinds or process equipment (though some think they are!), the problem has mainly been where to get the specialized knowledge that allows the best possible practice to be applied to the purchase of equipment. This applies to agitators as well as other engineered equipment [1].

In the 1960s and 1970s, many large CPI companies, and also a good many engineering firms, had staffs of specialists for a number of different kinds of process equipment. Examples include such items as heat exchangers, filters, evaporators, agitators, centrifuges, extractors, distillation columns, fractionation towers, dryers, and many other kinds of equipment.

In the 1980s, many CPI production companies began to regard such in-house expertise as a luxury they could not afford, so they reduced their staff size, and even eliminated whole departments in some cases. Yet they still needed such expertise to function optimally. Some got it from engineering firms, though many engineering firms also cut specialized staff during the same time period. Most were forced to rely on the expertise of equipment suppliers. Though the equipment worked in most cases, it did so because the leading equipment manufacturers had adequate, highly specialized, and experienced staff. Moreover, the end user did not have the expertise to determine if equipment was optimally sized, undersized or oversized. In addition, energy versus capital cost considerations were not always explored. Frequently, the purchase decision was based on trust in the vendor or its representative, rather than any quantitative basis for comparing bids. Equipment vendors, being well aware of this, cultivated relationships and sought to avoid direct bidding on an equivalent basis to their competitors, so as to avoid price being the basis of the purchase decision. Also, equipment vendors have never had enough staff to get involved early in the process development stages where it may be useful to look at alternative approaches or to aid in the design of experiments to assure adequate data are taken for scale-up.

During the late 1990s through the present, equipment purchases by the chemical process industries (CPI) declined on an absolute basis, and the price of the equipment fell when adjusted for inflation. The biofuel industry grew in the early 2000s, but diminished in the 2010–2020 time frame. Manufacturing efficiency gains assuaged some of the effect on equipment profit margins, but there was still a net decline in profitability. Equipment manufacturers responded by reducing the size and experience level of their Application Engineering and R&D departments, in some cases suspending R&D altogether for a while. So, the level of expertise that end users had come to expect from their equipment suppliers was no longer there Refs. [2] and [3].

Today, the typical end user engineer is faced with having to make purchase decisions on process equipment without extensive personal knowledge of the equipment, without a corporate staff of specialists, without available specialists at engineering firms, and with reduced experience and expertise at the equipment vendors. How can an engineer improve this situation?

One way is to consider the use of consultants. Several of the highly experienced engineers who were laid off during the staff reductions at equipment manufacturers have gone into business as independent consultants. They can fulfill the same role that inside specialist staff once did, as well as experienced vendor staff.

When one chooses the right consultant for the job, many benefits accrue. Examples include:

- Guidance of R&D efforts to assure proper data are gathered for scale-up.
- Assurance that process results are as expected from the equipment.
- Assurance of mechanical integrity standards.
- Objective bid specification to get all vendors on an equivalent basis ("apples to apples").
- Typically lower bid prices based on the above kind of specification (vendors must bid on an equivalent basis, rather than a different basis of their own which can obfuscate the decision process), saving much more than the consultant's fee.
- Objective bid evaluation to written criteria.
- Liaison between end user's needs and terminology and that of the equipment manufacturers.
- Much deeper involvement in the project than equipment vendor engineers have time for.
- Aid in looking at optimization or modification of equipment with little or no capital expenditure. Process equipment manufacturers' engineers have little interest in this.

Choosing the right consultant is, of course, a key ingredient in the success of a project. The right consultant is a part of the project team, not just a source of expertise.

Here are a few of the attributes that make for a good consultant to put on your team:

- Extensive experience with the kind of process equipment being purchased (typically 20 years or more), especially with manufacturers of that kind of equipment.

- Specific experience in the end user's industry or application area is a strong plus, though not always necessary.
- Flexibility of scheduling, to assure timely responses.
- Willingness to travel to jobsites or meeting sites as needed, especially on international projects.
- Adaptability to different corporate or national cultures.
- Personal reputation in the industry.
- Desire to optimize and save the customer money.
- Willingness to work on both small and large projects as needed.

A typical scope of activity for a consultant may include some or all of the following:

- Develop an experimental protocol targeted at the key process results required.
- If no such experimentation is planned, use published correlations to develop potential production designs on a feasibility basis.
- Scale-up and optimize the production design, taking into account any process constraints.
- Develop specifications for bid.
- Evaluate vendor bids.
- Train the customer staff on relevant design issues.
- Evaluate alternative vessel geometries and mass transfer rates versus volume required.
- Coordinate CFD simulations when required.

A typical range of fees for a US-based consultant, in 2022 dollars, would be about \$3000–30000, depending on the detailed scope of activity and deliverables. This is a typical range for biofuel and renewable chemical projects. The range would be higher for more complex processes, such as fermentation.

The main difficulty in finding a consultant specifically for fluid agitation is that there are very few of us, and we are getting old. There are probably no more than three to five working consultants with such qualifications in the United States as of this writing. This book is written partially to let the user make as informed a decision as possible without guidance from a consultant when none is available.

## Summary of Chapter

The choice of agitator supplier is an important part of overall process optimization. Equipment integrity, vendor service, and knowledge all pay a role in the decision process. In this chapter, we have endeavored to list as many relevant criteria as possible to evaluate vendors, though each end user must decide how to weight these factors. An experienced consultant can be of help as well, and we have identified key attributes and functions of such a consultant.

## A Supplier List

Here is a partial list of agitator suppliers believed by the author to have supplied agitators for the biofuel and renewable chemicals industry, mainly in the United States. It is not an exhaustive list; I am sure there are many others. It is not in any particular order. The inclusion of a vendor on this list is not an endorsement. Neither is an exclusion intended to disparage any vendor; it is just a function of the author's limited awareness of agitator vendors who have supplied to this field, as probably 90% of all fluid agitators have been supplied by the top four companies globally. Table 17.1 provides a list of things to check out when auditing suppliers.

| Name | Website | City, state, or province | Country |
|---|---|---|---|
| Fusion Fluid Equipment | http://www.fusionfluid.com/index.html | Michigan | United States |
| Fluidmix | https://www.agitadoresfluidmix.com/en/ | Madrid | Spain |
| Mixrite | http://www.jetmixing.net/ | Mumbai | India |
| Proquip | https://proquipinc.com/ | Ohio | United States |
| Cleveland Mixer | https://clevelandmixer.com/ | Connecticut | United States |
| Ekato | https://www.ekato.com/en/ | | Germany |
| Chemineer/NOV | https://www.chemineer.com/ | Ohio | United States |
| Lightnin/SPX | https://www.spxflow.com/lightnin/ | New York | United States |
| Mixel | https://www.mixel.fr/en/ | | France |
| Philadelphia Mixing Solutions | https://www.philamixers.com/; now part of SPX | Pennsylvania | United States |
| Hayward Gordon | https://haywardgordon.com/ | | Canada |
| GMM Pfaudler/ Mixion | http://www.gmmpfaudler.com | | India |
| MGT | https://www.mgt.co.il/ | | Israel |

**Table 17.1** Supplier audit checklist.

| | | | |
|---|---|---|---|
| Company name | | Principal products | |
| Founding year | | Number of employees | |
| Factory area, $M^2$ | | Factory employees | |
| Annual sales, USD | | Administrative employees | |
| Annual parts sales, USD | | Field service staff | |
| ISO certified? | | Engineers, <5 years | |
| ASME? | | Engineers, 5–10 years | |
| DIN? | | Engineers, 11–20 years | |
| JIS? | | Engineers, >20 years | |
| CSA? | | Code welders | |
| AGMA? | | Skilled machinists | |
| Other qualifications (fill in) | | CAD system used | |
| Top customers: Company name | Location | Contact person | E-mail or phone |
| | | | |
| | | | |
| | | | |
| | | | |
| | | | |
| Equipment history: Product type | First year sold | Cumulative quantity sold | Improvements made |

## References


**1** Benz, G. (2006). Team up with a consultant for equipment selection. *Chemical Engineering Progress* 102: 50–51.

**2** Ivona, O. (2009). The importance of tacit knowledge within the organization. https://www.semanticscholar.org/paper/THE-IMPORTANCE-OF-TACIT-KNOWLEDGE-WITHIN-THE-Orzea-Udrea/6fa14ec290e8a90d93eed743412b534b0856860d.

**3** Penya, A. (2013). Institutional knowledge: when employees leave, what do we lose? *HigherEdJobs Blog* (18 November 2013).

# 18

# Summary of Book

Though we probably have a sufficient supply of fossil fuel sources to last for quite a while, it is inevitable that the demand will outstrip supply at some point. Our long-term solution for energy must therefore be some kind of combination of renewable sources plus nuclear. And our long-term material supply for processes traditionally fed by petroleum will be renewable sources, which basically means solar energy converted into plant biomass.

In this book, we have endeavored to cover the fundamentals of agitation technology in a way that it will not become obsolete as new processes develop for renewable fuels and chemicals. Many current applications have been explored in detail, but the main thrust of this book has been to provide the reader with enough information to adapt agitation technology to new materials and processes as they develop. In many cases, lab testing will be required. Therefore, we have included examples of how to set up such a test and scale-up the results.

To our knowledge, no book of this kind has been written before. We hope the reader finds it useful for decades to come.

*Agitator Design Technology for Biofuels and Renewable Chemicals*, First Edition. Gregory T. Benz.
© 2022 John Wiley & Sons, Inc. Published 2022 by John Wiley & Sons, Inc.

# Appendix 1

## Generic Hydrofoil Pumping and Power Numbers (for Design Only)

| D/T = 0.25 | | | D/T = 0.3 | | | D/T = 0.35 | | |
|---|---|---|---|---|---|---|---|---|
| Reynolds number, $N_{Re}$ | Power number, $N_P$ | Pumping number, $N_Q$ | Reynolds number, $N_{Re}$ | Power number, $N_P$ | Pumping number, $N_Q$ | Reynolds number, $N_{Re}$ | Power number, $N_P$ | Pumping number, $N_Q$ |
| 1 | 32.4 | 0.3 | 1 | 32.4 | 0.27 | 1 | 32.4 | 0.25 |
| 2 | 16.2 | 0.31 | 2 | 16.2 | 0.28 | 2 | 16.2 | 0.26 |
| 3 | 10.8 | 0.31 | 3 | 10.8 | 0.28 | 3 | 10.8 | 0.26 |
| 5 | 6.5 | 0.31 | 5 | 8.25 | 0.28 | 5 | 8.25 | 0.26 |
| 7 | 4.63 | 0.31 | 7 | 5.98 | 0.28 | 7 | 5.98 | 0.26 |
| 10 | 3.3 | 0.31 | 10 | 4.27 | 0.28 | 10 | 4.26 | 0.26 |
| 12 | 3 | 0.31 | 12 | 3.61 | 0.28 | 12 | 3.6 | 0.26 |
| 15 | 2.8 | 0.31 | 15 | 2.95 | 0.28 | 15 | 2.93 | 0.26 |
| 20 | 2.3 | 0.3 | 20 | 2.29 | 0.28 | 20 | 2.27 | 0.25 |
| 25 | 1.91 | 0.3 | 25 | 1.89 | 0.27 | 25 | 1.87 | 0.25 |
| 30 | 1.64 | 0.3 | 30 | 1.62 | 0.27 | 30 | 1.61 | 0.25 |
| 50 | 1.1 | 0.3 | 50 | 1.09 | 0.27 | 50 | 1.08 | 0.24 |
| 70 | 0.85 | 0.29 | 70 | 0.86 | 0.26 | 70 | 0.85 | 0.24 |
| 100 | 0.73 | 0.29 | 100 | 0.71 | 0.28 | 100 | 0.67 | 0.25 |
| 200 | 0.53 | 0.34 | 200 | 0.52 | 0.31 | 200 | 0.48 | 0.28 |
| 300 | 0.47 | 0.37 | 300 | 0.45 | 0.34 | 300 | 0.41 | 0.31 |
| 500 | 0.41 | 0.43 | 500 | 0.39 | 0.38 | 500 | 0.36 | 0.35 |
| 1 000 | 0.36 | 0.5 | 1 000 | 0.35 | 0.43 | 1 000 | 0.32 | 0.41 |

*(Continued)*

*Agitator Design Technology for Biofuels and Renewable Chemicals*, First Edition. Gregory T. Benz.
© 2022 John Wiley & Sons, Inc. Published 2022 by John Wiley & Sons, Inc.

(Continued)

| D/T = 0.25 | | | D/T = 0.3 | | | D/T = 0.35 | | |
|---|---|---|---|---|---|---|---|---|
| Reynolds number, $N_{Re}$ | Power number, $N_P$ | Pumping number, $N_Q$ | Reynolds number, $N_{Re}$ | Power number, $N_P$ | Pumping number, $N_Q$ | Reynolds number, $N_{Re}$ | Power number, $N_P$ | Pumping number, $N_Q$ |
| 2000 | 0.34 | 0.54 | 2000 | 0.33 | 0.48 | 2000 | 0.3 | 0.44 |
| 5000 | 0.33 | 0.57 | 5000 | 0.32 | 0.5 | 5000 | 0.29 | 0.46 |
| 10000 | 0.33 | 0.58 | 10000 | 0.32 | 0.51 | 10000 | 0.29 | 0.46 |
| 20000 | 0.32 | 0.59 | 20000 | 0.31 | 0.52 | 20000 | 0.29 | 0.47 |
| 100000 | 0.32 | 0.59 | 100000 | 0.31 | 0.53 | 100000 | 0.29 | 0.47 |

| D/T = 0.4 | | | D/T = 0.45 | | | D/T = 0.5 | | |
|---|---|---|---|---|---|---|---|---|
| Reynolds number, $N_{Re}$ | Power number, $N_P$ | Pumping number, $N_Q$ | Reynolds number, $N_{Re}$ | Power number, $N_P$ | Pumping number, $N_Q$ | Reynolds number, $N_{Re}$ | Power number, $N_P$ | Pumping number, $N_Q$ |
| 1 | 32.4 | 0.23 | 1 | 32.4 | 0.22 | 1 | 32.4 | 0.21 |
| 2 | 16.2 | 0.24 | 2 | 16.2 | 0.23 | 2 | 16.2 | 0.22 |
| 3 | 10.8 | 0.25 | 3 | 10.8 | 0.23 | 3 | 10.8 | 0.22 |
| 5 | 8.25 | 0.25 | 5 | 8.25 | 0.23 | 5 | 8.25 | 0.22 |
| 7 | 5.98 | 0.25 | 7 | 5.98 | 0.23 | 7 | 5.98 | 0.22 |
| 10 | 4.24 | 0.25 | 10 | 4.23 | 0.23 | 10 | 4.23 | 0.22 |
| 12 | 3.58 | 0.25 | 12 | 3.57 | 0.23 | 12 | 3.56 | 0.22 |
| 15 | 2.92 | 0.24 | 15 | 2.91 | 0.23 | 15 | 2.9 | 0.22 |
| 20 | 2.26 | 0.24 | 20 | 2.25 | 0.23 | 20 | 2.24 | 0.21 |
| 25 | 1.86 | 0.24 | 25 | 1.85 | 0.22 | 25 | 1.84 | 0.21 |
| 30 | 1.59 | 0.23 | 30 | 1.58 | 0.22 | 30 | 1.57 | 0.21 |
| 50 | 1.06 | 0.23 | 50 | 1.05 | 0.22 | 50 | 1.04 | 0.2 |
| 70 | 0.84 | 0.23 | 70 | 0.83 | 0.21 | 70 | 0.82 | 0.2 |
| 100 | 0.68 | 0.24 | 100 | 0.66 | 0.22 | 100 | 0.68 | 0.21 |
| 200 | 0.46 | 0.26 | 200 | 0.46 | 0.24 | 200 | 0.46 | 0.23 |
| 300 | 0.4 | 0.28 | 300 | 0.39 | 0.27 | 300 | 0.4 | 0.25 |
| 500 | 0.35 | 0.32 | 500 | 0.34 | 0.3 | 500 | 0.35 | 0.28 |
| 1000 | 0.31 | 0.37 | 1000 | 0.3 | 0.35 | 1000 | 0.3 | 0.32 |
| 2000 | 0.29 | 0.4 | 2000 | 0.28 | 0.37 | 2000 | 0.28 | 0.35 |
| 5000 | 0.28 | 0.42 | 5000 | 0.27 | 0.39 | 5000 | 0.26 | 0.36 |
| 10000 | 0.28 | 0.42 | 10000 | 0.26 | 0.39 | 10000 | 0.27 | 0.37 |
| 20000 | 0.27 | 0.43 | 20000 | 0.26 | 0.4 | 20000 | 0.26 | 0.37 |
| 100000 | 0.27 | 0.43 | 100000 | 0.26 | 0.4 | 100000 | 0.26 | 0.37 |

| D/T = 0.55 | | | D/T = 0.6 | | | D/T = 0.65 | | |
|---|---|---|---|---|---|---|---|---|
| Reynolds number, $N_{Re}$ | Power number, $N_P$ | Pumping number, $N_Q$ | Reynolds number, $N_{Re}$ | Power number, $N_P$ | Pumping number, $N_Q$ | Reynolds number, $N_{Re}$ | Power number, $N_P$ | Pumping number, $N_Q$ |
| 1 | 32.4 | 0.2 | 1 | 32.4 | 0.19 | 1 | 32.4 | 0.18 |
| 2 | 16.2 | 0.21 | 2 | 16.2 | 0.2 | 2 | 16.2 | 0.19 |
| 3 | 10.8 | 0.21 | 3 | 10.8 | 0.2 | 3 | 10.8 | 0.2 |
| 5 | 8.25 | 0.21 | 5 | 8.25 | 0.2 | 5 | 8.25 | 0.2 |
| 7 | 5.98 | 0.21 | 7 | 5.98 | 0.2 | 7 | 5.98 | 0.2 |
| 10 | 3.5 | 0.21 | 10 | 3.5 | 0.2 | 10 | 3.5 | 0.2 |
| 12 | 3 | 0.21 | 12 | 3 | 0.2 | 12 | 3 | 0.19 |
| 15 | 2.83 | 0.21 | 15 | 2.83 | 0.2 | 15 | 2.83 | 0.19 |
| 20 | 2.27 | 0.21 | 20 | 2.27 | 0.19 | 20 | 2.26 | 0.19 |
| 25 | 1.87 | 0.2 | 25 | 1.87 | 0.19 | 25 | 1.87 | 0.18 |
| 30 | 1.6 | 0.2 | 30 | 1.6 | 0.19 | 30 | 1.6 | 0.18 |
| 50 | 1.07 | 0.2 | 50 | 1.07 | 0.19 | 50 | 1.07 | 0.18 |
| 70 | 0.85 | 0.2 | 70 | 0.85 | 0.19 | 70 | 0.85 | 0.18 |
| 100 | 0.68 | 0.21 | 100 | 0.68 | 0.2 | 100 | 0.67 | 0.19 |
| 200 | 0.48 | 0.23 | 200 | 0.48 | 0.22 | 200 | 0.48 | 0.21 |
| 300 | 0.41 | 0.25 | 300 | 0.4 | 0.24 | 300 | 0.41 | 0.24 |
| 500 | 0.36 | 0.29 | 500 | 0.36 | 0.28 | 500 | 0.36 | 0.27 |
| 1000 | 0.32 | 0.33 | 1000 | 0.32 | 0.32 | 1000 | 0.32 | 0.3 |
| 2000 | 0.3 | 0.35 | 2000 | 0.3 | 0.34 | 2000 | 0.3 | 0.33 |
| 5000 | 0.29 | 0.36 | 5000 | 0.29 | 0.35 | 5000 | 0.29 | 0.34 |
| 10000 | 0.28 | 0.37 | 10000 | 0.28 | 0.35 | 10000 | 0.28 | 0.34 |
| 20000 | 0.28 | 0.37 | 20000 | 0.28 | 0.36 | 20000 | 0.28 | 0.34 |
| 100000 | 0.28 | 0.38 | 100000 | 0.28 | 0.36 | 100000 | 0.28 | 0.35 |

| D/T = 0.7 | | | D/T = 0.75 | | | D/T = 0.8 | | |
|---|---|---|---|---|---|---|---|---|
| Reynolds number, $N_{Re}$ | Power number, $N_P$ | Pumping number, $N_Q$ | Reynolds number, $N_{Re}$ | Power number, $N_P$ | Pumping number, $N_Q$ | Reynolds number, $N_{Re}$ | Power number, $N_P$ | Pumping number, $N_Q$ |
| 1 | 32.4 | 0.18 | 1 | 32.4 | 0.17 | 1 | 32.4 | 0.17 |
| 2 | 16.2 | 0.18 | 2 | 16.2 | 0.18 | 2 | 16.2 | 0.18 |
| 3 | 10.8 | 0.19 | 3 | 10.8 | 0.18 | 3 | 10.8 | 0.18 |
| 5 | 8.25 | 0.19 | 5 | 8.25 | 0.19 | 5 | 8.25 | 0.18 |
| 7 | 5.98 | 0.19 | 7 | 5.98 | 0.19 | 7 | 5.98 | 0.18 |
| 10 | 3.24 | 0.19 | 10 | 3.24 | 0.19 | 10 | 3.24 | 0.18 |
| 12 | 3 | 0.19 | 12 | 3 | 0.18 | 12 | 3 | 0.18 |
| 15 | 2.8 | 0.18 | 2.82 | 2.93 | 0.18 | 15 | 2.83 | 0.18 |
| 20 | 2.27 | 0.18 | 20 | 2.27 | 0.18 | 20 | 2.27 | 0.18 |
| 25 | 1.87 | 0.18 | 25 | 1.87 | 0.18 | 25 | 1.87 | 0.17 |
| 30 | 1.6 | 0.18 | 30 | 1.6 | 0.17 | 30 | 1.6 | 0.17 |
| 50 | 1.07 | 0.18 | 50 | 1.07 | 0.17 | 50 | 1.07 | 0.17 |
| 70 | 0.85 | 0.18 | 70 | 0.85 | 0.17 | 70 | 0.85 | 0.17 |
| 100 | 0.68 | 0.19 | 100 | 0.68 | 0.18 | 100 | 0.68 | 0.18 |
| 200 | 0.48 | 0.21 | 200 | 0.48 | 0.2 | 200 | 0.48 | 0.2 |
| 300 | 0.41 | 0.23 | 300 | 0.41 | 0.22 | 300 | 0.41 | 0.22 |
| 500 | 0.36 | 0.26 | 500 | 0.35 | 0.25 | 500 | 0.35 | 0.25 |
| 1000 | 0.32 | 0.3 | 1000 | 0.32 | 0.29 | 1000 | 0.32 | 0.29 |
| 2000 | 0.3 | 0.32 | 2000 | 0.3 | 0.32 | 2000 | 0.3 | 0.31 |
| 5000 | 0.29 | 0.32 | 5000 | 0.29 | 0.32 | 5000 | 0.29 | 0.32 |
| 10000 | 0.28 | 0.33 | 10000 | 0.28 | 0.32 | 10000 | 0.28 | 0.32 |
| 20000 | 0.28 | 0.33 | 20000 | 0.28 | 0.33 | 20000 | 0.28 | 0.33 |
| 100000 | 0.28 | 0.33 | 100000 | 0.28 | 0.33 | 100000 | 0.28 | 0.33 |

# Appendix 2

## Generic Pitched Blade Turbine Pumping and Power Numbers (for Design Only)

| D/T = 0.25 | | | D/T = 0.3 | | | D/T = 0.35 | | |
|---|---|---|---|---|---|---|---|---|
| Reynolds number, $N_{Re}$ | Power number, $N_P$ | Pumping number, $N_Q$ | Reynolds number, $N_{Re}$ | Power number, $N_P$ | Pumping number, $N_Q$ | Reynolds number, $N_{Re}$ | Power number, $N_P$ | Pumping number, $N_Q$ |
| 1 | 50 | 0.38 | 1 | 50 | 0.34 | 1 | 50 | 0.32 |
| 2 | 25 | 0.38 | 2 | 25 | 0.34 | 2 | 25 | 0.32 |
| 3 | 16.7 | 0.38 | 3 | 16.7 | 0.34 | 3 | 16.7 | 0.32 |
| 5 | 10.5 | 0.38 | 5 | 10.5 | 0.34 | 5 | 10.5 | 0.32 |
| 7 | 7.5 | 0.38 | 7 | 7.5 | 0.34 | 7 | 7.5 | 0.32 |
| 10 | 6.14 | 0.38 | 10 | 6.14 | 0.34 | 10 | 6.14 | 0.32 |
| 12 | 5.53 | 0.38 | 12 | 5.53 | 0.34 | 12 | 5.53 | 0.32 |
| 15 | 4.88 | 0.38 | 15 | 4.88 | 0.34 | 15 | 4.88 | 0.32 |
| 20 | 4.2 | 0.38 | 20 | 4.2 | 0.34 | 20 | 4.2 | 0.32 |
| 25 | 3.76 | 0.38 | 25 | 3.76 | 0.34 | 25 | 3.76 | 0.32 |
| 30 | 3.45 | 0.38 | 30 | 3.45 | 0.34 | 30 | 3.45 | 0.32 |
| 50 | 2.71 | 0.39 | 50 | 2.76 | 0.36 | 50 | 2.76 | 0.33 |
| 70 | 2.3 | 0.4 | 70 | 2.38 | 0.38 | 70 | 2.42 | 0.35 |
| 100 | 1.92 | 0.45 | 100 | 2.14 | 0.42 | 100 | 2.1 | 0.4 |
| 200 | 1.75 | 0.55 | 200 | 1.75 | 0.5 | 200 | 1.75 | 0.45 |
| 300 | 1.6 | 0.6 | 300 | 1.6 | 0.54 | 300 | 1.6 | 0.5 |
| 500 | 1.46 | 0.66 | 500 | 1.46 | 0.61 | 500 | 1.46 | 0.56 |
| 1 000 | 1.37 | 0.73 | 1 000 | 1.37 | 0.66 | 1 000 | 1.38 | 0.64 |
| 2 000 | 1.37 | 0.79 | 2 000 | 1.37 | 0.72 | 2 000 | 1.38 | 0.68 |

*(Continued)*

*Agitator Design Technology for Biofuels and Renewable Chemicals*, First Edition. Gregory T. Benz.
© 2022 John Wiley & Sons, Inc. Published 2022 by John Wiley & Sons, Inc.

(Continued)

| D/T = 0.25 | | | D/T = 0.3 | | | D/T = 0.35 | | |
|---|---|---|---|---|---|---|---|---|
| Reynolds number, $N_{Re}$ | Power number, $N_P$ | Pumping number, $N_Q$ | Reynolds number, $N_{Re}$ | Power number, $N_P$ | Pumping number, $N_Q$ | Reynolds number, $N_{Re}$ | Power number, $N_P$ | Pumping number, $N_Q$ |
| 5000 | 1.37 | 0.83 | 5000 | 1.37 | 0.76 | 5000 | 1.38 | 0.72 |
| 10000 | 1.37 | 0.88 | 10000 | 1.37 | 0.8 | 10000 | 1.37 | 0.72 |
| 20000 | 1.37 | 0.88 | 20000 | 1.37 | 0.8 | 20000 | 1.37 | 0.72 |
| 100000 | 1.37 | 0.88 | 100000 | 1.37 | 0.8 | 100000 | 1.37 | 0.72 |

| D/T = 0.4 | | | D/T = 0.45 | | | D/T = 0.5 | | |
|---|---|---|---|---|---|---|---|---|
| Reynolds number, $N_{Re}$ | Power number, $N_P$ | Pumping number, $N_Q$ | Reynolds number, $N_{Re}$ | Power number, $N_P$ | Pumping number, $N_Q$ | Reynolds number, $N_{Re}$ | Power number, $N_P$ | Pumping number, $N_Q$ |
| 1 | 50 | 0.3 | 1 | 50 | 0.28 | 1 | 50 | 0.27 |
| 2 | 25 | 0.3 | 2 | 25 | 0.28 | 2 | 25 | 0.27 |
| 3 | 16.7 | 0.3 | 3 | 16.7 | 0.28 | 3 | 16.7 | 0.27 |
| 5 | 10.5 | 0.3 | 5 | 10.5 | 0.28 | 5 | 10.5 | 0.27 |
| 7 | 7.5 | 0.3 | 7 | 7.5 | 0.28 | 7 | 7.5 | 0.27 |
| 10 | 6.14 | 0.3 | 10 | 6.14 | 0.28 | 10 | 6.14 | 0.27 |
| 12 | 5.53 | 0.3 | 12 | 5.53 | 0.28 | 12 | 5.53 | 0.27 |
| 15 | 4.88 | 0.3 | 15 | 4.88 | 0.28 | 15 | 4.88 | 0.27 |
| 20 | 4.2 | 0.3 | 20 | 4.2 | 0.28 | 20 | 4.2 | 0.27 |
| 25 | 3.76 | 0.3 | 25 | 3.76 | 0.28 | 25 | 3.76 | 0.27 |
| 30 | 3.45 | 0.3 | 30 | 3.45 | 0.29 | 30 | 3.45 | 0.27 |
| 50 | 2.76 | 0.32 | 50 | 2.76 | 0.3 | 50 | 2.76 | 0.29 |
| 70 | 2.42 | 0.34 | 70 | 2.42 | 0.32 | 70 | 2.42 | 0.31 |
| 100 | 2.14 | 0.37 | 100 | 2.14 | 0.36 | 100 | 2.14 | 0.36 |
| 200 | 1.75 | 0.45 | 200 | 1.75 | 0.42 | 200 | 1.72 | 0.41 |
| 300 | 1.6 | 0.48 | 300 | 1.6 | 0.46 | 300 | 1.6 | 0.44 |
| 500 | 1.46 | 0.54 | 500 | 1.46 | 0.5 | 500 | 1.46 | 0.47 |
| 1000 | 1.41 | 0.59 | 1000 | 1.44 | 0.55 | 1000 | 1.45 | 0.51 |
| 2000 | 1.41 | 0.64 | 2000 | 1.44 | 0.62 | 2000 | 1.44 | 0.57 |

(Continued)

| D/T = 0.4 | | | D/T = 0.45 | | | D/T = 0.5 | | |
|---|---|---|---|---|---|---|---|---|
| Reynolds number, $N_{Re}$ | Power number, $N_P$ | Pumping number, $N_Q$ | Reynolds number, $N_{Re}$ | Power number, $N_P$ | Pumping number, $N_Q$ | Reynolds number, $N_{Re}$ | Power number, $N_P$ | Pumping number, $N_Q$ |
| 5000 | 1.4 | 0.67 | 5000 | 1.44 | 0.64 | 5000 | 1.43 | 0.58 |
| 10000 | 1.39 | 0.68 | 10000 | 1.44 | 0.65 | 10000 | 1.43 | 0.6 |
| 20000 | 1.39 | 0.69 | 20000 | 1.44 | 0.65 | 20000 | 1.43 | 0.61 |
| 100000 | 1.39 | 0.69 | 100000 | 1.44 | 0.66 | 100000 | 1.43 | 0.61 |

| D/T = 0.55 | | | D/T = 0.6 | | | D/T = 0.65 | | |
|---|---|---|---|---|---|---|---|---|
| Reynolds number, $N_{Re}$ | Power number, $N_P$ | Pumping number, $N_Q$ | Reynolds number, $N_{Re}$ | Power number, $N_P$ | Pumping number, $N_Q$ | Reynolds number, $N_{Re}$ | Power number, $N_P$ | Pumping number, $N_Q$ |
| 1 | 50 | 0.25 | 1 | 50 | 0.24 | 1 | 50 | 0.23 |
| 2 | 25 | 0.25 | 2 | 25 | 0.24 | 2 | 25 | 0.23 |
| 3 | 16.7 | 0.25 | 3 | 16.7 | 0.24 | 3 | 16.7 | 0.23 |
| 5 | 10.5 | 0.25 | 5 | 10.5 | 0.24 | 5 | 10.5 | 0.23 |
| 7 | 7.5 | 0.25 | 7 | 7.5 | 0.24 | 7 | 7.5 | 0.23 |
| 10 | 6.14 | 0.25 | 10 | 6.14 | 0.24 | 10 | 6.14 | 0.23 |
| 12 | 5.53 | 0.25 | 12 | 5.53 | 0.24 | 12 | 5.53 | 0.23 |
| 15 | 4.88 | 0.25 | 15 | 4.88 | 0.24 | 15 | 4.88 | 0.23 |
| 20 | 4.2 | 0.25 | 20 | 4.2 | 0.24 | 20 | 4.2 | 0.23 |
| 25 | 3.76 | 0.25 | 25 | 3.76 | 0.25 | 25 | 3.76 | 0.23 |
| 30 | 3.45 | 0.26 | 30 | 3.45 | 0.25 | 30 | 3.45 | 0.24 |
| 50 | 2.76 | 0.28 | 50 | 2.76 | 0.27 | 50 | 2.76 | 0.26 |
| 70 | 2.42 | 0.3 | 70 | 2.42 | 0.28 | 70 | 2.42 | 0.28 |
| 100 | 2.14 | 0.32 | 100 | 2.14 | 0.3 | 100 | 2.14 | 0.3 |
| 200 | 1.75 | 0.37 | 200 | 1.75 | 0.36 | 200 | 1.75 | 0.35 |
| 300 | 1.6 | 0.42 | 300 | 1.56 | 0.4 | 300 | 1.58 | 0.39 |
| 500 | 1.48 | 0.47 | 500 | 1.5 | 0.46 | 500 | 1.53 | 0.45 |
| 1000 | 1.47 | 0.54 | 1000 | 1.5 | 0.52 | 1000 | 1.52 | 0.51 |
| 2000 | 1.46 | 0.57 | 2000 | 1.49 | 0.55 | 2000 | 1.51 | 0.53 |

*(Continued)*

(Continued)

| D/T = 0.55 | | | D/T = 0.6 | | | D/T = 0.65 | | |
|---|---|---|---|---|---|---|---|---|
| Reynolds number, $N_{Re}$ | Power number, $N_P$ | Pumping number, $N_Q$ | Reynolds number, $N_{Re}$ | Power number, $N_P$ | Pumping number, $N_Q$ | Reynolds number, $N_{Re}$ | Power number, $N_P$ | Pumping number, $N_Q$ |
| 5000 | 1.45 | 0.59 | 5000 | 1.48 | 0.57 | 5000 | 1.5 | 0.55 |
| 10000 | 1.45 | 0.6 | 10000 | 1.48 | 0.58 | 10000 | 1.5 | 0.56 |
| 20000 | 1.45 | 0.6 | 20000 | 1.48 | 0.59 | 20000 | 1.5 | 0.56 |
| 100000 | 1.45 | 0.6 | 100000 | 1.48 | 0.58 | 100000 | 1.5 | 0.56 |

| D/T = 0.7 | | | D/T = 0.75 | | | D/T = 0.8 | | |
|---|---|---|---|---|---|---|---|---|
| Reynolds number, $N_{Re}$ | Power number, $N_P$ | Pumping number, $N_Q$ | Reynolds number, $N_{Re}$ | Power number, $N_P$ | Pumping number, $N_Q$ | Reynolds number, $N_{Re}$ | Power number, $N_P$ | Pumping number, $N_Q$ |
| 1 | 50 | 0.22 | 1 | 50 | 0.22 | 1 | 50 | 0.22 |
| 2 | 25 | 0.22 | 2 | 25 | 0.22 | 2 | 25 | 0.22 |
| 3 | 16.7 | 0.22 | 3 | 16.7 | 0.22 | 3 | 16.7 | 0.22 |
| 5 | 10.5 | 0.22 | 5 | 10.5 | 0.22 | 5 | 10.5 | 0.22 |
| 7 | 7.5 | 0.22 | 7 | 7.5 | 0.22 | 7 | 7.5 | 0.22 |
| 10 | 6.14 | 0.22 | 10 | 6.14 | 0.22 | 10 | 6.14 | 0.22 |
| 12 | 5.53 | 0.22 | 12 | 5.53 | 0.22 | 12 | 5.53 | 0.22 |
| 15 | 4.88 | 0.22 | 2.82 | 4.88 | 0.22 | 15 | 4.88 | 0.22 |
| 20 | 4.2 | 0.22 | 20 | 4.2 | 0.22 | 20 | 4.2 | 0.22 |
| 25 | 3.76 | 0.23 | 25 | 3.76 | 0.23 | 25 | 3.76 | 0.23 |
| 30 | 3.45 | 0.24 | 30 | 3.45 | 0.23 | 30 | 3.45 | 0.23 |
| 50 | 2.76 | 0.25 | 50 | 2.76 | 0.25 | 50 | 2.76 | 0.25 |
| 70 | 2.42 | 0.27 | 70 | 2.42 | 0.27 | 70 | 2.42 | 0.27 |
| 100 | 2.14 | 0.29 | 100 | 2.14 | 0.29 | 100 | 2.14 | 0.29 |
| 200 | 1.75 | 0.34 | 200 | 1.75 | 0.34 | 200 | 1.75 | 0.34 |
| 300 | 1.6 | 0.38 | 300 | 1.6 | 0.38 | 300 | 1.6 | 0.38 |
| 500 | 1.53 | 0.44 | 500 | 1.55 | 0.43 | 500 | 1.56 | 0.43 |
| 1000 | 1.53 | 0.5 | 1000 | 1.54 | 0.5 | 1000 | 1.55 | 0.5 |
| 2000 | 1.52 | 0.52 | 2000 | 1.54 | 0.52 | 2000 | 1.55 | 0.53 |
| 5000 | 1.51 | 0.54 | 5000 | 1.53 | 0.54 | 5000 | 1.55 | 0.54 |
| 10000 | 1.51 | 0.54 | 10000 | 1.52 | 0.54 | 10000 | 1.54 | 0.55 |
| 20000 | 1.51 | 0.55 | 20000 | 1.52 | 0.55 | 20000 | 1.54 | 0.55 |
| 100000 | 1.51 | 0.55 | 100000 | 1.52 | 0.55 | 100000 | 1.54 | 0.55 |

# Afterword

Much to my surprise, several years ago I began working on my first technical reference book, *Agitator Design for Gas-Liquid Fermenters and Bioreactors* which as published by Wiley InterScience in April of 2021. At the time I wrote that book, I had no plans to write another. However, I have been heavily involved in the biofuel and renewable chemicals industries as a consultant for many years. In doing that work, I noticed that there were virtually no books on agitation technology specific to those industries, and I felt I had something to contribute to the knowledge base. Though we have covered the fundamentals here and in my previous book, much of the information in this book has not been published before. It is my hope that people involved in these industries, as well as agitator manufacturers, will find this to be a useful reference. Thanks for reading it!

*Agitator Design Technology for Biofuels and Renewable Chemicals*, First Edition. Gregory T. Benz.
© 2022 John Wiley & Sons, Inc. Published 2022 by John Wiley & Sons, Inc.

# Index



*Agitator Design Technology for Biofuels and Renewable Chemicals*, First Edition. Gregory T. Benz.
© 2022 John Wiley & Sons, Inc. Published 2022 by John Wiley & Sons, Inc.

## g



## h



## i

Printed and bound by CPI Group (UK) Ltd, Croydon, CR0 4YY
17/09/2026
14964253-0003